BLACK
0.57

WHAT ON EARTH EVOLVED?

100 SPECIES THAT CHANGED THE WORLD

Christopher Lloyd

Illustrations by Andy Forshaw

B L O O M S B U R Y

LONDON · BERLIN · NEW YORK

Before Man page 9

On the impact of species that evolved in the wild

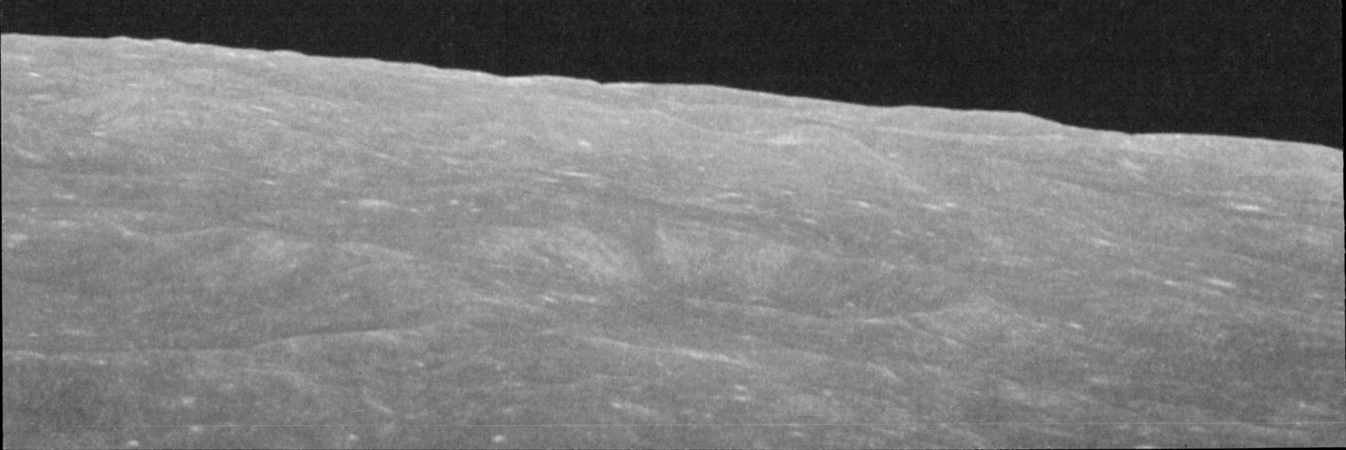

Introduction

WHAT IS LIFE? WHAT FORCES ARE IN CONTROL? Why have creatures evolved as they are? Where does humanity fit in?

This book is a jargon-free attempt to explain the phenomenon we call life on Earth. It traces the history of life from the dawn of evolution to the present day through the lens of one hundred living things that have changed the world. Unlike Charles Darwin's theory published 150 years ago, it is not chiefly concerned with the 'origin of species', but with the influences and impacts that living things have had on the path of evolution, on each other and on our mutual environment, planet Earth.

The first part surveys the mechanics of evolution before mankind, from the earliest replicating molecules to the rapid rise of mammals following the death of the dinosaurs, profiling fifty of the most successful life-forms (from now on, informally described as 'species') that emerged through *natural selection* up to the time when modern humans first trod the Earth.

The second part assesses how from *c.* 12,000 years ago humans introduced new evolutionary forces through the cultivation of plant and animal breeds to benefit human societies. The success of many creatures then came to depend on their capacity to impress and accommodate the needs of humans, which often replaced the traditional struggle for survival in the wild. What impact did these evolutionary forces have on the non-human world and on the direction of evolution itself? The second part profiles fifty of the most successful species that have developed through *artificial selection* up to the present day.

Finally, part three ranks the impact of the hundred chosen species – from the dawn of evolution to the present day – into a table of influence based on a set of simple, but arbitrary, criteria. Its intention is simply to stimulate thought, provoke comment and offer an alternative, sideways glance through the history of life.

The central purpose of this book is to cultivate a richer understanding of all history – not just as chronology – but as seen through the lens of the natural world itself. Its aim is not only to inform and entertain but, I hope, to stimulate debate about the place of mankind in nature and its pivotal relationship with non-human life and the Earth itself.

Before Man

From four billion to 12,000 years before present …

On the impact of species that evolved in the wild

1

On Viruses

How loose strands of genetic code swarmed across the early Earth, inserting themselves into all forms of life – past, present and future.

WHEN, WHERE AND HOW on Earth did life begin? The *when* bit isn't so hard – although recent estimates have pushed life's genesis further back into the Earth's early history to about four billion years ago, just 500 million years after the Earth was formed.

Where is easier. In the seas. Almost definitely. Chemicals required to support the construction of living things billions of aeons ago were found only in the oceans. The land was barren and the air thick with acrid, volcanic gas.

As to *how* – unfortunately there are almost as many theories as there are experts. They range from its delivery as a kit on a meteorite from outer space to the spontaneous fusion of inert chemicals at the bottom of the sea. So, to start our story, I am going to have to pick a theory. It is a recent one, as it happens, and not yet fully formed. But it is the one that, in the light of the latest evidence, seems to make most sense.

Life began as a *virus.*

At this point university professors and biology teachers will with one voice protest: how on Earth could life begin as a virus? Everyone knows that a virus itself is not a living thing!

Which is why at the start it was remiss of me not to ask the most salient and difficult question of all.

What is life?

Traditional definitions say living things must have internal stability, a form of energy consumption, be capable of growth, adapt to surroundings, respond to external forces and reproduce. Finally, all living things must comprise at least one or more jelly-like blobs called *cells*.

A virus fails all of these tests. On its own, it has no independent source of movement, it doesn't produce energy and it can't breed. Only when it accidentally brushes into contact with a suitable

Under attack: viruses inject their own genetic material (in blue) into a bacterial cell to hijack its copying machinery.

living 'host' cell does all this change. It then suddenly bursts into life, hijacking the cell's internal dynamics to make copies of its own kind. A virus is what scientists call a *genomic parasite* – a bundle of copying molecules (called genes) that germinate only when they infect another living thing. Like a seed, viruses have the potential for life, but only actually come alive when the conditions are just right.

As a result evolutionary biologists have generally dismissed the role of viruses. Instead, they have focused on trying to determine when the first living *cell* appeared – a primitive form of bacterium from which, they argue, all other life has descended. They call it the Last Universal Common Ancestor or LUCA for short (see page 29).

A simpler definition

It is now clear, thanks to computer simulations, that genetic evolution can actually come about thanks to only two simple conditions – not the traditional list of seven. The first is the presence of chemical 'replicators' that can combine and split up to make copies of themselves. The second is that, over time, some process must cause at least some of these copies to differ from one other in an unlimited number of small or large ways – that could happen via copying errors, mutation or some other process.

Provided these two conditions are satisfied then the process that Charles Darwin called 'natural selection' will emerge as predictably and as inevitably as water runs downhill. Those best-adapted replicators will breed fastest and will come to dominate the growing community of copying entities. But when environmental conditions change – for example the community gets more crowded, competition for raw materials increases or there is an adjustment in ambient temperature – the criteria for successful living will inevitably change.

Such a living system soon begins to resemble an arms race. Replicators best adapted to the system's current environmental conditions are statistically more likely to survive, and so they are 'selected', and evolve into the ecosystem's present champions. But as conditions change, others will take their place as a spiral of ever-increasing complexity develops. In much the same way, a small change in air or water temperature may naturally gather momentum, sometimes turning into an almighty hurricane or whirlpool.

A more satisfactory (albeit rather dry-sounding) answer to the question as to what life is may therefore be this: an increasingly complex evolutionary system composed of a variety of naturally occurring chemical replicators.

But how could viruses have begun the process of life when they always require the presence of other living cells to breed? Anyway, aren't viruses the cause of nasty diseases that have, at least over the course of human history, been responsible for countless millions of human deaths through illnesses such as **smallpox** (see page 24) and **influenza** (see page 16) and now **HIV/AIDS** (see page 18)? How could the credit for triggering the most momentous known event in all time and space, the birth of life itself, be awarded to a ruthless serial killer?

Clues as to why viruses may well have spawned life on Earth billions of years ago have recently been found buried deep within our very own genes …

The secret life of DNA

In 1953 biochemist James Watson and his colleague Francis Crick announced their discovery of a structure for deoxyribonucleic acid (DNA) – a complex string of chemicals that resides in the centre of living cells and is responsible for storing all the information required to build and operate a living organism from scratch.

Living things, be they as small as a single-celled amoeba, as large as an elephant, or as sophisticated as a human being, are made up of small, individual cells that contain this mysterious substance. DNA is composed of strings of chemicals called 'bases' that hang off a spindle that resembles a spiral staircase in the shape of a twisting double-helix. Long sequences of these bases are called genes and are organized in a way that is unique to each individual living creature.

These sequences are what make living creatures different, giving each its own individual identity, stamp or code.

The framework of DNA and its intricate genetic structure is known as a genome. A species is a very closely related group of living entities whose genomes are so similar that they can pair off and recombine during their reproductive processes to produce fertile offspring. The full spectrum of genomes within a species is called a gene pool.

The discovery of DNA in the mid 1950s gave scientists their first real glimpse into the evolutionary mechanics of life.

But it took another fifty years before scientists working on an initiative called the Human Genome Project, launched in 1990, were able to decipher the full genetic DNA code of human beings. Expectations of what secrets the Human Genome Project could yield ran high. Cracking the code was thought to be the big step towards directly manipulating the processes of life, offering scientists the prospect of switching on and off those genes specific to various human processes or behaviours.

Imaginations boggled at the potential for providing everything from immunity against crippling genetic diseases to a way to circumvent the ageing process and a new era of biological consumer choice at the point of conception. With enough genetic know-how perhaps designer babies could be cultivated with specific predetermined attributes ranging from eye and hair colour, to height, sexual orientation and breast size, catching the eye of pharmaceutical investors. Such prospects led to a bio-tech stock-market boom in the mid 1990s.

But the project had deeper implications. By decompiling human DNA, biologists hoped to see more clearly the evolutionary pathway that led from the beginnings of life to modern man. For example, 150 years ago Charles Darwin proposed that humans evolved from apes. But just how close is this relationship? And when did the momentous ancestral split between humans and our closest relatives, chimpanzees, actually occur?

More important still, what is the genetic *difference* between humans and chimpanzees? Could deciphering the human genome reveal which genes are responsible for intelligence, language and all those attributes we have come to associate with the 'higher' form of human life? Ultimately, by decoding human DNA it was hoped that one of the most taxing philosophical questions could finally be answered – *what is it that makes us human?*

Code-breakers

Watson, one of the prime movers behind gaining international public funding for the Human Genome Project, knew the challenge was vast – the human genome (our complete sequence of genes) was a highly sophisticated code comprising some *three billion* separate parts – all of them arranged in complex genetic sequences of just four bases that scientists have labelled A, C, G and T.

Researchers completed their work of deciphering the human genetic code thirteen years after the project's launch and after an investment of approximately $3 billion. The main findings were hugely humbling. The actual number of genes in the human genome was far fewer than expected – about 25,000, as opposed to estimates of up to two million at the start of the project. It means that the human genetic code is about the same size as that found inside the cells of a mustard plant. Smaller, simpler creatures such as the lungfish and even single-celled amoebas, it seems, have more genes than man.

Even more shocking was that of the 25,000 human genes at least 95 per cent turn out to have no apparent function at all. They are what is now described as 'junk DNA' – repetitive strings of inert gibberish. One sequence – called ALU – is repeated almost one million times, representing as much as 10 per cent of the entire human genome!

There is worse. It has proved almost impossible to work out, even for the relatively few genes that do control human biological processes, what each one actually does. There is no simple correlation between a gene and the production of a protein that affects bodily growth or behaviour.

Sometimes several genes are used in the process of making a single useful chemical, while others have multiple functions. The simplistic idea of an array of easy-to-control on/off switches has proved

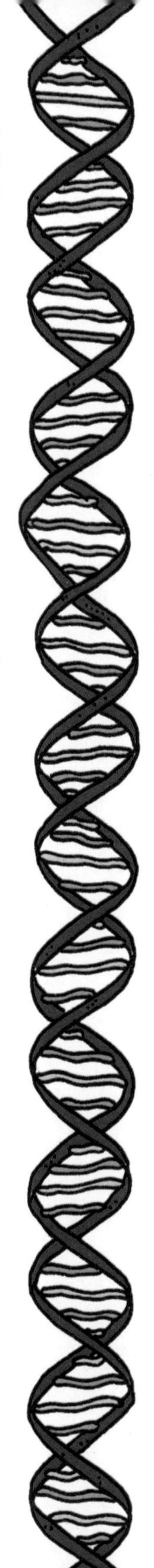

completely erroneous and now the task of working out what each gene actually does turns out to be a far greater challenge than the initial job of deciphering the code itself.

Investors grew impatient, bio-tech speculators lost money and many scientists have moved on. Not *everyone* was so disappointed with the results, however.

Viruses and humanity

Dr Luis Villarreal is an American professor of Mexican descent who just *loves* junk. Not usually of the food variety, although – like the students who attend his virology classes at the University of California – he is known to indulge in the occasional cheeseburger with fries.

Villarreal is one of the world's most highly regarded authorities on viruses. It has been his life's work to demonstrate that although Charles Darwin was right about the idea of relatedness between one species and another evolving over time, the process is far from what many of Darwin's recent followers have suggested – simply down to random copying errors in the genetic code of life.

Instead, Villarreal has come up with compelling evidence to suggest not only that viruses are the most ancient form of life, in fact the *original* form of life, but that all that DNA junk liberally spread around the human genome originated from viral genes that have survived over hundreds of millions of years of evolutionary time.

Human history is peppered with attempts to defy the scourge of death. Indestructible pyramids were built to house the tombs of Egyptian pharaohs so they could return to their earthly bodies in the afterlife and a humble Jewish carpenter's son called Jesus Christ sacrificed himself on a cross for the sake of his followers' spiritual immortality.

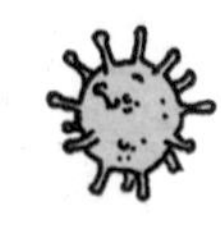

Viruses, the simplest forms of genetic code found in nature, also crave immortality. Unlike human beings, however, viruses aren't aware, intelligent or deliberate, so using language such as *crave* is meant in an entirely metaphorical sense – viruses behave only *as if* they crave immortality.

Viruses that copy their distinctive genetic sequences into other organisms are successful, while those that do not, die out. Even among purists who quibble about whether viruses are technically alive or not, everyone agrees that, once inside a living host, they are prolific replicators that absolutely obey the rules required for evolution to occur.

What gave Villarreal his biggest clue as to the huge impact of viruses on the evolution of other living things was the discovery in the 1970s of an enzyme called *reverse transcriptase*. This is a complex protein by which certain viruses are able to achieve that goal most sought after by the Egyptian pharaohs and others – immortality.

These viruses (called retroviruses, see **HIV/AIDS**, page 18) start by converting their own form of genetic code called ribonucleic acid (or RNA) into DNA, which they then splice into their host's genome. By this means they copy themselves through countless generations for millions if not billions of years. It transpires that most of the junk DNA that forms the vast majority of the human genome is generated by ancient viruses like these.

But even if we accept the rather bizarre fact that humans are genetically more virus than man, how come our ancestors managed to survive so many virulent infections?

The answer, says Villarreal, is that contrary to popular thought many viruses are harmless. Indeed, some are positively beneficial to the host organisms they infect (for examples, see pages 161 and 198). These viruses are called 'persistent' because their strategy is to copy themselves into the genome of a particular species and then let the hosts' own reproductive processes take over.

Persistent viruses work best in hosts that have small, widely dispersed populations. In theory a persistent virus could eventually spread throughout an entire species after infecting just one fertile individual. When that infected individual mates, the virus is passed to the host's offspring. The process then continues down the generations allowing the virus itself to attain immortality, accumulating, *ad infinitum*, as junk DNA.

Prehistoric humans were ideal hosts for persistent viruses because their populations were low and they did not generally live in close proximity to each other. One recent theory puts evolutionary differences between other mammals and man down to a series of ancient retroviral infections (see pages 337–8). According to Villarreal, such infections can often lead to the evolution of new species.

Not all viruses work by inserting themselves into their host's genomes, however. Some have succeeded by spreading their codes via coughs and sneezes (see influenza, page 16), virally infected saliva bites (such as rabies, see bats, page 163) or pustular sores (see smallpox, page 24). These viruses have affected humans in only relatively recent history – the past 10,000 years or so – since human social groups have grown dense enough for these 'acute' virus propagation strategies, which require close contact between hosts, to work effectively.

Viruses sometimes seek transporters (in the same way a flower recruits pollinators such as bees to spread its pollen. Aphids (see **potyvirus**, page 22) and bloodsucking insects such as mosquitoes (see page 134), fleas (see page 137), tsetse flies (see page 140) and ticks (see page 384) all act as carriers of viral disease to other species (including humans).

The trigger of life?

The properties of viral RNA have recently led some experts to suppose that the ancestors of today's viruses evolved either before or alongside the appearance of the first living cell.

Evidence is mounting because recent studies show that RNA viruses have the most diverse and varied combination of genes in the organic world. As many as 80 per cent of all RNA virus genomes are unique – which means they are unlikely to have evolved via genetic inheritance, in the way most life-forms do, from one generation to the next. For their diversity, viruses rely instead on a process of random gene transfer that operates via infection.

The business of splicing genes, copying them and experimenting *ad infinitum* with new genetic combinations is characteristic of the viral 'RNA' world that some experts believe may have started the whole saga of life on Earth. Over time, the most successful combinations survived, copied themselves and become dominant – until environmental conditions changed, giving other varieties a fresh chance.

As more complex forms of life evolved new opportunities opened up for these viruses to copy themselves by infecting not just each other but other living things such as cellular hosts. The fact that viruses today can survive only by infecting cellular life may simply be because, with so much cellular life about, what need is there now to infect another virus? The practice simply died out – like the limbs of whales (see page 166) – many aeons ago.

The influenza virus's spiky surface helps it to bind to host cells.

Viruses are today's most populous evolutionary entities. They reside in countless repetitious sequences within the genomes of almost every living organism, and even the oceans represent a viral cauldron of almost infinite capacity. As many as 10^{31} (that's a ten followed by thirty-one noughts) individual viral particles are thought to inhabit the seas, a mind-bendingly large number. So numerous are these genetic molecules that if placed end-to-end they would stretch across the entire known universe …

In the human world viruses will continue to have a lousy reputation thanks to the traumatic impact of a few devastating infections that have plagued humans throughout recorded history. It is ironic, though, to consider that were it not for the successful pursuit of immortality, perhaps begun by viruses at the dawn of life billions of years ago, human history may never have emerged at all.

Influenza

FAMILY: ORTHOMYXOVIRIDAE
SPECIES: INFLUENZAVIRUS A
RANK: 9

Historically one of humanity's biggest ever killers and still the largest threat to populations on Earth.

IN 1798 English economist Thomas Malthus predicted the imminent decimation of human populations by *'sickly seasons, epidemics, pestilence and plague'* that *'advance in terrific array, and sweep off their thousands and tens of thousands ... '*

This alarming prophecy was founded on the rapid rise of the human population. Not long after Malthus wrote his famous essay, the number of humans inhabiting the Earth rose past the one billion mark. Just over 200 years later, the human population stands at nearly seven billion, with approximately 211,000 extra bodies added every day.

Nothing could be better for viruses whose single concern is to copy their codes into as many other creatures (hosts) as possible. All living organisms from bacteria, plants and fungi to animals, fish and humans are vulnerable to infection by viruses – so great is their variety.

Influenza viruses infect a wide range of creatures including humans, pigs, birds, seals, mosquitoes, salmon and sea lice. At some point in the last few thousand years a variant of this virus family skipped across the species barrier and mutated, allowing it to spread from human to human. Its job was made much easier by the closer proximity between humans and animals that resulted from the rise of animal domestication and farming, beginning about 10,000 years ago (see page 192).

The influenza virus spreads by causing its host to erupt into severe bouts of coughing and sneezing. In dense populations of humans (such as military trenches, cities, trains or schools) copies of the influenza virus are easily spluttered from person to person in airborne droplets.

The human immune system is usually able to overcome these infections within two or three weeks. Old and young people are at greatest risk of severe illness. On average one person dies out of every one thousand infected.

But certain strains have had a much greater impact. Since Malthus wrote his essay, more than *one hundred million people* worldwide are thought to have died from various strains of flu. By far the worst outbreak came between 1918 and 1920 when between fifty million and one hundred million people fell victim. This was more than double the number of soldiers slaughtered in World War I (1914–18), which ended just as the influenza outbreak got under way.

The epidemic, which struck first in the United States, had a truly global impact. It spread throughout Europe and Asia, as far north as the Eskimo populations of the Arctic and west to the remote Pacific islands. Yet it came to be called 'Spanish flu' mainly because newspaper reports about its progress were least censored in Spain (which was not involved in the war effort and therefore had no ministry of propaganda) giving rise to the mistaken impression that the disease had Spanish origins.

Unlike in most flu outbreaks, the worst-affected victims were young adults. Experts believe this was because this particular strain (H1N1) causes the immune system to over-react, shocking it into producing too many immune cells and disease-busting chemicals, which badly damage body tissues – a phenomenon known as a 'cytokine storm'. Therefore people with the strongest immune systems – those in the prime of life – were most at risk. Death was gruesome. Massive haemorrhaging caused victims' lungs to fill with blood, and also suffer bleeding from the ears and internally.

Influenza viruses budding from the surface of an infected cell.

Today, scientists and politicians are especially anxious to avoid a repeat of this devastating pandemic. Sequencing the Spanish influenza virus was accomplished using preserved slivers of eighty-year-old lung tissue taken from soldiers who had died of the disease. In January 2007 results showed that the pandemic was a form of bird flu that had mutated in a way that enabled it to spread easily between humans.

A new strain of bird flu (H5N1) evolved in Asia between 1999 and 2002; however, it has had a limited effect on humans because this strain can be transmitted only from birds to humans, but not human to human, thereby restricting the risk of infection to those people who live or work in close proximity to birds. Nevertheless, an outbreak of another influenza variant (H1N1), originating in pigs, has since migrated throughout the world. Its easy spread from human to human caused this strain to be officially declared a global pandemic by the World Health Organization in the summer of 2009. International efforts to contain the spread of this so-called 'swine flu' were a conspicuous failure.

It is a worrying sign. Viruses have an extraordinary capacity for genetic diversity and change. The copying process used by influenza (which is based on RNA chemical replicators, see pages 14–15) makes errors far more often than the more sophisticated DNA genetic copying processes used in plant and animal cells. More errors mean more mutations, leading to the potential evolution of highly infectious lethal strains that can easily pass from human to human, turning each new victim into a deadly carrier of the disease. Mass immunization may or may not be effective, since precise counter-measures cannot be designed until it is known exactly how the virus has mutated – knowledge that can be gathered only *after* a new strain of the virus has struck. Mass production of a suitable vaccine would take at least three months by which time mutant strains of such an influenza virus could be wiping out tens if not hundreds of millions of people.

Malthus may have got his timing wrong, but in our overcrowded modern world and with highly infectious diseases like influenza, his prediction may be as pertinent today as ever.

HIV/AIDS

FAMILY: RETROVIRIDAE
SPECIES: HUMAN IMMUNODEFICIENCY VIRUS 1 & 2
RANK: 77

One of today's most perplexing and fastest-spreading diseases for which there is no sign of a permanent cure.

SOUTH AFRICA is one of the most beautiful countries on Earth. Its glorious coastline stretches for nearly 3,000 kilometres skirting the seas around the Cape of Good Hope to the east, south and west. For nearly 100,000 years different species of humans have shared the land with a rich ecosystem of wild game, exotic flowers and migratory birds. But then explorers from Holland settled near the southernmost tip in 1652, founding a refreshment station for merchant ships travelling from Europe in the West to the spice-rich markets of Asia in the East. Its strategic position close to the sea soon turned this bushy wilderness – now called Cape Town – into one of the richest, most sought-after cities in the world.

Today many people who live in and around Cape Town lead desperate lives. Their sorry story, which is repeated for the millions more who live in other Southern African cities, began thirty years ago when a devastating viral infection skipped across the species barrier from monkeys to humans. No one quite knows exactly how or when the HIV virus claimed its first human victim but it is thought that the incurable illness that stems from its infection, called AIDS, has so far claimed more than twenty-five million lives in twenty-five years, with at least another sixty million infected.

HIV is today's best known but least understood virus. As a member of the retrovirus family its ingenious tactics for copying its own genetic code into the genes of its human hosts are the product of billions of years of evolution (see reverse transcriptase, page 14). That's why, despite enormous advances in developing a vaccine for many of the most devastating human diseases, there is still no cure for HIV. Nor is there likely to be one any time soon.

Viruses like HIV replicate so fast and make so many copying errors in the process that they are always able to outwit vaccines. After infecting its host through bodily fluids such as blood, semen, saliva or breast milk, HIV mutates into so many different forms that the human immune system can't keep up. In many people the HIV infection can be contained by the body's defences for only about one year – although this can be greatly extended with the help of modern antiviral drugs. Eventually, however, the virus comes up with a new variant that the body's defences cannot circumvent. It then duplicates madly, wrecking what's left of its human host's disease-fighting systems. Death arises from other normally benign infections, such as the common cold, because the body has become too weak to fight them off.

Like all retroviruses, HIV converts its own RNA into DNA and splices it into its host's genes. It then multiplies at a ferocious rate and may be passed on to sexual partners or offspring. HIV represents something of a conundrum to many evolutionary biologists because retroviruses like HIV apparently defy the modern understanding of the laws of nature based on the survival of the fittest individual creatures (see page 60). Rather than countless millions of slightly different individual HIV-like retroviruses all competing with each other for success in attacking their hosts, they instead seem to collaborate as if working in a team (a phenomenon known as a quasi-species). Picture a group of like-minded thieves who all share the same objective of breaking into a bank vault. Instead of fighting against each other, every viral variant represents a fresh attempt by one potential thief at trying out a new combination on the safe's lock until at last one of them breaks the code. Once inside, the successful virus reproduces in massive numbers, overwhelming the body's defences.

A few people are naturally immune to HIV. A recent study has tried to explain why about 10 per cent of the European population benefit from a special type of mutation that is able to combat the spread of HIV. Research indicates that this mutation is present in human populations that suffered outbreaks of bubonic plague in the medieval period between *c.* AD 1000 and 1800. Viral haemorrhagic fever is first thought to have appeared about 2,500 years ago but infections became pandemic in Europe only since AD 1000, peaking with the Black Death in the mid fourteenth century. Unfortunately today's human population in Sub-Saharan Africa has no history of medieval bubonic plague. This infection has therefore gained its strongest grip in a place where natural human immunity is weakest. Experts also believe that for billions of years viruses like HIV may have been instrumental in triggering the appearance of new species.

It is interesting to speculate on what would happen to human populations in Africa in the absence of modern social, educational or technological intervention. Eventually the entire continent would become infected with HIV and all except for the few people with natural immunity would perish. It would be left to these immune survivors to repopulate the continent. But the new population would differ from the old one in genetic makeup by having a benign HIV signature in their genes, a signature that would effectively have been added to the ranks of these people's 'junk' DNA (see pages 13–14). Since human populations on other continents would not be able to breed with surviving African humans without themselves becoming infected, at least two isolated populations of humans would emerge – those with HIV in Africa and those without HIV in the rest of the world. The inability to interbreed between these populations means they would eventually 'evolve' into different human species.

Only about 1 per cent of African adults have so far been tested for HIV, so true infection rates are still largely unknown, making its future impact impossible to predict. In the absence of a vaccine or other cure, the disease continues to accelerate, devastating the social and economic lives of millions, the majority of whom live in the poorest parts of Africa and India. In 2007 an estimated 2.1 million humans died of AIDS, 330,000 of them children under the age of fifteen. South Africa has more than one million orphaned children, themselves HIV infected, thanks to the passage of the virus from mother to child via breast milk.

Meanwhile, in neighbouring Botswana average human life expectancy has plummeted from sixty-five years in 1988 to thirty-five years in 2007. Despite advances in modern medicine, the impact of HIV infection on Sub-Saharan Africa today puts it squarely alongside history's most devastating viral pandemics: avian influenza (see page 16) and smallpox (see page 24).

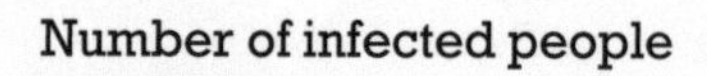
Number of infected people

1,000,000+

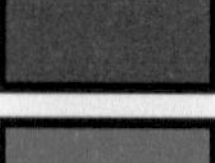
500,000–1,000,000

100,000–500,000

50,000–100,000
10,000–50,000
–10,000
No data available

The Spread of HIV/AIDS

How a virus that changes people's genomes is ripping through the world's human population.

Source: United Nations 2008 report on the global AIDS epidemic

Potyvirus

FAMILY: POTYVIRIDAE
SPECIES: POTYVIRUS SP.
RANK: 100

How a plant virus caused the first stock-market crash.

PLANTS are just as susceptible to infection by viruses as are humans, animals, fungi or bacteria. Less well known is how one virus had a significant impact on the birth of botany, horticulture and capitalism in the history of early-modern Europe.

Potyviridae is the largest of the thirty-four known plant virus families. It comprises about 180 members that together account for 30 per cent of all recorded plant viruses. These parasites are not experts in sneakily copying their own genetic code into the DNA of their hosts – nor do they spread by direct contact between one plant and another. Instead, these viruses employ the same technique for their reproduction as many plants use for pollination – they hitch a lift from a flying passer-by. The potyvirus uses the transportation of a species of insect called the green peach aphid (*Myzus persicae*).

Scuttling from plant to plant, these small creatures have transported potyviruses all over the world. They ride on the back of winds that blow them hundreds of miles across land and sea, and there is not a single region of the world that is not now affected by the viral infection they carry. Plants most at risk from damage include potatoes, turnips and plums and in most cases, once infected, their fruit or tubular production becomes seriously impaired.

However, Potyviridae's biggest impact on human history comes from one variant that ably demonstrates how viral infection can sometimes turn out to be highly beneficial for the survival prospects of its host species.

Tulips originated in Central Asia. They grow naturally in the mountainous regions of Turkey, Iran and Afghanistan. Long admired by Islamic sultans, these flowers were brought to Europe in the late sixteenth century by a Flemish doctor called Carolus Clusius (1526–1609) who was also responsible for running the imperial medicinal garden in Vienna. In 1593, when he moved to Leiden, southern Holland, he took with him the bulbs of some especially attractive tulips.

The name 'tulip' has the same origin as the Turkish word for turban. Not sweet-smelling, good to eat, nor useful as herbal medicines, these exotic plants had only one quality to attract the European merchants and travellers who took them home. As if by magic, some of the plants flowered with a highly distinctive flaming pattern that unfurled across their petals like marbled paper. No one knew why most tulips stayed a single, monotonous shade while a precious few broke into such a feast of whimsical streaks and stripes. The powerful allure of these exotic varieties among Dutch traders meant that, for a while, tulips became the most sought-after, valuable commodity money could buy.

So feverish did the Dutch appetite for tulips become that contracts for flowers due to be harvested in the autumn of 1635 were sold for more than the price of a house. The most celebrated was the vivid variety *Semper Augustus*, painted in all its majesty by famous Dutch artists of the day. A dozen or so samples owned by one tulip collector, Dr Adriaen Pauw, accounted for almost the entire world's supply, so rare was this stunning beauty. Despite numerous offers throughout the 1620s, Pauw refused to sell his collection at any price. When a single *Semper Augustus* bulb finally came up for auction, it sold for a record 6,000 florins (about £1 million today).

Prices like these were driven by scarcity. Even though tulip flowers are capable of producing up to 200 seeds a year, they hardly ever pass on their striking looks to their offspring (for an explanation see apple, page 344). The only way to ensure a further crop of similarly patterned plants is to take off-cuts from the original plant (a form of cloning) and, annoyingly for Dutch growers, it seemed that the more beautiful the tulip, the weaker and less numerous its bulbs from which off-cuts could be cultivated, partly explaining why prices shot through the roof.

It wasn't until the invention of the electron microscope in the 1920s that the cause of the tulip mutation that makes them develop stunning stripes and whorls was understood to be the work of a viral infection. Scientists discovered that the virus prevented the production of certain petal pigments, causing its colour to 'break'. Only then was the mystery of why some tulips had such stunning looks resolved, and the link between virus, aphid and tulip finally pieced together.

Semper Augustus, the marbled tulip, as painted by seventeenth-century Dutch artist Pieter Holsteijn II.

Modern consumer marketers could not have done a more impressive job than did this virus in boosting its host's worldwide appeal. By playing to humankind's inherent attraction towards brightly coloured, intricately variegated exotica (see On Beauty, page 336), this parasitical infection inadvertently caused its host tulips briefly to become the most likely plants to be cultivated and traded by humans. And, in a masterstroke of scarcity-marketing genius, the smaller number of bulbs produced by its host meant the value of infected tulips, driven by unsatisfied demand, increased beyond all reasonable measure.

The tulip speculators' bubble finally burst in the spring of 1637. Gamblers who had promised to pay thousands of florins for that year's harvest were left with nothing and those who had bought tulips the previous year for vast sums saw their fortunes evaporate overnight. Some historians claim it took years for the Dutch economy to recover from 'Tulip-mania'. Others say its impact has been over-hyped. From the 1920s onwards the virus was systematically eradicated by Dutch farmers anxious to increase the yields of their tulip crop. By producing hybrids of other species, similar visual effects to those originally produced by the virus could be cultivated for less cost, time and effort.

Yet it is a fascinating and significant historical fact that a plant virus was instrumental in creating the first ever speculative financial crash. 'Tulip mania' was an early forerunner of the 'South Sea Bubble', and more recently the 'dot-com boom' and 'sub-prime mortgage' crisis. More significantly, though, this virus-induced episode firmly established the Netherlands as a world centre of botany and horticulture and shows, however inadvertently, how a virus can sometimes help the survival prospects of its host species. Today, more than *three billion* tulip bulbs are exported every year from Holland and the tulip, largely thanks to some truly brilliant and entirely natural tactics in viral marketing, has become one of the world's most highly propagated and exported flowers.

Smallpox

FAMILY: POXVIRIDAE
SPECIES: VARIOLA VERA
RANK: 63

A highly infectious virus that fundamentally altered the course of human and evolutionary history.

JANET PARKER was an exceptionally unlucky woman. Her tragic death on 11 September 1978 marked the passing of a long and inauspicious era in the history of human disease. Only six weeks earlier, she had been infected by a rogue strain of the smallpox virus *Variola major* that had escaped through an air vent from an unregulated research laboratory beneath her photographic studio in a university campus building in Birmingham, England. Parker, aged forty, was the last victim of the smallpox virus. Within only two years of her death, the World Health Organization officially declared the disease extinct in nature – the first and only case of a natural infection being totally eradicated by collective human endeavour.

Despite its recent extinction, smallpox has been one of humanity's most lethal viral killers. This disease has had a significant impact on the course of human history on every continent of the world. The disease passes easily through the air between one human and another whenever they are about two metres or less apart. Typically, about 30 per cent of those infected subsequently die, although in some outbreaks the rate increases to nearer 100 per cent. Cumulatively this disease has been responsible for the deaths of hundreds of millions, if not billions, of people.

Variola infects only humans. It began as a mutant strain that originated in another member of the Poxviridae family, cowpox (*Vaccinia*). Cowpox and monkeypox are non-lethal infections that can affect both humans and other animals. At some point after humans started keeping farm animals, about 10,000 years ago, a variant jumped the species barrier from cows to humans and evolved into a lethal new strain.

Poxviridae do not hijack the nucleus of a host's cell to replicate. Instead, they use their own double-stranded DNA as copying machinery, only helping themselves to their host cell's cytoplasm as a source of body-building protein. This unique behaviour has led some virologists to hypothesize that at a very early stage in the evolution of life on Earth a pox-like virus invaded simple bacterial cells (Prokaryotes), seeding them with a nucleus of DNA. If so, the effect that the ancient ancestors of the smallpox virus had on life on Earth were monumental. Their vital impact may therefore have been to provide the essential DNA genetic copying machinery that has made all forms of higher life possible (see page 45).

Death by smallpox was extremely unpleasant. Signs of trouble emerged about twelve days after infection. Huge numbers of infected viral particles then spread through the bloodstream to all parts of the victim's body. High fever, muscle pain, headaches and vomiting gave way to a pernicious rash, usually inside the mouth, on the tongue, down the throat and all over the face. Rupturing blisters oozed copious doses of viral infection into the saliva. Two days later the rash spread all over the body, usually bursting out into large leaking pustules that eventually scabbed over. If the disease itself failed to kill its host, other diseases such as bronchitis or pneumonia often took over. The most fortunate survivors – like Queen Elizabeth I of England – were usually left with lifelong scars (in her case covered up with thick dollops of white lead make-up). The less fortunate survivors went permanently blind.

Understanding the enormous impact that this disease has had on human society is only possible where good historical records survive. One of the most ancient accounts is from the Greek historian Thucydides who described a 'Plague of Athens' that decimated the city's population in 430 BC. From his account it seems almost certain that a smallpox epidemic contributed significantly to the Spartan victory in the famous Peloponnesian Wars (431–404 BC).

By 250 BC clear evidence emerges of smallpox infection in China, possibly spread by Central Asian Huns. Frequent outbreaks followed in succeeding centuries and by the sixth century it had spread from Korea to Japan via Buddhist monks. Urbanization in Japan from 710 (with the building of the imperial capital of Nara) helped the disease spread, where it wiped out the ruling Fujiwara family and, from *c.* 735, became endemic (permanent in the society). One result was that Japan's Buddhist religious fervour grew more intense in a desperate attempt to relieve the Empire of this appalling pestilence.

In Africa the disease long plagued the dynasties of ancient pharaohs, claiming the life of Ramesses V in 1157 BC, whose mummy has recently been examined revealing a rash of raised pustules. It spread from there to the Middle East via the Hittite empire in Turkey. By *c.* AD 550 a mostly Christian army from Ethiopia invaded Yemen, reaching as far as Mecca in AD 570 – the year of the birth of the Muslim Prophet Mohammed. Then smallpox struck. The vastly outnumbered but infection-free Arabs were easily able to fight off the diseased Christians in a conflict known as the Elephant War. As Mohammed's warriors galloped unimpeded, north, east and west, the disease spread further. What, if anything, may have become of Islam had this disease not struck those early Christians when and where it did? The virus travelled with Muslim warriors who pushed on into Spain, through the Middle East and even across the hot, dusty Saharan deserts into the plains of West Africa.

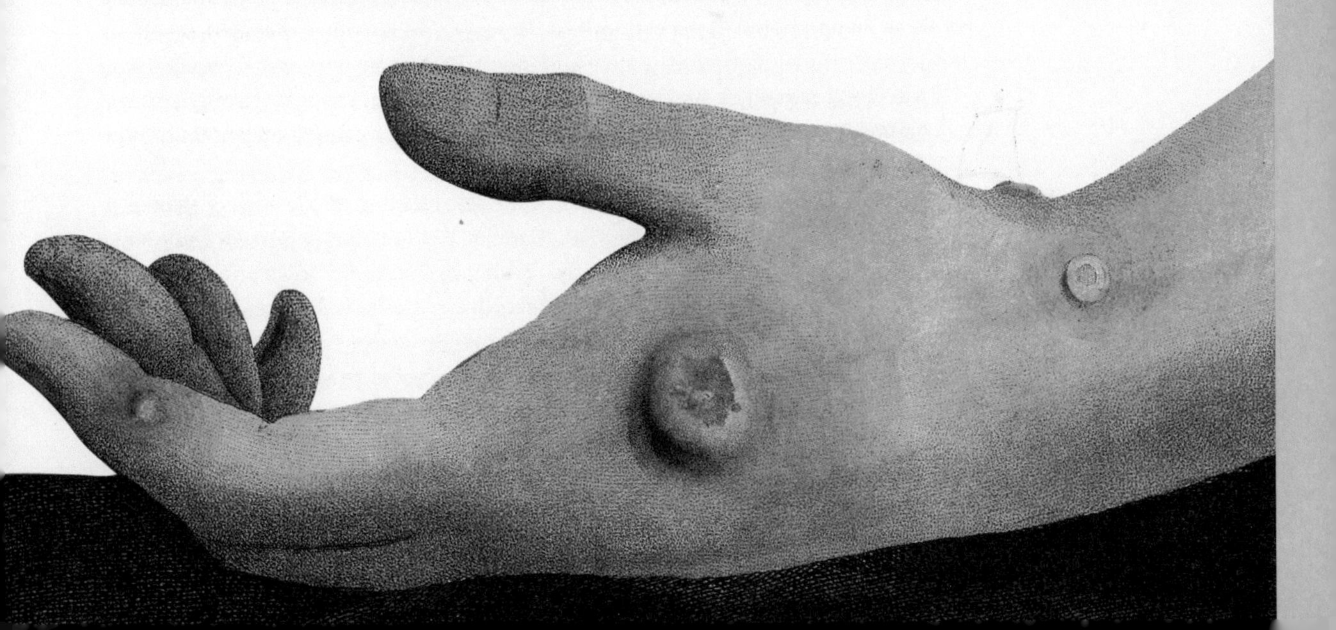

Pustulating smallpox sores on a human hand, etched by British artist William Thomas Strutt (1777–1850).

Smallpox had its greatest impact several hundred years later, carried by sixteenth-century newcomers to the Americas – the Spanish conquistadors. They were soon followed by West African slaves, shipped across the Atlantic to work in the plantations of the New World. Both waves of immigrants brought smallpox to the Americas, virtually annihilating the Native American populations who had no immunity to the disease because, traditionally, none had lived in close proximity to farm animals.

It is thought that a single African slave in Spanish invader Hernán Cortés's entourage brought the disease to Mexico in 1521. Within a year the Aztec Empire and its glorious capital city of Tenochtitlan had been subjugated by an army of no more than 500 Spanish soldiers – almost entirely because disease had killed so many natives. According to one observer the Indians '*died in heaps like bed-bugs*'. From an estimated population of about thirty million before the Spanish arrived, the disease systematically reduced the population of Mexico to a low point of about 1.6 million just one hundred years later.

Further south, the Inca 'Empire of the Sun' collapsed for the same reason. Smallpox killed its emperors in quick succession (Huayna Capac and then his designated heir Ninan Cuyochi) throwing the empire into a desperate five-year civil war. So weakened were these people by war and disease that cunning Spanish opportunist Francisco Pizarro successfully conquered them with just over one hundred men, a handful of horses and few primitive cannons. Spanish rule in the Americas only exacerbated the disease with the introduction of more slaves from Africa who brought with them new waves of fresh infection. Jesuit missionaries who rounded up natives into camps unwittingly facilitated the spread of the virus among the natives who, in the words of one historian, *'might as well have been in a slaughter house'*.

One hundred years after the fall of the native Central and South American empires, the disease struck North America, this time spread by French settlers in Nova Scotia. An epidemic starting in 1617 wiped out much of the Native American population along the east coast within just a few years. By the time the Pilgrim Fathers arrived from England on the *Mayflower* in 1621 *nine-tenths* of the indigenous east coast population had died. Not that the European setters were totally immune; about twenty of those on the *Mayflower* also succumbed to the disease.

Thanks to the domestication of farm animals, smallpox had a clear preference for some human populations over others. However, unlike other diseases, smallpox never discriminated between rich and poor. European imperial dynasties and royal families such as the Austrian Habsburgs and English Stuarts both had royal claims (Spanish and British respectively) shattered by the deaths of their heirs from smallpox.

The first glimmer of a human fightback came in 1717, largely thanks to an enlightened English feminist. While living in Turkey, Lady Mary Montagu, wife of the British ambassador to Istanbul, wrote a series of letters describing life in and around the court of the Ottoman Emperor. In one of them she described a method of 'inoculation' against smallpox. It is thought to have been pioneered in ancient India under the auspices of Hindu mystics who had their own goddess of smallpox, *Shitala Mata*. The process involved some powdered scab taken from a previously infected person who had recovered from the less virulent form of the disease, *Variola minor*. This was inserted into a small incision made in the skin of an uninfected person, provoking a mild form of the disease. Usually the process provided lifelong immunity, although it wasn't without risk. Occasionally such inoculation resulted in the full-blown disease and, since it involved transmission of the live smallpox virus, these treatments were also extremely dangerous to administer. But Lady Montagu was determined to lead the way and by 1721 she had both her children inoculated, the first such procedure that was ever performed in England.

Inoculation against the smallpox virus, although not a perfect remedy, had its own major impact on human history. Later that century, George Washington had all the troops under his command inoculated during the American Revolutionary War (1775–83), a decision that was instrumental in keeping America's dreams of independence alive: *'I know that it is more destructive to an army in the natural way than is the sword'* wrote Washington to the Governor of Virginia in 1777. American aspirations to wrest Canada from British control were also scuppered by the disease – as John Adams, a future US President, recalled: *'The smallpox is ten times more terrible than the British, Canadians and the Indians together. This was the cause of our precipitate retreat from Quebec.'*

Lady Mary Montagu was Europe's earliest advocate of smallpox inoculation.

While inoculation provided some protection from the disease, it wasn't until 1796 that a genuinely safe process was discovered by which immunity could be conferred without risk to doctor or patient. By injecting pus from a cowpox blister into a healthy human, Edward Jenner, a doctor from Gloucestershire, England, found a new method of immunization. He called it vaccination (from the Latin *vacca* for cow). Cowpox causes a mild and temporary skin infection in humans that the body's immune system has no problem controlling. If someone who has already had cowpox is later infected by smallpox, the body can successfully attack the disease because the viruses are so similar that it is tricked into thinking that the smallpox infection is a recurrence of cowpox. Immunity is almost always assured.

About 175 years passed between Jenner's discovery and the death of Janet Parker. In that short period, hundreds of millions of people continued to die from the disease, until its eventual eradication, certified in 1980. Today only two laboratories in the world are officially permitted to keep live samples of the virus for research in case of future outbreak – although to what extent other samples are held in secret by countries that were once part of the former Soviet Union remains unclear.

Despite its demise, most people alive today would be highly vulnerable to smallpox infection were it ever to recur. Since the early 1970s, mass vaccination programmes have become too costly to be worth bothering about, so low is the risk of infection thanks to human herd immunity. Most people below the age of thirty-five have never been vaccinated while for others the vaccine's efficacy will have worn off after approximately ten years. A biological weapon, in the form of a new smallpox outbreak, is now reckoned to be one of the most lethal and plausible bio-terrorist options. Extinct as this virus is now in nature, for humans at least, its past and future potential impact remains as awesome as ever.

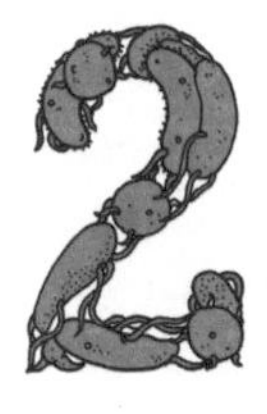

On Simple Cells

How versatile single living cells established new patterns of evolutionary behaviour, filling every available niche with life.

SOME PEOPLE dream of seeing a new world. The story of Dorothy, whisked away by a tornado into the wonderland of Oz, has enthralled children and adults for over a hundred years. More recently, TV addicts have been able to digest more than 700 episodes and ten feature films following the adventures of Captain Kirk and his intrepid crew on board the Starship Enterprise, fearless explorers charged with boldly going where no man has gone before. But fantasy and science fiction aside, the man who made history's most remarkable discovery of a new world actually lived in Holland nearly 400 years ago.

Antonie van Leeuwenhoek (1632–1723) was a Dutch draper who pioneered a way for humans to see a world they previously had no idea existed. His lifelong passion was to craft glass lenses into magnifying instruments that could be used to view objects otherwise too small to see.

Van Leeuwenhoek constructed hundreds of microscopes during his lifetime, some of which are thought to have been capable of magnification of up to 500 times.

What he saw as he peered through his lenses at everything from spit to sperm triggered a complete transformation in our understanding of life on Earth. Far beyond the gaze of the naked human eye, van Leeuwenhoek discovered a new world of microscopic life-forms ranging from bacteria to blood cells. So bizarre and so small were these tiny spectacles, like swimming fish, that he called them *animalcules*.

In his excitement, van Leeuwenhoek sent numerous letters to the English Royal Society, which began publishing them in 1673. One letter (17 September 1683) describes his observations on the sticky stuff called plaque gouged out from between his teeth (and those of two of his lady friends) which he then analysed under one of his many instruments:

'I saw, with great wonder, that in the said matter there were many very little living

animalcules, very prettily a-moving. The biggest sort … had a very strong and swift motion, and shot through the water (or spittle) like a pike does … The second sort oft-times spun round like a top … and these were far more in number … Moreover, the other animalcules were in such enormous numbers, that all the water seemed to be alive.'

What van Leeuwenhoek saw was humanity's first view of formerly invisible organisms, tiny creatures that transformed evolution by developing living cells – the building blocks of life as we know it today, the stuff of you and me.

Life's first citadels

The emergence of life in its conventional form of organisms that harbour their own living cells had a massive impact not just on the evolution of other living things, but also on the overall composition of the Earth's biosphere – the land, sea and sky.

Living cells are like mini castles – microscopic power stations in which replicators (genes) can manufacture energy and make copies of themselves from within the privacy and comparative safety of their own walls. While not immune from attack by other rogue genetic molecules (such as viruses), cellular citadels have become the basic building blocks of life. The most ancient and simplest types of living cells still thrive today and represent the earliest recognized kingdom of living things. They are called Prokaryotes.

Arguments rage over whether or not all cellular life is descended from a specific ancestral cell that is thought to have emerged in the seas sometime about 3.5 billion years ago. Scientists call her LUCA (short for Last Universal Common Ancestor) and some hold on to the notion that this first, original single-cellular being is like an ancient seed, and that all life today is in some way distantly related to her.

In reality, there is absolutely no reason why there may not have been several LUCAs appearing in different places and employing different cell designs in a competitive bid to create energy, mobility and defence. Indeed, the idea that cellular life may have begun independently several times was given a big boost about thirty years ago after it was discovered that bacteria, which were thought to comprise the only members of the Prokaryote domain, are far from alone. Recently, as a result of DNA sequencing, other types of single-celled organisms have been found and placed in a separate domain, called Archaea. What's more, when another single-celled entity called the mimivirus was recently sequenced, its unique genetic structure gave rise to the idea that it may be the living descendant of a third single-celled domain, about which scientists currently know very little.

While it is still impossible to prove whether or not all life today is derived from a single ancient cell, a clearer picture of how evolution was transformed from primitive virus-like replicators to independently mobile, well-defended living cells is at last beginning to emerge.

A few steps on from the pioneering instruments used by van Leeuwenhoek, advanced electron microscopes today allow scientists to peer so deeply into the microscopic world that they can observe an object in incredible detail down to the level of a single atom. It is entirely thanks to such instruments

Antonie van Leeuwenhoek, pioneer of microscopes and discoverer of previously invisible life (left) and his own imaginative hand-drawings of human sperm (below).

fig: 2.

fig: 3.

fig: 4

that Professor Helen Hansma and her team at the University of California have recently proposed the most compelling theory yet of how the first living cells evolved. According to Hansma, cellular life began in a kind of embrace between the Earth and primitive genetic replicators right where it seems most natural – between the sheets.

Mica is an extremely thin, delicate and glittery type of silicon rock found mostly in India, China, the eastern USA and Canada. It is so intricately assembled through natural crystallization that in a piece just one millimetre thick there are as many as *a million separate sheets* of silicate layered one on top of another, each one separated from the next by a tiny microscopic space.

Hansma has spent most of her professional life using sheets of mica as a flat surface on which to deposit materials for analysis under an atomic force microscope (AFM). One morning in 2007, a wave of curiosity overcame her when she placed a fresh piece of mica under her scope.

As Hansma peered through her eyepiece she could see that the mica sheet was slightly soiled with organic material. DNA molecules had been ingrained into tiny ridges on the rock's surface.

> *'As I was looking at the organic crud, it occurred to me that this would be the perfect place for life to originate – between these sheets that can move up and down as temperatures rise and fall and that are permeable to sea water tides – it was a perfect environment for providing the mechanical energy needed to make and break chemical bonds.'*

Like evolutionary incubators, mica sheets may have provided just the right structures within which cellular life could evolve. They also happen to be rich in life-enhancing chemicals such as oxygen, potassium, magnesium and iron. When bathed in sea water, the additional shot of sodium would complete the range of ingredients necessary for the construction of cellular life.

Hansma can see the image clearly in her mind:

> *'I see all the molecules of life evolving and re-arranging among mica sheets before budding off with cell membranes and spreading out to populate the world.'*

Amongst the earliest single-celled organisms were life-forms that learned to feed near nutrient-rich volcanic vents under the seas. They are called methanogens and represent the earliest-known forms of archaea.

Now able to convert energy from one form to another from inside their walls, some of these cells developed their own distinctive forms of propulsion. An extraordinary array of microscopic tails, wheels, cogs, screws and propellers emerged that still powers many of the world's bacteria and archaea from one place to another. In a relentless search for food, these motile structures, called flagella, can rotate at speeds of up to 1,000 revolutions per minute. They are also unique in being the only occurrence of wheels found in nature.

Natural extreme machines

What possible major impacts could invisible organisms like these have on the big picture of evolution in which creatures like fish, plants, trees, animals and humans seem to dominate?

For a start, these creatures have extraordinary survival skills, regardless of what nature throws at them or however difficult the living conditions.

Plants and animals are limited to just two basic modes of energy production (respiration and

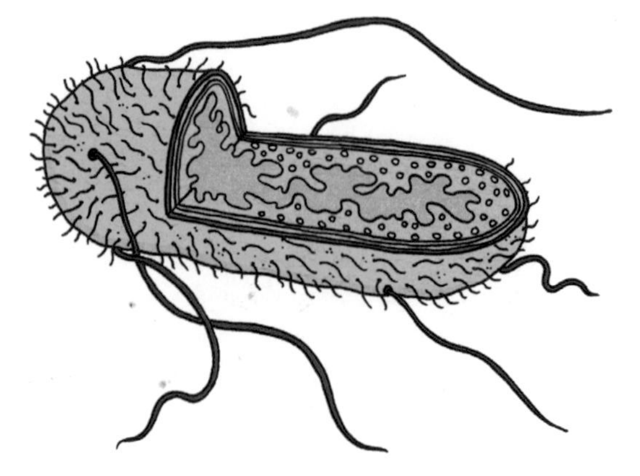

Cross-section through a Prokaryote with its cell wall (shaded green). Inside reside loose strands of genetic DNA (pink) surrounded by cytoplasm (turquoise). String-like flagella (red) are attached to the cell wall.

photosynthesis), yet Prokaryotic bacteria boast at least twenty, allowing them to feed on molecules as diverse as sulphur, carbon dioxide, nitrogen, water, hydrogen, oxygen, iron, ammonia, methane and sugar. Diverse tastes in food have made these creatures by far the most robust in nature, allowing them to thrive in places where others fear to tread. Modern scientists have been astonished by their resilience.

When experts examined a sample of water taken from a stagnant pool deep inside a South African gold mine, they didn't expect to find any signs of life. In this place, over three kilometres below the Earth's surface, ingredients normally essential for sustaining life such as natural light and oxygen have been absent for millions of years. But here communities of single-celled, endolithic (within rocks) bacteria have been reproducing without interruption for aeons. *Desulforudis audaxviator* has adapted to the extreme nature of its habitat by living on hydrogen, cast off as a waste gas from the process of natural radioactive decay.

No less astonishing is the story of creatures that thrive beneath the impossibly cold conditions of Antarctica. Ace Lake is located in the Vestfold Hills near the Antarctic Research station of Davis, built by the Australian government in 1957. Here, pools of trapped sea water have remained undisturbed since ancient glaciers gradually withdrew after the last great ice retreated more than 12,000 years ago. When curious scientists scooped out their first samples of water, taken from the freezing lake bed twenty metres below the surface, much to their astonishment they discovered thriving communities of methane-producing microscopic archaea. Research continues into how these species (*Methanogenium frigidum* and *Methanococcoides burtonii*) can survive such bitterly cold temperatures in an environment totally devoid of the three conditions that most life requires: heat, oxygen and light.

Such resilience helps to explain how life survived at least two highly dangerous epochs in geological time when the Earth was plunged into such a severe ice age (each time lasting many millions of years) that it resembled a giant snowball (see also page 65). These cryogenic (ice-creating) eras were caused, it is thought, by the arrangement of the Earth's tectonic plates. If at some point in the future the Earth was plunged into such a deep freeze again, these bacteria from the Antarctic lakes would be sure to continue the processes of evolution whatever else happened to other life during such a time of such environmental stress.

At the other extreme, some Prokaryotes are most at home living in temperatures found in a boiling kettle. Anything higher than a hundred degrees centigrade is just fine for the aptly named *Pyrococcus furiosus*, a species of hyperthermophile archaea that feeds on naturally occurring sulphur in geothermal geysers and hot-underwater marine sediments. No wonder the collective name for these super-tolerant micro-organisms is 'extremophile' – lovers of extreme conditions.

Numerous varieties of Prokaryote life-forms have been found in the Atacama Desert, in Chile, one of the driest places on Earth where it rains only once in every twenty to fifty years (see page 40).

Perhaps nowhere could be more extreme than living in the human stomach – an environment so fiercely acidic that until recently it was thought that no organism could possibly last long there. Yet again, single-celled Prokaryotic life has confounded all expectations. As recently as 1981 scientists began to realize that stomach ulcers were not generally caused by stress or spicy foods, as was widely believed, but by an infectious bacterium called *Helicobacter pylori* that makes its living by using its six flagella to drill into the lining of the human stomach where it can feed on partly digested food.

Nature's 'Terminator'

It would be wrong to think that there are no conditions in which Prokaryotes cannot survive – a complete lack of any suitable food supplies, total desiccation or excessive ultraviolet radiation must eventually affect all forms of life. Still, some Prokaryotes have found an ingenious way of ensuring they never have to give up their zest for survival – even in the event that conditions get too tough. Firmicutes are bacteria that can construct emergency nuclear shelters designed to keep them safely out of harm's way.

The process – called sporulation – takes only about eight hours to complete from start to finish. A bacterium, such as **anthrax** (see page 36), which is often fatal when inhaled by humans, makes a copy of its all-important genetic code, which it then surrounds with a new cell wall that is encased in a toughened outer coat. Finally, after separating from its dying parent, this new entity, called an *endospore*, is designed to be blown about by the wind until it eventually settles in a place conducive to new life.

Boiling, cooking, baking, freezing, starving and x-raying won't harm these almost indestructible spores. Even total immersion in highly corrosive chemical agents or being blasted into the absolute cold and vacuum of outer space usually makes little difference. Barring total conflagration, all the obvious strategies used to sterilize life are ineffectual against such persistent survival machines. When environmental conditions finally become favourable, these dormant life-forms effortlessly reactivate themselves as fully-functioning bacterial cells, once more able to reproduce and continue life afresh.

Awareness of Prokaryotic hardiness has recently led to a renewed interest in the search for extraterrestrial life. Astrobiology is a branch of science not geared towards the search for a man in the moon or little green Martians. Rather, it is on the lookout for methane- or carbon-dioxide-rich atmospheres on other planets or moons that could conceivably support the survival of microscopic life. Microbes that live on Earth are so robust that it is believed some could survive (or may once have lived) on planets as apparently hostile and dead to higher forms of life as Mars or Venus – or even as far as the moons of Jupiter. Space missions to find out more are already under way.

The impact of bacteria

Prokaryotes called **cyanobacteria** (see page 34) are microbes that, beginning about 2.5 billion years ago, pioneered a radical new way of life. They learned to use carbon dioxide in the atmosphere, water and the energy in sunlight to make food. During this process, called photosynthesis, a high-energy waste gas, oxygen, was produced that literally changed the world because without it the evolution of all plants and animals – which rely on supplies of freely available oxygen – could never have taken place.

The impact of other microscopic Prokaryotes on land-life has been just as profound. Plants that humans depend on – legumes such as peas and beans, for example – have co-evolved with bacteria from a group called ***Rhizobia*** (see page 40). These are microbes that live within nodules on plant roots, performing a critical function converting or 'fixing' nitrogen from the atmosphere into chemicals such as ammonium, essential for plant growth.

Other bacteria in the soil underpin the ongoing organic cycle of birth, life, death and renewal. It has been estimated that there are more than *forty million* bacterial cells in any one gram of soil. *Bacillus thuringiensis*, a close cousin of deadly anthrax, is a soil-dwelling bacterium used today as a natural pesticide whose endospores kill moths, butterflies, flies, mosquitoes and beetles.

Harnessing the power of microscopic nature to help solve current environmental issues on Earth gives Prokaryotes a potentially vital role in the future. **Pseudomonas** (see page 38) is a soil-dwelling bacterium that has the potential to turn industrial pollution into biodegradable plastic. Meanwhile, methanogenic archaea already play an essential part in the world's increasingly important renewable energy mix. These organisms are key ingredients in anaerobic digesters that convert organic waste into methane gas – fuel that can be used for everything from powering cars to cooking food and heating homes.

But these are all *indirect* impacts, changing the parameters for other life by stirring the Earth's mixture of organic materials through the production of oxygen, nitrates or methane gas. Just as important are impacts that directly diverted the course of evolution itself.

Life's first communities

Prokaryotes built the world's first communities. Biofilms are thin but visible bacterial mats that hang about on the surface of stagnant water or are artificially cultivated in sewage treatment works to help break down human waste. They comprise trillions of Prokaryotes that have formed into tightly knit groups for mutual benefit.

Research shows that many bacteria are at a distinct survival advantage when they link up. Several human bacterial infections, such as cystic fibrosis pneumonia, are caused by biofilms that are resistant to traditional antibiotic treatments and to the immune system.

A group of organisms can only be stronger than the sum of its parts if co-operation exists between its individual members – a condition which requires some form of communication. The world's first de-centralized decision-making process can be traced back to these primitive living communities in the Prokaryote kingdom. Bacteria that live together in tightly packed groups use chemicals, called pheromones, to help assemble and then assess the size of their community. Upon reaching a critical mass, new messages are passed along in a chain from one individual to another inducing each cell to start manufacturing chemicals that favour the success of the entire group. This phenomenon is called 'quorum sensing'.

Vibrio fischeri is a highly community-minded bacteria whose group behaviour also helps out its host organism – the Hawaiian bobtail squid – in a most extraordinary way. Individual bacteria pass along chemical messages causing the community to secrete a special substance that shines in the dark, lighting up the surroundings so that the squid can more easily locate food in the pitch black depths of the sea. Social species, such as ants, bees and termites, that evolved hundreds of millions of years later, have adopted the same basic system for group communication and decision-making (see also slime mould, page 50).

Why should these bacteria bother wasting energy making a luminous chemical for a deep sea squid? Such mutual collaboration means that when the bobtail squid finds food, the bacteria also gain a meal as it is passed through its host's gut. Scientists called this two-way relationship *symbiosis*.

Evolutionary biologists have spilled plenty of ink in recent years trying to prove that, in the wild, all species act only according to the individual selfish interests of their genes in an unfettered competitive battle for the survival of the fittest. Less well appreciated is how normal it is for one species to align itself with the interests of another. Even more surprising is how often such co-operative 'symbiotic' behaviour has profoundly altered the course of evolution (see On Symbiosis, page 43).

The ability to build intimate relationships between different species was a life-skill pioneered by single-celled Prokaryotes. Some prove beneficial to both parties (mutual); others benefit one side but harm the other (parasitic), see page 129; and some benefit one party more than the other, but cause no serious damage (commensal). From these basic beginnings, relationships between all living things began to evolve along a spectrum – from occasional helpfulness to such tightly bound co-operation that neither party could survive without the help of the other (a type of relationship known as obligate). Occasionally, what were once separate species would even fuse together at the genetic level through mutual necessity and new species would arise (for an example, see page 64).

Since the dawn of cellular life, symbiosis pioneered by Prokaryotes has become such a significant agent of evolutionary change that in the words of one leading expert it *'plays the major creative role in the genesis of new species'*.

Cyanobacteria

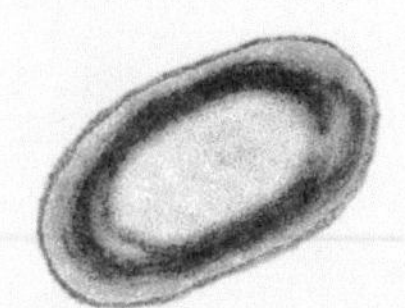

FAMILY: SYNECHOCOCCACEA
SPECIES: PROCHLOROCOCCUS MARINUS
RANK: 3

A photosynthesizing bacterium that provided the essential breath of life.

IMAGINE A WORLD with no plants, trees or animals. The land would be a barren blend of rock and silt, the seas would be bereft of fish and the air silent except for the wind. This is a world devoid of a vital ingredient on which almost all forms of 'higher' life – fish, birds, insects, plants and animals – depend. It is a world without oxygen.

Until about fifty years ago no one imagined our planet Earth as a place without oxygen because throughout recorded human history (about 10,000 years) the amount of oxygen in the air has remained constant at a steady 21 per cent. Now we know differently. The Earth's first atmosphere probably contained only trace supplies of oxygen that quickly disappeared from the atmosphere after reacting with other elements, such as iron, to form rocky ores of iron oxide. Over the last few thousands years these rocks have been extensively mined by humans to extract the iron for crafting tools, making weapons and building shelters.

Plants, trees and animals all owe their existence to the presence of oxygen in the atmosphere and oceans, supplies of which were originally established by a certain type of Prokaryotic bacterium. Indeed, these single-celled life-forms have excreted so much oxygen that surplus supplies have accumulated in the air. The culprits were cyanobacteria – so called because of their blue-green (cyan) colour – which evolved a feeding process now called photosynthesis. They break down carbon dioxide into its chemical constituents, carbon and oxygen, using sunlight, and combine the carbon with hydrogen in water to produce sugars, giving off oxygen as a waste gas in the process.

Prochlorococcus, a genus of cyanobacteria alive today, is thought to be responsible for as much as *20 per cent of all global oxygen supplies* and is reckoned by some experts to be the most populous creature on Earth with as many as one hundred octillion (10^{29}) individuals alive in the seas.

Stromatolites are rock-like structures that jut up along the shoreline in places like Shark Bay in western Australia. They are sticky to touch because their surfaces are covered in colonies of slimy cyanobacteria that absorb sunlight and give off oxygen. Although they are quite rare today, the world was once peppered with these structures, which first appeared about 2.8 billion years ago, contributing to the massive injection of oxygen into the Earth's atmosphere.

Most microscopic life that thrived in the ancient seas avoided all contact with oxygen as its reactive nature was highly toxic. Prokaryotic life that evolved in an oxygen-free world therefore either had to adapt to an oxygenated environment or find places where levels of oxygen were sufficiently low to ensure their continued survival. Some organisms adapted, learning to feed on oxygen themselves (such as the genus *Rickettsia*). Others linked up, using oxygen as an ingredient for proteins that could glue cells together, to make more complex, multi-cellular creatures that were better adapted for life in oxygenated environments (see page 46). Others went into hiding – either at the bottom of the sea, hugging close to hydrothermal vents full of sulphur-rich nutrients (such as the methanogens) or, as other life-forms evolved, deep within the guts of creatures where levels of the gas are low.

Life's original oxygen-hating Prokaryotes live on today inside the oxygen-free innards of animals such as cows and humans, where they feed on ingested food, often providing essential services in return. Colonies of bacteria that live in cows' stomachs break down cellulose, plant material that cows cannot digest themselves. In humans, more than 500 oxygen-hating species of bacteria populate our guts, some of which synthesize valuable chemicals such as vitamin K – essential for blood clotting. Others produce global-warming gases such as methane as a by-product of their own digestive processes, which is why creatures like humans and cows frequently break wind.

The rise of energy-rich oxygen presented awesome opportunities for the evolution of new types of creatures, triggering the genesis of an entirely new domain, the Eukaryotes (see page 43). All forms of higher life – from plants and fungi to fish, birds, mammals and man – are Eukaryotes, whose existence is dependent on adequate supplies of oxygen. Therefore without cyanobacteria, and the photosynthesizing process they originally established, it is hard to see how the creation of a visible, multicellular living world could ever have taken place.

Stromatolites, rock-like structures formed by the secretions of algae, that host communities of photosynthetic cyanobacteria. These are found in Shark Bay, western Australia.

Anthrax

FAMILY: BACILLACEAE
SPECIES: BACILLUS ANTHRACIS
RANK: 70

A virtually indestructible life-form that will survive as long as life itself.

BOCA RATON is an affluent city in south-east Florida. Finely manicured, automatically watered golf courses meander through its plush estates where exclusive houses and hotels are painted pink and their roofs tiled in terracotta, like Spanish villas. Its character is a cultural throwback to the first European rulers, conquistadors from Spain, who landed on Melbourne Beach on 2 April 1513, about 160 kilometres to the north. They arrived during *Pascua Florida*, the Spanish for Easter, from where today's sunshine state derives its modern name.

In September 2001, the United States of America was reeling from the shocking trauma of the 9/11 terrorist attacks in New York and Washington in which 2,974 victims died, with another twenty-four missing, presumed dead. This was a news event like no other in the history of the modern media, as Bob Stevens, an experienced sixty-three-year-old photo editor at *The Sun* newspaper, based in Boca Raton, knew all too well. What Stevens couldn't possibly know was that, just one week after Islamic terrorists ploughed two fully laden passenger planes into the World Trade Center and a third into the Pentagon, an innocent-looking letter would arrive in his newspaper's mailroom that, for Stevens, was going to prove just as fatal.

All he did was open an envelope. A mixture of brown and white endospores – looking like mixed up salt and pepper – spilled on to his keyboard and puffed into the air. Chances are Stevens thought nothing of it. Maybe he never even noticed the fine powder during what was easily the most hectic week in the history of broadcast news. Ten days later this journalist was in appalling shape. Violent vomiting and high fever were accompanied by severe disorientation and complete loss of speech. On Tuesday 2 October Stevens was admitted to hospital. By Friday he was dead.

Anthrax is a frequently fatal disease caused by the inhalation of endospores from the soil-dwelling bacterium *Bacillus anthracis.* Thanks to modern antibiotics and careful disposal of infected animals that have contracted the disease from spores in the soil, it doesn't kill so many people these days. In 2009 a drum-maker died after inhaling spores carried in imported hides. But only in malicious attacks, such as the yet-to-be-solved biological terrorism of September 2001, are large-scale incidents likely to occur. Eight letters containing spores were sent to US senators and media professionals, killing seven and seriously injuring seventeen. Unlike viral smallpox, this is a disease that cannot easily be passed from one human to another.

Bacillus anthracis matters because it was this strain of bacterium that ultimately led to the discovery by humans that microbes were the source of many deadly infections. In 1875, German medic Robert Koch (1843–1910) published a new theory as to why animals and humans sometimes suffered spontaneous death. After a series of experiments, he was able to reveal the lifestyle of *Bacillus anthracis*. Too much exposure to today's highly oxygenated atmosphere causes this bacterium to become distressed, triggering the process of sporulation in which it shields itself in an almost indestructible microscopic shell that can survive in the ground or air for thousands of years (see pages 31–2).

Koch's experiments showed that ingestion or inhalation of these spores by humans or grazing animals, such as cows, horses and sheep cause this bacterium to reawaken in the more favourable, less oxygenated environment inside a host's body. Once in the lungs, the endospores are attacked and engulfed by phagocytes, cells that form a vital part of the body's immune system (see pages 49–50). But these spores are so resilient that once inside the phagocyte they begin to awaken, germinating into living bacteria that feed off the phagocytes' cytoplasm. Eventually the immune cell explodes, spreading swarms of newly fashioned bacteria throughout the host's body, infecting tissues and causing severe bleeding that rapidly leads to death. Exposed to oxygen in the air through wounds in the host's carcass, sporulation is triggered once gain, allowing a vastly increased population of *Bacillus anthracis* to lie dormant in wait, *ad infinitum*, for its next lucky break.

Having proved his germ theory, Koch and his pupils went on to identify different strains of bacteria that were found to be the cause of many lethal diseases including tuberculosis, cholera, diphtheria, typhoid, pneumonia, gonorrhoea, meningitis, leprosy, bubonic plague, tetanus and syphilis. Their research led directly to the development of vaccines and antibiotics that unleashed humanity's own campaign against microbes, dramatically improving human life expectancy. Success in this war has been a major factor behind the exponential rise in human populations from a total of two billion people in the early twentieth century to nearly seven billion individuals today – with all its associated ecological impact (see page 370).

A British scientist pictured on the toxic Scottish island of Gruinard, which the British government deliberately contaminated with anthrax during World War II.

Being able to bunker down and survive the most extreme conditions through sporulation makes *Bacillus anthracis* rank highly among organisms most likely to survive in the future, regardless of human or natural catastrophe. These microscopic spores are so difficult to destroy that decontaminating a few US government buildings after the 2001 attacks took more than five years at a total cost of $1 billion. On the Scottish island of Gruinard, used by the British military during World War II to test biological weapons based on anthrax, the clean-up took forty-eight years.

Pseudomonas

PHYLUM: PROTEOBACTERIA
SPECIES: PSEUDOMONAS PUTIDA
RANK: 94

A bacterium that promises to help solve the problem of plastic waste.

KAMILO BEACH should be one of the most picturesque places on Earth. Reaching out towards the southern tip of tropical Hawaii's Big Island, this shoreline is a natural mecca for sun-worshippers. But, today, no right-minded tourist would willingly go there because over the last fifty years it has become the most polluted beach in America, if not the world. The reason, in a word, is plastic.

In the summer of 1997, Captain Charles Moore, heir to a successful Los Angeles oil family, took the short way home after participating in a prestigious trans-Pacific yacht race. Instead of heading along the normal sea routes that track the trade winds, he motored his fifteen-metre catamaran directly through the middle of the Northern Pacific Gyre, a slowly swirling oceanic circulatory system that rotates clockwise in a huge circle from Alaska in the north down the west coast of the United States and finally across to Asia.

What Moore didn't count on was spending an entire week sailing through thousands of miles of plastic junk that has been collecting in the middle of this vortex for more than fifty years.

> *'This part of the ocean is like a giant toilet bowl that never flushes ... Waste that washes up on to Kamilo Beach goes down a foot deep. At one time there were toothbrushes, pens, cigarette lighters, plastic bottles and caps, but now it's been ground down by the weather into plastic sand.'*

These tiny plastic particles wash into the sea as microscopic fragments, with long-term health risks to both the marine food chain and other predators, like humans, who feed on fish.

After returning from his yacht race and witnessing the scale of the impact that modern man is having on the marine environment, Moore decided to devote his life and his wealth to raising awareness about the problems of plastic pollution. Researchers at the Algalita Research Foundation, established by Moore, are now exploring the full extent of 'The Great Pacific Garbage Patch', as the oceanic vortex of plastic waste off Hawaii is often called. Their latest findings indicate that the dump covers an area twice the size of Texas, and contains more than *100 million tonnes* of non-biodegradable plastic. Most of it has been washed into the sea from landfill sites by the natural process of erosion and now swirls around in a never-ending cycle. The problem is getting worse: worldwide plastic production is increasing so quickly that Algalita researchers say the Garbage Patch is growing at a runaway rate, expanding tenfold every three years.

Plastic is an artificial polymer made from crude oil that does not easily biodegrade. Man-made objects are generally designed around the needs of humans, not nature, and so plastic is durable, not degradable. It is a remarkable thought that all the plastic ever created by humans still exists – somewhere. Some of the floating junk in the Pacific Ocean can even be dated back to the first major phase of plastic production during World War II. In 2006, a piece of plastic was found in an albatross's stomach that bore a serial number tracing it to a World War II seaplane shot down in 1944. According to a United Nations Environment Programme published in July 2004, more than one million seabirds are killed by plastic debris each year and, on average, every 2.6 square kilometres of ocean contains 46,000 pieces of floating plastic junk.

What's to be done? Kevin O'Connor is a microbiologist at the University of Dublin who thinks he may have an answer to a large part of the problem. For the last four years his team of scientists has been studying the behaviour of a group of Prokaryotes called Proteobacteria and, in particular, a variety called *Pseudomonas putida.*

Bacteria are well known for their diverse taste in food (see pages 30–31). O'Connor believes that with sufficient encouragement, and in the right conditions, this species of bacterium could help solve the issue of plastic pollution by producing a form of biodegradable plastic called polyhydroxyalkonate (PHA for short), which can be used to make everything from plastic wrap and plates to knives, forks and cups. What's more, the food these microbes use to make their own form of organic plastic is styrene, a highly toxic by-product of the manufacture of synthetic rubber (see page 259).

O'Connor's ultimate goal is to cultivate, and, if necessary, genetically modify this bacterium to make the manufacturing of bio-plastics an efficient, economic process. His aim is to turn the fifty-seven million kilograms of styrene waste annually dumped into landfill sites in the United Sates into thirty-four million kilograms of biodegradable plastic – as much as Americans buy in plastic goods every year. O'Connor believes his vision could soon make economic sense. In future, companies could sell styrene instead of paying others to dispose of it, reducing the cost of bio-plastic production to about one dollar a kilogram compared with two dollars for a kilogram of artificial plastic today.

The prospect of bacteria reducing mankind's damaging impact on the natural environment is not altogether surprising. Prokaryotic organisms have underpinned the essential processes of all 'higher' forms of life (plants and animals) for hundreds of millions of years. But natural evolutionary adaptations typically take place in geological not human timescales. The dramatic addition of so much artificial waste in the last fifty years – as graphically highlighted by Moore and his researchers at Algalita – has been too quick for most bacteria naturally to adapt.

Yet, as O'Connor and his team have demonstrated, species like *Pseudomonas putida* do seem capable of performing the biological equivalent of medieval alchemy. Although their full impact lies in the uncertain future, the potential benefit they offer is awesome: saving the living world from drowning in oceans of non-biodegradable, toxic, man-made plastic junk.

Rhizobia

FAMILY: RHIZOBIACEA
SPECIES: RHIZOBIUM LEGUMINOSARUM
RANK: 4

A bacterium that largely accounts for the successful establishment of human civilizations.

BAQUEDANO railway junction sits next to a derelict mining town in the scrubs of the Atacama Desert in northern Chile. Jutting out from its old engine sheds, which surround a giant disused turntable, are the impressive hulks of several enormous steam trains, perfectly intact. They have been petrifying there, in the scorching sun, after being abandoned more than one hundred years ago.

This is like no other place on Earth. There is so little water that humidity actually reaches zero. A few poor locals still manage to get by, squatting in zinc-roofed houses built by British engineers at the end of the nineteenth century when the town was at the centre of a mighty international battle for the control of some of the world's most precious natural resources. Most coveted of all was saltpetre, a natural ingredient for manufacturing explosives and fertilizer vital to European nations such as Germany and Britain during the build-up to World War I (1914–18).

Saltpetre is the by-product of a Prokaryotic process. Over millions of years, symbiotic bacteria from the Rhizobiales order have teamed up with certain plants to convert nitrogen in the air into water-based ammonium and nitrates. These are chemicals that can be absorbed by plants to make proteins, the basic ingredients for cellular life. In return, these plants provide a high-energy diet of

Whistle-stop: the locomotive cemetery at Baquedano junction, Chile, a petrified relic of the saltpetre age.

carbohydrates for the bacteria, which is essential for the nitrogen-fixing process and their general welfare.

Plants have long since disappeared from the Atacama Desert. The land is so dry that none could grow here now. Yet the bacteria that these long-gone plants once depended on still remain. Researchers have recently found them surviving in the arid soil alongside millions of tonnes of saltpetre nitrates, by-products of these bacteria's former lives when vegetation was all around. Like the old steam trains, these precious chemicals – which in most places would have long since dissolved – have been preserved for countless aeons entirely thanks to this region's complete lack of moisture.

Rhizobium leguminosarum is one of the few organisms that is capable of 'fixing' atmospheric nitrogen into soluble nitrates that fertilize the soil so that plants and trees can thrive. It evolved over hundreds of millions of years in association with the legume family of plants, many of them vegetables that have become essential to animal and human diets, such as spinach, beans, peas, lentils, soya and clover. Chemicals, called flavonoids, secreted by the roots of these plants, induce *Rhizobium* to infiltrate. Once established, the bacterium lives in nodules that jut out of the plant's roots, where they assiduously convert nitrogen from the atmosphere into nitrates that are absorbed by the plant.

Roman farmers were the first to note the importance of legumes for nourishing the soil – not that they had any inkling that symbiotic bacteria were literally the root cause of their agricultural prosperity. Luckily, legumes don't just fix nitrogen for themselves. After harvesting, the bacteria in their root nodules leave substantial supplies of excess ammonium and nitrates in the ground, which act as natural fertilizers to support the growth of subsequent crops. Centuries later, Arab agriculturalists, who transported food produce across the vast Islamic empire from China and India to Spain (see page 194), established the technique of crop rotation in which legumes such as spinach and aubergines were alternated with other crops to ensure the soil was kept rich in nitrates.

Medieval European farmers developed their own system of crop rotation using the legume clover to provide essential fertilizer. In the civilizations of Central America, a planting system known as the 'Three Sisters' ensured other legumes, such as beans, were planted alongside gourds and maize, once more ensuring the ground was filled with nitrogen-rich nutrients.

But in the early twentieth century the natural bargain between bacteria, plants and humans began to break down. Two German scientists, Fritz Haber and Carl Bosch, pioneered an industrial process that could artificially 'fix' nitrogen without the need for plants, bacteria or saltpetre. They found that by heating nitrogen and hydrogen gas under high pressure they could synthesize large quantities of ammonium, so that by 1915 a new age of artificial fertilizers was born (see page 198).

Now any country could make its own supplies of gunpowder and fertilizers using nothing more than fossil fuels to generate the heat required to run the process. Imperial dependency on saltpetre from the driest deserts of the world (India and Chile were the main sources) was abandoned – along with the steam trains of Baquedano that once transported huge volumes of the stuff from inland mines to ports on the coast.

The historic impact of nitrogen-fixing *Rhizobium leguminosarum* and its several closely related Prokaryotic varieties is beyond doubt. Without them, the Neolithic revolution from which human agriculture has its origins (see On Agriculture, page 191) would most probably have stalled through soil exhaustion, throwing into doubt the long-term viability of settled human civilizations.

Its other big impact began in the mid ninth century when Taoist monks in China discovered how to manufacture gunpowder from naturally occurring nitrates (saltpetre). The knowledge spread westwards with the rise of the Mongol Empire (1206–1368) that began with Genghis Khan, eventually reaching fractious Europe where cannons were first used in anger during the Hundred Years War between France and England (1337–1453).

But future prospects for a revival in widespread organic nourishment for the world's soil now look brighter. Scientists and governments alike are aware of the environmental side-effects of modern food production systems on the Earth's climate and the eventual depletion of fossil fuel supplies. The Haber process alone uses as much as *5 per cent of the world's total production of natural gas.* How much more safe, efficient and cost effective are nature's bacterial systems that use none! Artificial fertilizers also acidify the soil and release additional greenhouse gases such as ammonia and nitrogen dioxide (NO_2) into the atmosphere.

Traditional organic methods that rely on nitrogen-fixing bacteria and their legume hosts are the only known long-term source of soil fertilizer. All organic farming depends on them. Therefore the story of *Rhizobia*, the nitrogen-fixing bacterium, is likely to have a future just as important as its past – even if, for a few eccentric years, it went briefly out of fashion when the world went fossil-fuel mad.

On Symbiosis

How genes and cells converged, establishing a new set of evolutionary rules that led to multicellular life.

GEORGIA'S MOUNTAINS are covered in trees. Deep in her forests, in the bitter winter of December 2001, two woodmen found some rusting metal drums that had been dumped in a clearing. What really aroused their curiosity was that the metallic containers were hot to touch. It seemed like the perfect gift from God – a constant source of heat that never seemed to need refuelling.

The canisters contained radioactive waste discarded from a former Soviet power station. As the woodmen dragged the drums back to their makeshift camp, disaster struck. One of the drums exploded, crippling both workers with deep burns that soon turned septic. Days later the men were close to death in a hospital in Tbilisi. Doctors were in despair. No amount of conventional antibiotics seemed able to cure the sores that covered them. It seemed they were heading inexorably towards an agonizing death.

Yet the men lived. An old Soviet cure, long since forgotten in most of the rest of the world, miraculously saved the day. The woodmen's wounds were covered with biodegradable patches impregnated with specially cultivated viruses called bacteriophages that naturally destroy bacteria. Within days the men had recovered and their wounds were healed.

Evolution's locks and keys

About 2.7 billion years ago, long before familiar organisms such as plants, animals or fungi had evolved, life in the seas was perpetually changing. It was dominated by just two basic types of evolutionary life-forms, both of which still populate the oceans today. The first were viruses that swarm about chaotically (see pages 11–12). The second were Prokaryotes, such as bacteria, with cell walls and a fully independent source of power, movement and reproduction (see page 28).

Viruses use specialist equipment for breaking through the tough cell walls of Prokaryotic bacteria.

After securely attaching themselves, they plunge a syringe-like rod deep inside a captured cell down which they inject their own genetic code (see picture on page 10). Within minutes the Prokaryotic cell falls victim to its predator and dutifully starts to manufacture new generations of its viral master's offspring. Infection such as this is nature's most ancient form of biological contact.

As a defence, bacteria have evolved a wide range of protective walls with oddly shaped spikes, nodules and bumps to try to prevent viruses from being able to cling on. But successful viruses have mutated into different shapes, like elaborate keys, that can sometimes evade these lock-like defences. However, so intricate are the bacteria's designs that usually only one type of viral key ever fits a particular bacterium's lock. It means that each strain of bacterium has its own specific viral enemy (called a bacteriophage or 'phage' for short) and each phage its own special Prokaryotic target.

The Georgian woodmen owed their lives to this ancient evolutionary protocol. Their medicinal patches contained a virus that had evolved the specific key that enabled them to attack and destroy the very bacterium that had penetrated their wounds. The virus produced no unwelcome side-effects since its phage is useless against all other forms of life and hence is safer even than traditional antibiotic treatments.

Bacterial locks and viral keys evolved as a result of a process of genetic change that scientists call 'horizontal gene transfer'. When bacteria and viruses come into contact in the seas, they constantly trade genes for the purposes of attack and defence. In this way evolution remains in a state of flux, which is how life on Earth was for at least a billion years after the first signs of cellular life.

What an Alice-in-Wonderland world! *Evolution is not hereditary* and *variety is constant.* There were no fixed species in life's primordial soup because no living thing was stable for long enough to become established. There were also no common ancestors because living things were mostly the random results of a game of genetic mix 'n' match.

The reason this sounds so different from the conventional view of evolution is because from about 2.7 billion years ago a profoundly different type of biology began to emerge. Textbooks usually focus on the story of life from this time onwards, relegating evolution's origins to a footnote of history. Even Charles Darwin didn't deal with the machinations of primordial microscopic life because, when he first published his work *On the Origin of Species* 150 years ago, so little was known about it.

Today all that has changed and so has our understanding of the profound change of direction that evolution took in the seas just over a billion years after life's first viral twitch.

Selfishness versus collaboration

Lynn Margulis is an American biologist who truly knows what it is like to face rejection. In 1966 she wrote a scientific paper in which she proposed a new origin for life after the evolution of bacteria: *'The paper was rejected by about fifteen scientific journals,'* she recalls. *'They said it was flawed, also it was too new and nobody could evaluate it!'*

Margulis said that the origin of complex 'Eukaryotic' cells – the constituent parts of all higher forms of life such as animals, plants and fungi – came from a novel collaboration between Prokaryotic bacteria that learned to live together *inside* the same cellular space. She called this coming together of previously separate individuals 'endosymbiosis'. Evolution was, she said, as much to do with the acquisition of genomes via *co-operation* between living things as it was to do with a Darwinian competition for survival between them.

The reason Margulis faced such hostility was because at about the time her paper was published the fashion among biologists was to try to combine Charles Darwin's theory of natural selection with the more recent discovery of the role of genes in evolution. Experts such as Richard Dawkins, who published his famous book *The Selfish Gene* in 1976, described the evolution of animals, plants and fungi in terms of replicators (individual genes) and survival machines (multicellular organisms such as animals and plants).

He claimed that the behaviour of all complex life was ultimately determined by genes acting according to the interests of their own selfish desires for immortality. Plants and animals were simply genetic 'survival machines' whose biological function was to pander to the insatiable appetites of their genes to achieve immortality. According to this point of view instinctive behaviour in animals and plants is best understood as strategies to improve the chances of their genes' long-term persistence into future generations.

For Dawkins and other 'Neo-Darwinists' evolution is seen as an elaborate process of random mutation in which genes collaborate only if their chances of survival are well served by doing so. Altruism, sacrifice, charity, even unnecessary risk-taking are never found in nature unless they serve a gene's individual selfish prospects. Underlying any such behaviour is an instinctive and complex 'cost-benefit analysis', which always seeks to increase the chances of individual genes copying themselves into the future. In the political climate of the 1970s and 1980s – with its resurgence of Thatcher-Reagan style free-market capitalism – explanations of such a state of nature became very popular indeed.

Unfortunately, Margulis's ideas represented a direct and rather awkward alternative. She said that genes couldn't just be selfish agents predetermining the behavioural instincts of all higher forms of life, because the foundation of all complex life was itself built on an intrinsically collaborative not competitive framework.

Margulis claimed that at some point after the emergence of Prokaryotic life, individual organisms gave up their individual independence for the greater good of a new co-operative. The highly complex Eukaryotic cellular environment came about through the fusion of many separate Prokaryotic bacteria into new, larger cells.

From citadels to cities

If Prokaryotic bacteria have castle-like walls (see page 29) then Eukaryotic cells resemble enormous medieval cities. Inside their boundaries are everything from power stations for generating energy (mitochondria) to warehouses for storing food, waste products or proteins (vesicles) and a central library for archiving architectural blueprints on how to make and copy a complete new city from scratch (nucleus).

They also contain tubelike scaffolding (micro-tubules) and bendy cross-beams (filaments), which give the cell its shape, allowing it to yield flexibly without causing structural damage. In this way a Eukaryotic cell is able to engulf food (such as a bacterium) that can be stored and digested. *Chloroplasts* – the photosynthetic elements of some Eukaryotic cells – originated from the union of a cyanobacterium with another Prokaryotic cell (see page 34). *Mitochondria* – a cell's respiratory machinery – were once purple bacteria that fed off atmospheric oxygen. Eukaryotic *flagella* – whip-like structures that drive some cells along – were originally rod-like Prokaryotic spirochaetes (see also tick, page 384) that move with a twisting motion.

Today Margulis's theory is widely accepted. Verification came from the sequencing of genes, beginning in the 1980s, when analysis conclusively revealed that buried inside all plants, fungi and animals are the genes of once unrelated bacteria that teamed up billions of years ago to create different, far more complicated types of living cell.

An infectious surprise

However, a big curiosity has remained concerning the Eukaryotic cell's nucleus, the part that contains its all-important reproductive apparatus. This is so totally different from the internal workings of any known Prokaryotic bacteria – which have no nucleus, just loose strands of DNA – that the origin of the cell nucleus has remained an elusive evolutionary mystery. What bacterium could possibly have contributed this novel genetic storage and copying device when there is no record of a Prokaryote ever having had a nucleus in the first place?

Recently, it has been suggested that the Eukaryotic nucleus had its origins in another type of symbiosis, this time between a *virus* and a bacterium. Copying machinery used by certain DNA viruses (such as the Pox virus, see page 24) employs similar techniques to those found in

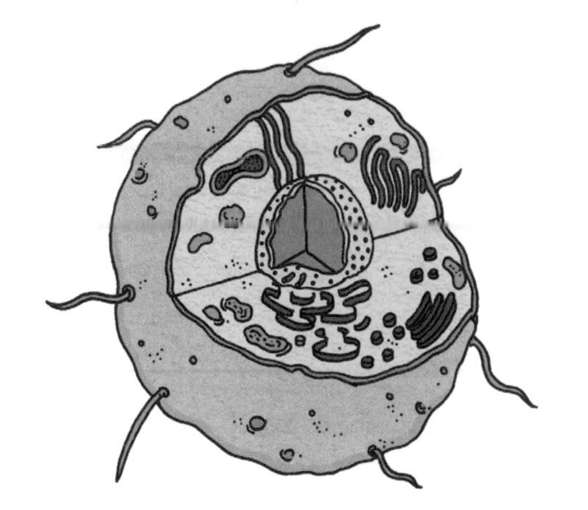

Cross-section through a Eukaryotic cell with a central nucleus (orange), which contains the DNA, and is surrounded by its own thin membrane (yellow). Microtubules (red) keep the cell's structure intact.

Eukaryotic cells, leading some experts to conclude that the most critical innovations leading to the first Eukaryotic cell may well have been derived from ancient viral infection.

Such evidence adds yet more weight to Margulis's initial idea that the foundations for complex life were laid through symbiotic co-operation, not hostile competition.

Safety in numbers

Collaboration between microscopic genetic replicators may have seemed rather attractive on Earth roughly 2.7 billion years ago for a variety of reasons. Increasing levels of highly reactive atmospheric oxygen released by photosynthesizing stromatolite cyanobacteria (see pages 34–5) would have given some bacteria cause to seek refuge inside the safe harbour of another's cell wall as it would have given them additional protection against overexposure to this toxic gas.

Safety from attack by an attacking virus (phage) was another possible incentive. Cyanobacteria that now inhabit Eukaryotic plant cells as chloroplasts may initially have been seeking shelter from cyanophages (viruses that attack cyanobacteria) by merging with another cellular organism.

Other organisms may have benefited from viral infection. Bacteria are vulnerable to genetic breakdown if overexposed to ultraviolet light. Infection from DNA viruses might confer a significant survival advantage for such bacteria, helping them repair broken genetic fragments more quickly. Such symbiosis may have given rise to the first Eukaryotic nucleus.

Escape from atmospheric toxins, viral attack and the incorporation of new genetic repair kits gave ample incentives for individual organisms to collaborate. Once assembled, however, these genetically disparate entities needed to work out a way of living together in collective harmony. How did they reorganize themselves for mutual benefit, precluding any one of them from establishing a way of securing individual advantage? Without new rules of engagement, conglomerate cells would most likely fail through the greed or appetite of their strongest constituent part. Successful Eukaryotic cells were therefore only those that were able to establish an entirely new regulatory set-up. Their success resulted in the evolution of all higher forms of life – animals, plants and fungi – and the first signs of natural collective intelligence.

Life's new rules

When investors put money into a business, in return for their cash they receive shares and get to become part-owners of the new collective. It is taken for granted that a well-defined and fair set of rules determines exactly how proceeds from the business should be shared out. Every company has its own set of rules, called 'articles of association', to make sure no individual investor can secure an unfair advantage. Investors also demand sufficient reassurance that their money is safe in the collective and that the business is stable enough to survive.

Modern business corporations mirror rather closely the organizational dynamics of life's first successful complex cells. Eukaryotic life succeeded (probably after a great deal of trial and error) because it eventually worked out a new evolutionary rubric. Three rules of collaboration emerged that profoundly impacted the evolution of all higher life-forms that followed.

1: Fair play

The first concerned fairness. Solitary genetic replicators in any evolutionary system are primarily concerned with copying their codes *ad infinitum* into the future. Therefore, stability in any such collective necessarily relies on an equitable system for genetic duplication.

The solution that evolved in Eukaryotic life is called 'mitosis'. It is the cellular copying system that underpins all life in plants, animals and humans. Inside the collaborative Eukaryotic cell, individual genes are stitched together to form structures called chromosomes. During cell duplication, the process of mitosis ensures that identical copies are made of *each and every gene*. As the cell divides into two parts, each part receives its own complete set of genes (a chromosome copy).

Therefore, when a Eukaryotic cell divides in two (a process called *cytokinesis*) every gene is certain of a place in both parent and child cells. Like shareholders who insist on fair treatment over future business transactions, mitosis means genes are guaranteed their own future through a system of scrupulously equitable cell division and reproduction.

Mitosis became a founding design principle for all Eukaryotic organisms. It means that inside every individual cell there resides an organism's *complete genetic code*. Human brain cells therefore don't just contain instructions relating to thinking and/or remembering. They also contain the genetic code for every other form of cell in the human body from the liver to the lungs (there are roughly 250 different types of cell in the human body), even though brain cells don't ever use them.

Nature is usually super-efficient when it comes to conserving energy. But, in this case, the absolute need for genetic fairness has overridden the demand for efficiency savings. *Every* cell in the body has to carry *all* the genes needed to build and operate an entire organism from scratch. This excess baggage is the unavoidable cost of doing business in a Eukaryotic enterprise. Without this 'mitotic bargain', which ensures fair play between different genes, the foundations for complex life (and indeed the advantages to humans from processes such a grafting in horticulture or cloning in genetics) would have proved impossible. Its impact is truly immense (for an example, see apple, page 344, or grape, page 296).

2: Stability

Alongside mitosis, a second rule evolved that substantially increased the survival chances of these new, more complex cells. It is a system we humans are still very much addicted to more than a billion years after it first emerged – we call it sex.

Sexual reproduction increases the *stability* of a species. Without it many creatures alive today – especially those that have remained unchanged for millions of years – would have long since become extinct (for an example, see velvet worm, page 72).

Sexual reproduction is a system of natural encryption pioneered by Eukaryotic organisms as a life insurance policy against being wiped out by disease or severe changes in environmental conditions. It works by jumbling up the genetic codes of two (or more) organisms to produce offspring that have their own unique genetic sequence. If every individual is slightly different then at least a few should be less susceptible to lethal infection or adverse climate conditions (for an example, see water mould, pages 52–3).

If every member of a Eukaryotic species had *identical* genes they would be equally vulnerable. This would put the species at grave risk that a single infection, a new predator or cataclysmic environmental event could consign its members to genetic oblivion.

Sex is therefore a powerful incentive for genes to collaborate within a Eukaryotic framework since the security of its *species-wide gene pool* increases each individual gene's chances of persistence into future generations.

'Meiosis' is the name of the process that randomly jumbles up the genes of separate Eukaryotic individuals (usually one male and one female). 'Fertilization' is the process that fuses them back together again to create a new organism (or embryo) whose genetic sequence is different from either parent.

Sexual reproduction introduced the evolutionary concept of inheritance and common descent where genetic characteristics are passed down through generations. It is thanks to the stability of sexual reproduction that the basic designs of many creatures have survived almost unchanged for hundreds of millions of years (for example, see shark, page 78).

Evidence that sex works as a system for species preservation can even be found in human history. Were it not for sexual reproduction it is possible that humans would not have survived the onslaught of smallpox when it first jumped the species barrier from animals to humans *c.* 10,000 years ago (see page 24). Thanks to genetic sexual encryption at least some individuals had natural immunity, avoiding complete catastrophe for our species.

With assurances of fair play in copying and the better chances of survival within a species-wide pool of genes, it is not surprising that Eukaryotic lifestyles proved popular. Thanks to the success of these two new rules evolution powered into overdrive. New species evolved that still defy classification today. Ancestors of organisms such as **water moulds** (see page 51) represent some of the most primitive forms of Eukaryotic life. Neither animals, nor plants, nor fungi, they are now lumped together by biologists into a kingdom called the Protoctista. Meanwhile, co-operative instincts reached new levels of sophistication. Some species linked up to create life's first multicellular creatures, sometimes toggling between joining together and splitting part, depending on the needs of their kind. Such behaviour illustrates the first signs of Eukaryotic collective intelligence (see **slime mould**, page 49).

Other complex cells teamed up to become the most primitive group of animals whose descendants (see **sponges**, page 58) still survive in the seas today.

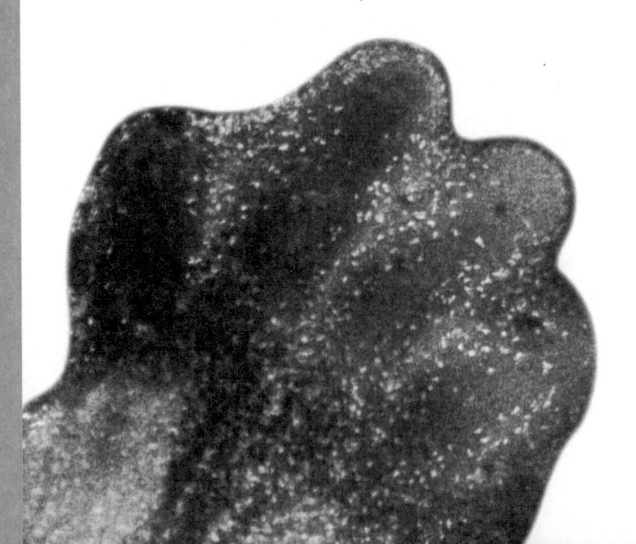

Apoptosis: cells of a developing mouse embryo captured on camera in the process of dying off (see bright green areas) to help form separate digits on one of its feet.

The oldest multicellular fossils in the world are plants that date back at least 1.2 billion years. They belong to ancestors of today's **algae** (see page 54), species of photosynthesizing Eukaryotes that formed the world's first seaweeds.

3: Self-sacrifice

However, for Eukaryotic cells to succeed in multicellular collectives a third rule for collaborative behaviour became increasingly necessary for survival. 'Apoptosis' is the process of planned cell death – a kind of cellular suicide that is fundamental to the survival of multicellular species.

When a human embryo forms in its mother's womb it is the programmed *die off* of individual cells that allows fingers and toes to grow and separate. In the average human, up to *seventy billion* cells die prematurely every day in order to keep the body fit for life. If cells divide faster than they die off, cancerous tumours result. Apoptosis is therefore essential for the effective functioning, maintenance and wellbeing of any multicellular being. Apoptosis in a tree means healthy cells deliberately dying off to form the hardened structures necessary to make a sturdy wooden trunk.

This is perhaps the best example of why co-operation between individual evolutionary entities became such a powerful characteristic of the natural world. Competition between creatures continued, of course, but each and every multicellular creature's chance of life has its origins in the profound spirit of co-operation between different genetic entities triggered by the Eukaryote revolution *c.* 2.7 billion years ago.

New rules of fair play, gene-pool stability and self-sacrifice meant that life *c.* 2.7 billion years ago stood at its most important crossroads. Collaboration, fairness and species stability created the foundations for multicellularity and a new biological fitness. Thanks to Eukaryotic collaboration and its new articles of evolutionary association, wonderful *visible* life began to fill up the seas.

Slime Mould

PHYLUM: MYCETOZOA
SPECIES: DICTYOSTELIUM DISCOIDEUM
RANK: 74

The surprising origins of intelligent, self-organizing systems.

IT MUST HAVE BEEN one of the most bizarre announcements in the history of scientific research. In the autumn of the year 2000, Japanese scientist Toshiyuki Nakagaki told the world how a slime mould could successfully navigate its way through a maze by the shortest possible route. It was, he claimed, a demonstration of the most primitive form of Earthly intelligence.

Working with a team of researchers at the Biomimetic Control Research Centre in Nagoya, Nakagaki dotted several pieces of slime throughout a three-centimetre by three-centimetre maze constructed on top of a plate of agar. Over the course of a few hours the blobs grew tube-like structures called pseudopodia that caused the whole maze to fill with a long, sticky column of slime.

There were four possible routes through the maze. Food was placed at both exits. After eight hours the column of slime had shrunk so that its 'body' filled only the parts of the maze that formed the shortest route from one piece of food to the other. Remarkably, it had worked out how to organize itself into the most efficient possible shape for foraging for food and in the process solved the puzzle of finding the quickest and most direct passage through the maze.

Slime moulds are living descendants of some of the earliest forms of Eukaryotic life. Since their modern discovery in a North Carolina forest in 1935, scientists all over the world have been fascinated by their extraordinary behaviour. Slimes are so puzzling that they have variously been classified as plants, animals and fungi. The consensus today is that while these creatures share some characteristics with all three, they rightfully belong in a separate kingdom altogether, called Amoebozoa. Recent genetic sequencing of the slime mould *Dictyostelium discoideum* confirms their evolutionary significance. They split off from the Eukaryotic tree very early on, just after the initial division between plants and animals and before the first sign of fungi.

The peculiar life cycle of the slime mould brilliantly demonstrates how versatile life became once symbiotic cells (Eukaryotes) had evolved new rules for collaborative living. As the slime mould maze experiment shows, the emergence of what humans understand as basic intelligence starts right here.

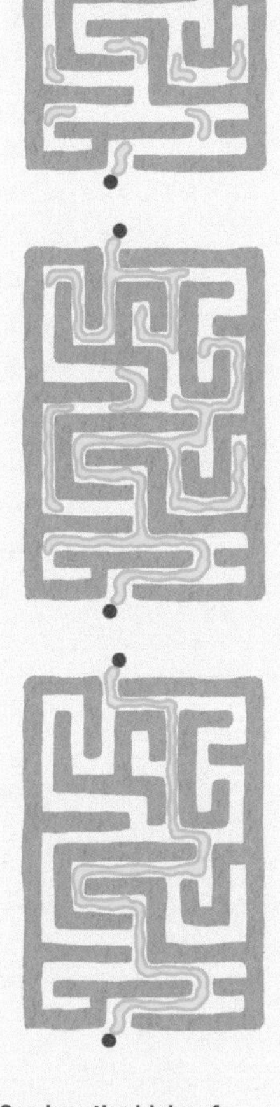

See how the blobs of slime join up into a single slug-like creature to find the shortest route to the food (marked in pink).

The slime life cycle begins when spores hatch out on moist ground, growing into single-celled Eukaryotes that graze on bacteria in the soil or on rotten logs. Their flexible cell structure allows them to surround their prey, engulfing and then digesting it as food. This process, called phagocytosis, is of special interest to modern science as it represents the earliest appearance in nature of tactics later employed by vertebrate animals, including the human immune system, to fight disease (see shark, page 78). White blood cells in our bodies behave just like individual slime cells, engulfing and digesting bacterial invaders to neutralize any threat they pose to the body. Recent genetic sequencing of *Dictyostelium discoideum* shows that the genes responsible for attacking disease in humans are essentially the same as those found in this much more ancient slime mould.

Slime's real eccentricity becomes apparent when food runs short. Individual cells secrete a chemical (adenylyl cyclase) that acts like an alarm informing other nearby slime cells that since food is scarce, genetic survival depends on collaboration. They do not engage in a battle to fight each other in a war over scare resources. Instead, they group together in an orderly manner to form a completely new multicellular creature – a type of slug – from scratch. For years scientists thought there must be some master-cell co-ordinating the process – if only they could observe this bizarre behaviour closely enough. But it is now clear that slime cells represent a natural self-organizing system. The collective actions of many individuals following a simple set of collaborative rules result in intelligent group behaviour.

As soon as a cell emits the chemical alarm, its neighbours either move towards the chemical or amplify its effect by secreting more alarm signals. In this way, moving in highly regulated six-minute pulses, up to 100,000 cells can rapidly swarm to form a single body of multicellular slime. The aggregate organism substantially increases the chances of successfully finding food, as ably shown by the slime in the Japanese maze, since as a collective its size, reach and powers of movement and detection are so much greater than those of any one individual.

When conditions are just right, the slime mould begins to metamorphose once again. Within about ten hours it has turned itself into a fruit. Some cells then sacrifice themselves (see apoptosis, page 48) to form a stalk made of cellulose that raises up other cells at the front that have turned into balls of hardened spores ready to be blown away by the wind. Like the bacterium anthrax (see page 37) these well-protected cells can survive for years, waiting for a favourable moment to begin germination, starting the slime's life cycle all over again.

The significance of slime today lies in the insight this extraordinary life-form has given into the workings of evolution and emergent, self-organizing systems. Intensive study of *Dictyostelium discoideum* since the 1960s and the sequencing of its genome (completed in 2005) has yielded a fresh understanding of how complex intelligent systems arise without the need for centralized direction (see page 162). Examples are everywhere, from the formation of fruiting slime moulds to the construction of embryos in the womb and the human immune system. In each case, the three basic Eukaryotic conventions of collaboration in terms of fairness (mitosis), stability (meiosis) and self-sacrifice (apoptosis) apply.

In the last fifty years, humans have begun to mimic these systems and a new branch of science called 'artificial life' has emerged as a direct result. Open-source software development, evolutionary computer games, the self-organizing maintenance of complex communications networks – even the rise of the Internet – are just a few examples of the recent human application of natural systems that first evolved in slimy moulds, hundreds of millions of years ago.

Water Mould

CLASS: OOMYCETES
SPECIES: PHYTOPHTHORA INFESTANS
RANK: 30

How man unwittingly helped a microscopic, unicellular protoctist reshape the course of modern human history.

PROTOCTISTA, which includes such diverse groups as amoeba, algae and slime, is one of the most ancient and least well known divisions of Eukaryotic life. Yet the influence of some of its species during recent human history (especially for those of Irish descent) has been especially profound. Of particular note are the Oomycetes, microscopic unicellular creatures, more commonly known as water moulds.

Talented French botanist Heinrich Anton de Bary first described the parasitic creature *Phytophthora infestans* in 1861. Among his many weird fascinations was mixing tiny spores, removed from the leaves of potato plants, into drops of water, which he placed under a microscope. He was astonished to see that the spores germinated and multiplied, each growing their own tails and swimming about as if they were miniature tadpoles, just hatched out from eggs.

The more he watched, the more de Bary realized these were no normal vegetative fungi. When he saw how these miniature animals behaved on the surfaces of potato leaves he was amazed. Tubes seemed to grow out of the organisms, punching their way through the leaf's surface. A creature would then squeeze its way through the opening, leaving an empty, dead layer of skin behind. Once inside, it would swell, elongating and branching its way from cell to cell, helping itself to the plant's nutrients until the leaves were completely consumed. Eventually, the mould would fruit by thrusting its roots back up through the pores of the stricken leaf (see picture on page 53), so that its spores could be borne by the wind towards another nearby plant.

De Bary quickly realized the significance of what he had observed – which is why in 1861 he named these microscopic creatures *Phytophthora*, meaning 'plant-eater' in Greek. Just sixteen years earlier, the most devastating famine in modern times had struck the Western world. De Bary had at last worked out its cause.

In mid-nineteenth-century Europe potatoes were still not accepted as respectable food on the dinner tables of many countries, associated as they were with dubious origins among the 'savage' natives of Central and South America (see potato, page 206). But this was not the case in Ireland. Laws passed by the ruling English parliament prohibited natives from owning more than a few acres of land each, so high-yielding potatoes had become a key part of poor people's diets. Tubers from these plants provided everything someone with only a little land could wish for. They were simple to grow, easily stored and full of proteins and vitamins B and C; a diet of potato mash with milk proved sufficient to feed a large family, and all its livestock, throughout the year. It was largely thanks to the potato that, within 100 years of its introduction, the population of Ireland rocketed from three million to eight million.

Meanwhile, more than six thousand miles away in the mountains of Peru, the little protoctist *Phytophthora infestans* was enjoying its own population boom, this time feasting on potato plants grown by the indigenous agriculturalists of South America. Wisely, the ancestors of these industrious farmers cultivated hundreds of different potato varieties. It meant that if any one type succumbed to disease, other more resistant varieties could still be successfully farmed in their place. Echoing the natural benefits of sexual reproduction, their pursuit of biodiversity provided an insurance policy against devastating infection (see page 47).

In the early 1600s, when Spanish conquistadors subdued the Incas of South America, it seemed necessary to export only a few popular types of vegetables as curiosities back to Europe. Little did these opportunists realize just how valuable these weird-looking, knobbly tubers would eventually become. Nor did they appreciate the consequences of exporting only a handful of varieties, such as 'Lima', 'Peruviennes' and 'Cordilieres'.

It wasn't until the early 1840s that diseased tubers crossed the Atlantic Ocean in ships from North America to Europe, triggering outbreaks of late blight in Belgium, Germany and Holland, beginning in 1845. No country suffered quite like Ireland, so dependent were its people on their simple diet of potatoes and milk. After the failure of three harvests (1845, 1847 and 1848), the country was devastated. More than one million of its people are thought to have died directly of starvation and disease brought on by malnutrition. The stench of rotting potatoes filled the air. Black, spotted leaves would appear overnight, turning once healthy crops into a foul-smelling mush. Politics and prejudice heaped on yet more misery, as British overlords continued to export grain and livestock from Ireland in a bid to protect their incomes. Desperate to escape the appalling conditions of the worst famine to affect Europe since the time of the Black Death of 1348, millions of Irish fled, driven off the land by laws that made anyone who owned more than a quarter of an acre ineligible for aid.

Most went to America, although some settled in Canada and Australia. The impact on the Western world has been immense. The United States now has thirty-six million Irish Americans, 12 per cent of its entire population. Not that *Phytophthora infestans* was the only cause or indeed the initial trigger of Irish emigration to the Americas, which began soon after the Napoleonic Wars (1804–15) yet its arrival in 1845 massively increased the flow.

The effects of blight and its associated human emigrations have been felt ever since, from the global spread of Catholicism (two-thirds of Irish immigrants were Catholics), to an enduring bitterness between many English and Irish people over the failure of the British government to aid its own people, a crime that some have since called genocide. The blight triggered demands for Irish independence, leading to the Irish War of Independence (1919–21), the partitioning of the country into North and South and the founding of an independent Ireland, ratified by the Anglo-Irish treaty of 1921, a status quo that persists to this day.

The political aftermath of this crop disease looked set to outlive the disease itself, thanks to a French student of de Bary's called Pierre Millardet. During the 1860s, only a few years after *Phytophthora infestans* devastated European potato crops, a close Oomycete relative, *Phylloxera*, was also given a free ride on board American ships bound for Europe.

This parasitic protoctist attacked French vine leaves, devouring them every bit as ferociously as its potato-loving cousin. Only by grafting American vines on to French roots could France's world of wine-making be saved from almost certain extermination (see grape, page 296). But, by 1885, Millardet found an easier solution – literally – by mixing copper sulphate crystals in water and quicklime. The liquid – called Bordeaux Mixture – could be sprayed on to vines (and potato plants), killing the water moulds. In the process, Millardet helped launch the first systematic human war against plant diseases, ushering in an age of pesticide science that has since become a cornerstone of modern agriculture.

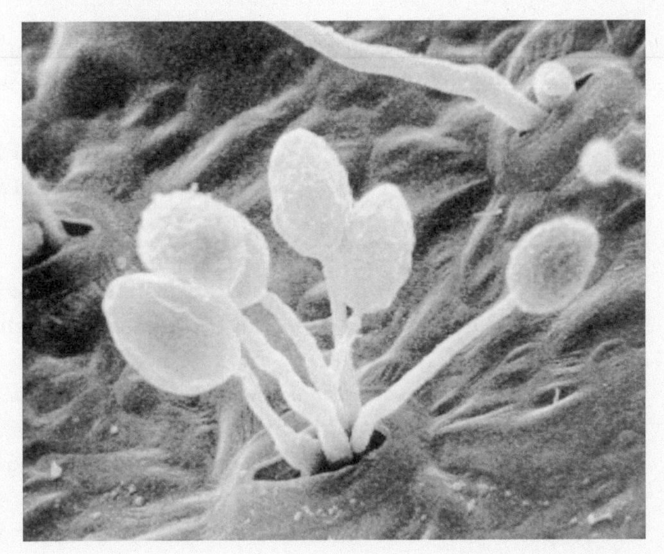

Sporulating branches of *Phytophthora infestans* punching upwards through the stomata of an infected potato plant.

Late blight in New Zealand in 1905 was one of the last water mould epidemics, thanks to the effectiveness of Bordeaux Mixture, although Chinese potato plants remained unprotected as late as the 1950s. With the birth of systemic pesticides following World War II, it seemed the story of potato blight – *Phytophthora infestans* – had finally come to a close.

That is what everyone thought. But never underestimate the power of Eukaryotic sexual reproduction to secure the survival of a species. In the 1980s, a resurgence of new varieties of *Phytophthora infestans* gave the farming communities of America and then Europe a nasty shock. Thanks to sex, new strains of water mould (they are capable of both sexual and asexual reproduction) began to appear with sufficient genetic diversity to benefit from at least some natural immunity to even the most persistent onslaught from artificial pesticides. In 1984 a sexual variant (called A2) that originated in Colombia was spotted in Switzerland and has since been adventuring around Europe. To the utter dismay of the potato farming community, it reached Ireland in 1989. During its progress, *Phytophthora infestans* has been mating with other strains, producing genetically unique specimens of enormous diversity, some of them offering resistance to traditional cures for late blight disease.

So blight is back. *Phytophthora infestans* is once again being described as 'the world's worst agricultural disease', costing farmers billions of dollars a year to contain. A new fightback is now under way by genetic scientists who are splicing genes taken from a wild South American potato plant *Solanum bulbocastanum* that has, over an unknown number of generations, developed its own natural immunity to these persistent protoctists. Genetically modified, protoctist-resistant potatoes may eventually become a reality, if only the world's seed stock can be comprehensively updated. But, in true Eukaryotic tradition, how long will it be before another sexual deviant saves its species from destruction by humans?

Algae

FAMILY: DICTYOSPHAERIACEA
SPECIES: BOTRYOCOCCUS BRAUNII
RANK: 2

How algaculture – the farming of pond scum that thrives in murky water – could be the aviation industry's best long-term biofuel bet.

DAVID DAGGETT is a man with a mission. His goal is no less than to save the worldwide aviation industry from collapse. Volatile fuel prices and global concern about carbon emissions are causing people and governments alike to think twice before they fly. Even for those reluctant to kick their globe-trotting habits, aviation faces an uphill struggle. Industry experts are being forced to prepare themselves for a new age where demand for oil continually outstrips supply. Witness the beginning of a global trauma, predicted for years by some, of a phenomenon called peak oil.

As Technology Leader for Boeing Aerospace's energy and emissions product development group, Daggett's task is to identify a new long-term source of aviation fuel that's plentiful, cheap to produce and kind on the environment. Thanks to his own passion for growing microscopic algae at home, Daggett believes he may just have found the solution.

Algae are among the oldest and most diverse organisms on Earth – like slime and moulds, they are Eukaryotes and are classified today in the kingdom of Protoctista. Early-nineteenth-century scientists divided algae into groups based on their different colours. Green algae are the ancestors of all land plants and trees. Red algae range from large seaweeds to microscopic varieties that attach themselves to the skeletons of coral fish, turning their reefs pink (see page 66). Brown algae, the third major group, form vast marine forests of seaweed called kelp that grow as fast as half a metre a day. Like coral reefs, these twisting underwater thickets, which stretch out as long as eighty metres each, are a vibrant part of the marine ecosystem, providing excellent habitats in which many aquatic species can lay their eggs safely and hide away from hungry predators.

Algae are prodigious producers of oxygen, having gained a capacity for photosynthesis from an ancient symbiosis with cyanobacteria (see page 45). Without marine-based algae today global oxygen supplies would quickly dwindle. Collectively they produce a far greater quantity than all the land plants and trees combined.

Without the countless varieties of microscopic algae, larger forms of sea life would never have been able to evolve, let alone survive today. Small filter-feeding marine animals such as krill, clams, mussels and oysters depend on hundreds of thousands of different species of microscopic algae that float through the water, as do larger creatures such as flamingos, basking sharks and even the enormous baleen whale. Plankton, the collective name for many of these algae, lies at the base of the entire aquatic food chain. Land-life would have been just as impoverished without marine algae. Ancestors of the lungfish, creatures that masterminded the escape from the seas, would have found precious little to feed on in their newfangled terrestrial habitat without land plants – all of which are descended from ancestral green algae from the seas (see page 81).

Like many of the earliest Eukaryote life-forms, algae are highly collaborative creatures. Lichens, corals (see page 66) and sponges (see page 58) have survived for hundreds of millions of years chiefly thanks to the services of photosynthesizing algae that provide them with plentiful supplies of oxygen and sugars.

Algae have played a key role in human history, too. Various types provided ancient human civilizations with a source of highly nutritious food. Edible kelp called Kombu (*Laminaria japonica*) has been eaten for thousands of years by cultures in the Far East – today it is still a vital source of food in countries such as Japan where it is used to add flavour and nutritional value to soups and broths. Other varieties of kelp have been used in the relatively recent past to make potash, a necessary ingredient in soap and glass manufacture. In the nineteenth century, Britain found itself bereft of wood as a source of industrial fuel, owing to extensive deforestation. British manufacturers turned instead to potash made from kelp harvested by Scottish fishermen. Dried and burned, the kelp produced enough potash to enable them to continue to make essential commodities such as glass, textiles, soap and paper.

In other places algae has become an important fertilizer. Maerl is a type of red algae found on the seabed off the coast of Cornwall, now used as an organic soil fertilizer in place of artificial nitrates.

Other types of microscopic red algae called coccolithophores and foraminifera form hard outer shells by absorbing carbon dioxide from the atmosphere and turning it into calcium carbonate (see page 400). As part of this process they

Giant kelp (*Macrocystis pyrifera*) is a species of marine algae found in the Pacific Ocean that can grow more than 45cm per day and is used in everything from cosmetics to glassmaking.

have a direct influence on the Earth's all-important water cycle, a process which has made land-life and the evolution of humanity possible. A waste gas produced by these algae, called dimethyl sulphide, helps seed clouds by providing perfectly sized surfaces around which water vapour can condense back into liquid droplets. Apart from delivering fresh water supplies around the world, clouds also play an essential role in helping to cool the Earth, making it suitable for life. Much of the Sun's infrared radiation reflects off the tops of clouds back into space, so helping to regulate global temperatures.

Indeed, the impact of this diverse family of creatures on human history goes further. When these tiny organisms die, they sink to the ocean floor and, over millions of years, get crushed into a type of sedimentary rock called limestone, which has been one of the most popular building materials used by humans ever since the rise of the Egyptian pyramids more than 5,000 years ago.

Easy to carve, chisel and cut into blocks, limestone is still a traditional favourite for construction, especially in Europe and America. During the nineteenth and early twentieth centuries, banks, railway stations and monuments in cities such as London were frequently made out of limestone (Portland limestone is one example), a product of dead red algae mixed with various quantities of clay, silt and sand all crushed together by the tectonic processes of geological time.

It is the future bounty of particular types of algae that experts including Dave Daggett think warrant closest attention. Some, including *Botryococcus*, have the potential to help save the modern world from a global warming catastrophe by weaning us off our dependency on fossil fuels such as coal, oil and gas. In the process, these plants absorb some of the excess carbon dioxide currently heating up the atmosphere and causing such damaging environmental effects.

A report entitled *Alternative Fuels for use in Future Aircraft*, presented by Boeing's Daggett to the American space agency, NASA, in April 2008, leaves little doubt about the potential for bio-diesel produced by certain species of algae as a sustainable fuel for the future needs of aviation. One of its great benefits, says the report, is the high yield compared to other biofuel crops:

'Algae could produce 150 to 300 times more oil than a crop of soybeans ... eighty-five billion gallons of biojet fuel could be produced on a landmass the equivalent size of Maryland (8.5 million acres)' – that's less than one seventh of the land area in the US currently dedicated to growing corn. Moreover, if biojet fuels were fully compatible with legacy aircraft, the report concludes that *'this would be sufficient to supply the present world's fleet with 100 per cent of their fuel needs well into the future'*.

Part of the reason why businesses and governments are getting so excited about the possibilities of harvesting fuel from algae is because certain species of green algae, such as *Botryococcus braunii,* have recently been found capable of producing quantities of oil with a mass of up to 40 per cent of their dry body weight. With a constant supply of carbon dioxide, microscopic algae like these can be farmed in tanks and then separated from water, dried and crushed to extract their oil.

'Algaculture', the practice of farming algae for oil production, has ancient natural origins. Kerogen is a mixture of organic compounds buried deep underground that is made up of decomposed marine algae, cyanobacteria and other protoctists. When it is heated to the right temperature in the Earth's crusts it releases crude oil or natural gas. Just imagine being able to accelerate that same process which normally takes hundreds of millions of years into real-time production – that's the potential of algaculture.

As a result governments, energy companies and opportunistic entrepreneurs are now showing a great deal of interest in the potential of this sustainable and environmentally friendly energy source. Something of a green gold-rush is unfolding with speculators large and small flooding into the market, chasing the prospect of potentially limitless biofuels without the drawbacks seen with crops that require large amounts of land, water and artificial fertilizer. Some companies are even experimenting to see whether pumping the carbon dioxide emissions from power-station smoke stacks directly into tanks of algae can convert this greenhouse gas into materials that can be harvested as bio-diesel. Others are turning to genetic engineering, tinkering with the creation and patenting of algae species that are faster-reproducing, higher yielding and easier to harvest in order to help make the mass-production of bio-diesel more profitable.

Meanwhile, other prospectors are focusing on algae micro-production, targeting potential home-growers. Instead of having to farm traditional labour-intensive bio-crops such as corn or soya, algae can be grown on special translucent trays. These contain enough algae to produce upwards of forty-five litres of bio-diesel a week – more than enough to satisfy a family's needs for transport and heating oil.

Another possibility is to use algae as an environmentally friendly way of producing hydrogen – a potential fuel source for planes, trains and automobiles. When certain species of algae, such as *Chlamydomonas reinhardtii,* are deprived of sulphur they switch from producing oxygen as a waste product to producing hydrogen, and in such volumes that, according to one recent scientific paper, it could make the process commercially viable.

In other trials, species such as *Sostera marina* are being harnessed by Japanese scientists in an experiment to construct giant carbon sinks that can absorb vast quantities of carbon dioxide, turning this problematic climate-changing gas into organic fertilizers and fuel. A pilot project, begun in 2005, envisages one hundred vast nets full of algae, each measuring ten kilometres square that will float in the oceans. Each net will grow to a weight of 270,000 tonnes, absorbing huge quantities of carbon dioxide from the atmosphere to be harvested for use as biofuel twelve months later.

In a newspaper interview project leader, Dr Masahiro Notoya, from the University of Tokyo, put the case for the impact of algae, past, present and future rather succinctly:

> *'It's actually thanks to seaweed that we're here at all. When the world was young, it was the little blue-green algae and other seaweeds that, over the years, converted so much of the carbon dioxide in the air into oxygen and eventually pushed it up to the levels it is at today. Now that the balance is being thrown off, it's time for the seaweed to come and help again.'*

Sponge

PHYLUM: PORIFERA
SPECIES: MICROCIONA PROLIFERA
RANK: 72

A primitive animal survivor that pioneered the art of multicellular construction.

MOST PEOPLE naturally associate a sponge with having a bath. Not so Henry Wilson, an early-twentieth-century professor of biology at the University of North Carolina, who instead used a sponge to conduct one of the most illuminating and bizarre biological experiments of all time.

The azure vase sponge (*Callyspongia plicifera*) growing on the sea floor off the Caribbean island of Grenada.

Sponges are classified as primitive animals. Rooted to the sea floor, they use tiny internal hairs to beat sufficient water through their mostly hollow bodies to extract the microscopic algae and plankton on which they live. Simple and motionless they may be, but to Wilson sponges were his complete *raison d'être*. He spent his whole life studying them.

Modern science owes a huge debt to Wilson's obsession. To the great surprise of his biologist colleagues, in the early summer of 1907 Wilson demonstrated conclusively that the apparently simple sponge can perform a rather remarkable conjuring trick. His demonstration went as follows:

First, Wilson minced up a silicate sponge (*Microciona prolifera*) by ramming it through a sieve until all that was left was a pile of individual, crumb-like cells. Next, he placed the sponge cells in a dish of water and stained them red so that he could see what, if anything, would happen.

Then he watched.

To begin with, the cells just seemed to jiggle about independently at random. Gradually he saw them begin to gather together forming a red cellular clump. Then the cells began to take on various shapes and patterns. Finally, the cells reorganized themselves into an exact replica of the sponge that Wilson had previously crumbled through the sieve.

What Wilson witnessed was the extraordinary process by which nature builds multicellular bodies. The natural emergence of self-organizing biological systems is the process by which an entire organism is built out of independently operating cells from scratch – and all without the need for any form of central command. Some people say this is the basis of all 'intelligent' forms of life.

After seeing Wilson's experiment, scientists have never been so dismissive about sponges again. Indeed, this classic sponge experiment has been instrumental in the development of a specialist branch of body-building biology called embryology, which now provides some of the most important insights into the overall foundations of evolutionary science.

In-depth analysis of the sponge experiment shows that just *three* basic processes lie at the heart of what makes multicellular life possible. First is the need for some kind of glue that helps individual cells bind together. Plant and animal cells secrete special proteins that do this job. Collagen is one such, also found extensively in humans. It helps to keep the *two trillion individual cells* in our body stuck firmly together. Secondly, cells have different types of shapes and surfaces by which they attach themselves to each other. These rivet-like structures are key to making sure individual cells organize themselves correctly, latching on to other cells in the right order. Finally, given that there is no overall conductor or chief in charge of the operation, it is essential that each cell can communicate with its neighbours, passing on instructions such as whether to (a) die, (b) divide, or (c) issue further instructions to other nearby cells. Again, they secrete special proteins to carry out these necessary communication functions.

A sponge may seem an absurdly simple creature compared to a reptile, bird or human, but in fact the cellular routines that it uses to build and (if sieved up) rebuild itself are similar to those used by a human embryo in its mother's womb. All multicellular organisms (which includes all plants and animals) use the same basic self-organizing system as a sponge, allowing their cells to collaborate in a way that maximizes the creature's overall chances of survival and reproductive success.

While sponge-like creatures have been detected in the fossil record as far back as 600 million years ago (about fifty million years before creatures in the sea developed bones, shells and teeth, see pages 62–3), there still remains the question of why it took so long for multicellular life to emerge. After all, the first traces of Eukaryotic life appear about *one billion years* before the first seaweed and sponge-like fossils.

The reason for this delay is now thought to be down to levels of atmospheric oxygen. Collagen, and other proteins that provide the glue that help cells stick together, can be manufactured only in oxygen-rich environments. Therefore only when levels of oxygen increased sufficiently (due to photosynthesizing cyanobacteria, see page 34) could multicellular bodies begin to develop. By 600 million years ago visible life-forms were finally filling up the oceans – seaweeds, jellyfish, sponges and worms being among the first – leaving their shapes for posterity as occasional fossils in the rocks.

Once creatures like sponges adopted the habit of linking up their cells into giant collectives – of growing big – it is safe to say that life on Earth was never the same again.

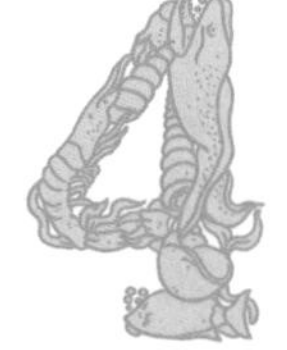

On Sea Life

How biological variety and constantly changing environments caused an explosion of new species to evolve in the seas.

ABOUT 550 MILLION years ago life in the seas was in the throes of another dramatic transformation. Exotic fossils found in 1909 in the Burgess Shale, Canada, show vividly how at that time the seas were brimming with vast numbers of visible species, many of them strange-looking animals with spiny backs, grasping arms and eyes mounted on stalks.

Before about 600 million years ago, however, there is scant evidence of large, multicellular animal life in the seas at all. One or more events must have occurred that caused an explosion of species to radiate into the seas. Evolution, it seems, was given a jump-start, causing modern scientists to herald this time as the beginning of a new era – the Phanerozoic – a word deriving from ancient Greek meaning *visible life*.

But why should so many forms of life suddenly explode into the fossil record as if from nowhere? Investigating changes in the form and function from one species to another and understanding their historical relationships is what lay at the heart of Charles Darwin's *On the Origin of Species*. Its investigation continues to this day, reaching back more than 500 million years, when the first evidence of animal and plant life appeared in the fossil record. Darwin had a name for the evolutionary process that he believed lay behind such variety:

> *'This preservation of variations, and the destruction of those which are injurious, I have called Natural Selection, or the Survival of the Fittest.'*

But Darwin was modest enough to confess to not knowing exactly how it worked. He explained in a letter to his friend, the botanist Sir Joseph Hooker (see also page 258), that rather than the work of a divine creator he thought natural selection had *'appeared by some wholly unknown process'*.

Still today, despite huge advances in the scientific understanding of genetics and heredity, there is much debate among scientists as to exactly what causes variation in nature and why it happens.

Random mutation

It is a fashion among many modern biologists to account for the rise of variety in nature through the *random mutation* of living things from one form into another. As genes copy themselves from one generation to the next, it is inevitable, they say, for mistakes in the process occasionally to occur, creating new types of organisms. If the affected creature survives, the law of averages says that it may *by chance* be better adapted to its living environment. In that case it will thrive, passing on its genetic changes and spawning a successful new species.

Although it is clear that random genetic mutations do happen in nature, it is also apparent that mutations in some genes matter a great deal more than in others.

Thanks to the recent genetic sequencing of many different living creatures, a particularly important set of genes has been discovered that is almost identical within the genomes of all types of animals. These genes – called Hox genes – represent a kind of ancient design 'toolkit' that is responsible for determining where different parts of the body are located. For example, Hox genes ensure that eyes appear on the head not legs, and legs jut out from the sides of the body and not through the mouth.

Genetic analysis suggests that about 600 million years ago a subtle mutation in the function of these genes eventually resulted in a dramatic divide that prised the animal kingdom apart into two great branches, which are commonly known today as vertebrates and invertebrates.

Flatworm-like creatures living at the bottom of the sea separated into 'protostomes' (creatures whose embryos develop a mouth first) and 'deuterostomes' (creatures that first develop an anus, then a mouth later on). The former evolved into invertebrates, such as millipedes, crabs, insects and spiders that have segmented bodies, jointed legs, an exoskeleton and a nervous system that runs along the bottom or belly of the animal. The latter eventually became vertebrates, creatures like fish, amphibians, reptiles, birds and mammals, that have an internal skeleton held together by a central backbone and a nervous system that runs along their topside or backs.

In this case, a small, chance change in a highly specific part of a worm-like creature's genetic code seems to have had huge consequences for the evolution of the rest of life on Earth.

When bigger is better

Take a quick look inside a *large* multicellular creature and it will quickly become clear why certain design features in animal species had to evolve as they did. Starting from about 600 million years ago, after the Earth emerged from a prolonged ice age, the *physics of size* became a powerful evolutionary force that began to determine how and why some creatures adopted different shapes and forms.

In certain circumstances, being large made survival sense. Provided there were sufficient supplies of food to feed a big body, larger organisms could move faster to catch prey or flee faster to avoid danger. Being large depended on multitudes of individual cells specializing into different tissue types that could carry out various functions, usually resulting in species that varied considerably from their smaller ancestral relatives.

The blue swimmer crab (*Portunus pelagicus*) is a protostome with a segmented body, jointed legs and exoskeleton.

Food transport is vital in a larger organism to ensure all its cells receive sufficient supplies of nutrients. *Caenorhabditis elegans* are nematodes, a form of **roundworm** (see page 68) whose ancestors evolved a primitive digestive system able to deliver food throughout their bodies. A cavity that runs the length of the worm takes food in through one end (the mouth) and expels waste through the other (the anus). As nutrients pass along the tube, specialist endoderm cells protrude in columns, increasing the surface area of the creature's gut for absorbing nutrients into the body proper. Other mesoderm cells then separate the digestive tract from the outer layer of the organism providing an internal space called the coelom which, in more complex creatures, is where specialist organs such as a heart, liver and lungs can safely reside. A third basic cell type – ectoderm – forms a protective layer of outer scaly skin around the surface of the roundworm. All forms of animal life (vertebrate and invertebrate) have subsequently evolved around this basic triploblastic (three-layer) body design. Humans are, in terms of evolutionary design, elaborate worms.

Despite their cellular specialization roundworms are still relatively small, thread-like creatures. To grow larger, animals needed other systems for carrying essentials around their growing bodies, such as iron molecules to transport oxygen for respiration. Networks of vessels appeared in the ancestors of larger worms called annelids – the phylum to which today's hugely important earthworms belong (see page 97). These contained a complex protein (haemoglobin) that incorporated iron in order to carry oxygen throughout the body.

Cross-section through a roundworm's triploblastic body design: a digestive endoderm layer lies at the centre; an open space for the nervous system, eggs and ovary sits in the coelom; and a layer of ectoderm cells form the outer muscles and skin.

Specialist digestive and circulatory systems freed creatures from the limitations of staying small. Oxygen, which could otherwise reach inside a body only through natural diffusion, could now be delivered in sufficient quantities to make growing large a matter of evolutionary choice.

Oxygen and food transport systems were intrinsic designs that shaped the form and variety of new species. Features we humans take for granted about our bodies – a digestive tract, a circulatory system, blood, a mouth for feeding and an anus for excretion – were all designs pioneered by species of increasingly complex worms living at the bottom of the sea some 600 million years ago. They grew larger by evolving repeated segments in their bodies, some of which later evolved to have other specialist functions such as antennae and small stubby legs (see **velvet worms**, page 72).

Special measures

Once cells began the habit of dividing up tasks between them, other tactics for living successfully soon began to emerge. Some of the earliest recognizable animal fossils were jellyfish-like creatures that belonged to a highly successful phylum called Cnidaria, which contains more than 9,000 surviving species today. Among the first to evolve were sea anemones and **stony corals** (see page 66). Common to all Cnidaria are specialist stinging cells that paralyse passing prey. Some species of coral evolved to secrete calcium carbonate, surrounding their soft-bodies in a hard, protective outer-skeleton, the remains of which have accumulated into enormous structures called reefs.

Other creatures that were successful in the evolutionary fast-moving times *c.* 600 million years ago developed rival systems of attack and defence also based on the concept of secreting calcium-based compounds that could be shaped into hard shells (for defence) or teeth (for attack). Bite marks in tiny cone-shaped creatures called cloudinids, which thrived *c.* 550 million years ago, provide some of the first fossil evidence of species that incorporated calcium-based materials into their body designs. Thanks to the evolution of hard body parts, dead creatures began to leave a record of the shape and form of their bodies in rocks that fossilize, and so from this time the fossil record of life on Earth truly begins.

Hard body parts became very popular indeed. Ancestors of molluscs, familiar today in the form of oysters, clams, and mussels, were mostly peaceful creatures that stuck motionless on to the surfaces of rocks, filter-feeding off algae or plankton in the water. These creatures were masters in the art of defence, surrounding their soft, slimy bodies in bi-valvular shells that snapped shut at the first sign of danger.

Other specialist cells were those that became sensitive to light, probably evolving shortly before the mutant vertebrate–invertebrate split. Clear underwater vision was first achieved by the **trilobites** (see page 70), invertebrate crustaceans that became experts at seeing all round. It helped them become one of the most successful species of their time, radiating into countless numbers, thriving for hundreds of millions of years and peppering the world's rocks with fossils that have since become the prized possessions of amateur and professional palaeontologists alike.

Sensory awareness

Some experts believe that the emergence of senses such as sight dramatically changed the direction of all animal evolution. Hunting from a distance, adapting skin colours to provide effective camouflage, mimicking other creatures as a means of warning potential predators away and evolving exotic looks as a form of sexual attraction all date from the advent of sight in the seas, about 550 million years ago. Such innovations were once again down to the specialization of collaborative cells that divided up different tasks between them, taking on functions such as lenses, receptors and nerves that, in the case of eyes, formed life's first clear pictures of the outside world.

Nervous systems that incorporated other senses, such as feeling and hearing (for example, in bony fish), were equally profound. An ever-increasing sensory awareness became the primordial pivot around which Darwin's process of Natural Selection began to twist.

Cnidarians, like this giant sea anemone (*Condylactis gigantea*), host an impressive array of stinging tentacles to capture their prey.

An alternative modus vivendi, explored by other types of ancient sea creatures, involved the practice of relocation. Larvae from sedentary animals that were attached to the sea floor, ancestors of today's **sea squirts** (see page 76), developed a stiff internal rod-like structure called a notochord to help them swim to a new location in a bid to find safe, fertile places to grow. Their descendants took the innovation much further, thickening the notochord into a backbone to support ever larger bodies. With the additional support of bones and an internal skeleton, other cells could turn themselves into muscular tissue for swimming ever faster, either to catch prey, or to flee from danger.

By 400 million years ago, the sea was populated with grasping invertebrate arthropods such as **sea scorpions** (see page 74), vertebrate **sharks** (see page 78) and placoderms that had biting power equivalent to a *Tyrannosaurus rex* (see page 125), challenging the upper physical limits of life's potential for size, strength and speed.

Their success provoked other species, such as the lobe-finned fish (see page 112), into dabbling with even more radical lifestyles, such as breathing air through an experimental lung in a determined effort to make the most of the comparative safety of a life spent half in the water and half on land.

Merger and acquisition

The extraordinary richness and diversity of species such as those that exploded in the seas between *c.* 600 million and 400 million years ago can also be explained through that ancient practice of genetic fusion called symbiosis (see pages 44–5). Recent research indicates that new life-forms were sometimes created by the *merging* of different types of organisms. Sexual contact between distantly related creatures may even have resulted in the formation of new fertile species that comprised a genetic mixture of both (see also chicken, page 226).

Such a theory explains how different types of invertebrate creatures, such as sea urchins and starfish, have nearly identical larvae, whereas those of much more closely related species are sometimes very different. Velvet worms (see page 72), creatures thought to have been among the first to crawl out of the seas, may well have contributed at least some of their genetic traits to later land creatures such as caterpillars, moths and bees, accounting for these creatures' worm-like larval phases.

The stinging cells of the Cnidaria phylum of corals and jellyfish may once have been free-living fungus-like organisms called microsporidia whose genes were symbiotically acquired by comb-jellies that subsequently passed their traits on to future generations (see also lichens, page 82).

Many of the most important grasses that have been domesticated by humans over the last 10,000 years were derived from similar genetic fusions (see wheat, page 201; rice, page 230; and maize, page 234).

Viral infections

Viruses have made other important evolutionary contributions. Biologists tend to ignore the effect of viral infections in their accounts of life on Earth because many of them regard these genetic agents as more dead than alive (see page 11). Yet rather than a footnote to biological history, past viral infections have been powerful agents of both evolutionary stability and change.

For example, one of the most significant innovations in animal evolution emerged with the development of the adaptive immune system in bony fish (Osteichthyes) and sharks (see page 78) about 440 million years ago. This sophisticated mechanism for fighting infection evolved in vertebrates (fish, amphibians, reptiles, birds and mammals). Specialist blood cells identify, attack and remember invading pathogens (such as bacteria and viruses), providing ongoing immunity against future infections.

Strategies used by viruses to infect and colonize their hosts are directly analogous to those used in the vertebrate immune system to attack invading pathogens. Such insights have led some experts to conclude that the molecular machinery needed to operate the adaptive component of the vertebrate immunity actually originated in the genomes of a series of persistent retroviruses that infected bony fish hundreds of millions of years ago.

Snowball Earth

Inanimate environmental factors also played a big part in determining the evolution and variation of species, which ushered life into the visible Phanerozoic age.

About 575 million years ago, just as creatures in the seas began to leave their first traces as fossils in the rocks, the Earth was emerging from a prolonged and severe ice age, now often dubbed 'Snowball Earth'. Debates still rage as to what effects (if any) such global warming may have had on the explosion of species variety that soon followed in the seas. One possible scenario is that as the ice melted, higher levels of sunlight penetrated through the oceans providing conditions that encouraged specialist cells in small multicellular trilobite-like creatures to evolve the world's first eyes.

Such an innovation could have triggered a barrage of changes as sea creatures found themselves in a competitive race to adapt to a world in which some species had developed sight.

Geological events have also had profound effects. Mass extinctions have regularly punctuated the fossil record since visible life first appeared in the oceans *c.* 600 million years ago. The most catastrophic occurred 252 million years ago. What is sometimes called 'the mother of all extinctions' or the Great Permian Extinction, destroyed up to 96 per cent of all marine species including the last of the trilobites and sea scorpions. Following this extinction, as in all others, life gradually recovered and new species emerged to fill the biological void left by those that had disappeared.

More often than not these episodes have been caused by random events that have upset the ecological balance of the planet (such as massive meteorite strikes) or by the movement of the Earth's continents brought about by plate tectonics. Events like these can cause dramatic changes to global temperatures, the composition of gases in the atmosphere and the acidity of the oceans.

Episodes of this sort led to the extermination of certain species simply because their bodies were not well adapted to surviving such environmental changes. Others thrived because, by chance, their bodies were blessed with what turned out to be better-suited designs. In any analysis, life survived such traumas thanks only to its spectacular diversity of species stimulated by a powerful cocktail of genetic mutation, cell specialization, the physics of size, symbiosis, viral infection and geological chance.

Snowball Earth was a prolonged ice age period (*c.* 600 million years ago) in which it is thought practically the entire surface of the Earth froze up.

Stony Corals

ORDER: SCLERACTINIA
SPECIES: POCILLOPORA DAMICORNIS
RANK: 7

One of nature's oldest and most colourful species, which was founded on a mutual collaboration.

'Multiply, vary, let the strongest live and the weakest die.' So said Charles Darwin as he boldly tried to wrap up his life's work by boiling the behaviour of the natural world down to a single general law.

Observers wishing to challenge this claim would be wise to consider the case of corals – organisms that evolved near the start of the saga of multicellular life about 600 million years ago. Cnidaria, the phylum to which jellyfish, anemones and corals belong, shows up clearly in the earliest animal fossils.

One leading naturalist recently declared that corals are ecologically speaking *'surely among the most significant of all animals'* chiefly owing to their capacity to build underwater mountains. More than one million square kilometres of coral reefs are thought to exist in the world's oceans. Each reef, standing in waters up to fifty metres deep, is mostly composed of the dead skeletons of tiny organisms called polyps, related to jellyfish, which secrete hard calcium carbonate coatings around their soft-bodies as a simple form of protection.

Although these reef-building coral polyps can feed using tiny stinging tentacles (like their relatives the jellyfish), most of the time their energy is supplied through a symbiotic collaboration with a genus of microscopic red algae called *Zooxanthella*. More than 90 per cent of the polyp's energy requirements come from the ingestion of these algae that then reside inside the polyp's body tissue turning sunlight into energy via photosynthesis. In return, the polyp protects the algae and supplies the carbon dioxide it needs to photosynthesize.

When polyps die, the deposition of their calcium skeletons, layer upon layer, creates huge reefs, and eventually enormous underwater mountains, providing an essential habitat for both living polyps and algae. This ensures that the algae remain as close as possible to the sea's surface where the light from the Sun can easily penetrate. The red algae also secrete calcium carbonate in layers, adding their own contributions to the structure of a successful, thriving coral reef. Such close interdependent co-operation between species is known as mutualism.

The Great Barrier Reef complex off the coast of north Australia is easily the world's most massive structure made by living organisms. More than 2,900 individual reefs

and 900 islands stretch across 2,600 kilometres of open, shallow seas, covering an area of approximately 350,000 square kilometres – that's nearly twice the size of Great Britain ...

Scale and size aside, the impact of the symbiotic union between red algae and these tiny jellyfish-like animals that first appeared in the Cambrian Period (542–488 million years ago) has been immense. Repeated reef-making eras throughout geological history have harboured the richest diversity of marine species. Over 4,000 different species of fish inhabit today's healthy reefs, making their homes in cracks and crevices that provide perfect breeding grounds for all manner of marine creatures. It's not just fish that benefit. Reefs also provide excellent homes for sponges, jellyfish, worms, crustaceans, molluscs, starfish, sea squirts, sea turtles and sea snakes.

Until recently, mammals rarely paid them a visit, except, perhaps, for a few curious but friendly dolphins. While aboriginal Australians have regularly fished from the reefs off Queensland for the last 40,000 years, the modern threat to coral reefs comes from rising levels of carbon dioxide in the air, which dissolves in the sea making it more acidic. Polyps eject symbiotic algae from their systems in response to sea temperature rises and ocean acidification, a process that leads to coral bleaching. Feeding carnivorously using their stinging tentacles is an alternative but temporary survival strategy. Unless conditions revert, the reefs eventually die.

As many as 10 per cent of the world's reefs are thought to have perished in the last hundred years because of the impact of human-induced climate change. Further damage to these jewel-like marine ecosystems has come from the introduction of modern artificial fertilizers that run off farmland into the seas, increasing nutrient levels in the water, causing the wrong type of algae to grow. This has the effect of blocking out sunlight from the water's surface, smothering a reef to death. Tourism is another hazard. Since the end of World War II, coral islands, with their attractive climates, beautiful turquoise-blue seas and exotic marine life, have become some of the world's most popular holiday resorts (for example, in the Maldives, along the Red Sea, Indonesian and Mexican/Belize coasts).

The process of evolutionary mutualism that gave birth to life's extraordinary coral reefs does not mean Darwin was wrong in his claim that the modification of species (evolution) occurs through a perpetual competitive struggle for survival. It's just that, as the symbiosis of coral fish and red algae shows, in certain environments relationships between different species are just as likely to be founded on mutual benefit rather than competitive pressure (see also oak, on page 143, for an example of mutualism on the land). Coral reefs are powerful places for the natural conservation and co-operation of species, which is largely why the lifestyles of polyps and their symbiotic *Zooxanthella* partners have survived so well to this day.

An artist's impression of life at the bottom of the seas *c.* 450 million years ago, dominated by now-extinct rugose corals, predecessors of today's stony corals.

Roundworm

PHYLUM: NEMATODA
SPECIES: CAENORHABDITIS ELEGANS
RANK: 20

A super-successful, super-terrestrial life-form that has stretched the art of survival to its elastic limits.

WHEN SPACE SHUTTLE *COLUMBIA* broke up on 1 February 2005 some sixty-one kilometres above the Earth's surface, no one expected search teams to find any signs of life in the wreckage. Amazing as it sounds, despite the inevitable and tragic loss of human life, we now know that they did.

A team of scientists headed by Dr Catherine Conley from a NASA research centre in California successfully recovered five canisters of *Caenorhabditis elegans*, a species of roundworm that had been taken on board the mission (ST-107) as part of an experiment designed to discover more about the effects of zero gravity on organisms living in space. After painstakingly sifting through the wreckage for seven weeks, five out of seven canisters were recovered. In all but one container the worms had survived their highly traumatic re-entry in the Earth's atmosphere as well as a colossal and mostly unprotected impact with the ground, crashing into the Earth at speeds estimated at more than 600 miles per hour.

This extraordinary story underlines the incredible resilience of certain small animals, such as these nearly microscopic worms. More than 80,000 different species of roundworm are currently known to science. Most feed on microscopic algae and fungi or dead organic matter, decomposing and recycling nutrients in the soil and seabed. A few (about 15,000 species) live as parasites in other hosts, sometimes causing diseases in humans, farm animals, cats, dogs and plants.

Roundworms are among the most numerous and least fussy creatures on Earth, adapting to life in almost every conceivable habitat from the Arctic to the tropics

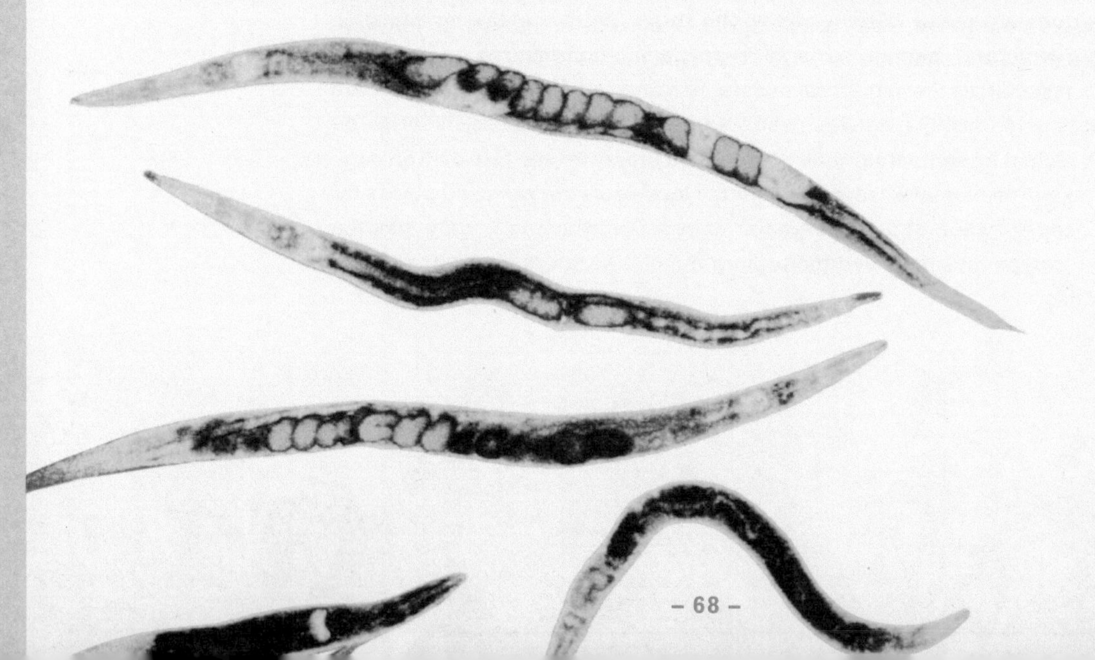

Space-faring nematode: *Caenorhabditis elegans* is stained and magnified under a microscope to reveal its soon-to-be-laid eggs.

and from the tops of mountains to the deepest mid-ocean trenches. Some tens of thousands of individual species have been described by scientists to date. The number undescribed is thought to run into the millions. Creatures belonging to this phylum represent a staggering *90 per cent of all life found on the sea floor*. On land, literally billions are contained in just one acre of topsoil, and as many as 90,000 roundworms are said to be able to occupy a single rotting apple.

Today's roundworm ancestors were among the first creatures to establish a fully-developed three-layer (triploblastic) system of body design (see page 62). Unlike jellyfish and corals (cnidarian) which have only two layers of cells (diploblastic), more sophisticated creatures later exploited the three-layer design to support the emergence of internal organs, circulation, muscular activity and a skeleton without disrupting the all-important and ongoing processes of internal digestion.

As well as being the first roundworm species to survive a space crash, *Caenorhabditis elegans* has another special significance. It was chosen by Nobel-Prize-winning biologist Sydney Brenner as a model organism for detailed scientific research into how animal bodies work and how their cells specialize to take on different tasks and functions (roundworms have a primitive nervous system as well as a digestive tract). He chose well. *Caenorhabditis elegans* has proved ideal for intensive scientific study, being small, simple and easy to cultivate. Its CV also includes coming fully back to life after being frozen indefinitely in liquid nitrogen. Once thawed, on it goes, feeding, excreting, breeding more of its kind. So venerated has this species become in research fields that it was awarded the honour of becoming the first creature whose genome was fully sequenced. Scientists published their results in 1988. The subsequent publication of the Human Genome in 2003 showed rather surprisingly that humans and roundworms share approximately the same number of genes (about 20,000–25,000).

The biological resilience and survival capacity of *Caenorhabditis elegans* following its remarkable journey into space and back has recently boosted theories for the extraterrestrial origins of life – a branch of science called astrobiology (see also page 32). If small animals such as roundworms can successfully travel through space and land on a planet at more than 600 miles per hour, then who knows what could have arrived on the Earth on meteorites hundreds of millions of years ago, seeding life or at least shuffling its course into new, unexpected directions?

Trilobite

PHYLUM: ARTHROPODA
SPECIES: OLENELLUS SP.
RANK: 31

A family that switched on the lights of life.

CHARLES DARWIN was always anxious to remind his readers of the old dictum that 'nature never moves in leaps' (*natura non facit saltum*) – indeed his whole theory of natural selection was built on the idea that gradual, incremental changes from one generation to another are what have caused, from so simple a beginning, *'endless forms most beautiful to have been evolved ...'*.

It is still with some courage that modern scientists challenge Darwinian orthodoxy. One American naturalist, Stephen Jay Gould, famously tried to destroy the idea of incremental evolutionary change in a detailed study of the extraordinary fossils from the Burgess Shale in Canada (see page 60). His book *Wonderful Life*, published in 1989, proposed that chance rather than fitness was the major agent of evolutionary change and that exotic creatures etched into the Burgess Shale were often more various and therefore more 'evolved' than those that have followed.

A big debate still rages today between scientists as to whether evolution is a gradual or a more punctuated process. Australian biologist Andrew Parker has recently made a powerful case for at least one profoundly punctuated moment in the evolution of animals which, if correct, puts the impact of one type of extinct creature right at the top of nature's most significant species.

Trilobites were marine arthropods (invertebrate creatures with jointed legs) that looked rather like large, heavily armoured woodlice. Although they have been extinct for more than 250 million years, their arrival on to the stage of life at the beginning of the Cambrian Period 542 million years ago appears to have caused a major evolutionary upset.

Before trilobites, most animals were worm-like, although a few, like coral polyps, had started to experiment with sponge and jellyfish-like forms. At about the time the last 'Snowball Earth' ice-age periods retreated (575 million years ago), sea creatures began to grow larger probably because of rising oxygen levels in the atmosphere. According to Parker the increased intensity of sunlight that penetrated through the melting ice became a significant selective pressure in favour of creatures that could determine light from shade. Swimming towards the light meant floating upwards nearer the surface of the sea where photosynthesizing food sources, such as algae (see page 54) were most abundant. Mobile creatures sensitive to the light therefore thrived.

Parker believes this increasing luminescence was responsible for the birth of vision. *'With our eyes open suddenly we see the world very differently ... The light switch was turned on for the first and only time – and it has been on ever since.'* Once one type of organism learned how to see, through myriad calcite lenses connected to a series of internal nerves, evolutionary pressures in the sea were affected all

A trilobite fossil with semi-circular ridges around its head, etched out by the sockets of all-seeing eyes.

around. Trilobites quickly became supreme rulers of the marine world. Fossil evidence makes it hard to tell exactly which of the 6,000 trilobite species so far identified by palaeontologists were life's first visual pioneers, but one reasonable candidate is Olenellus, a common and highly successful early Cambrian trilobite. Its fossils show that it definitely had eyes.

Within just a few million years, almost instantaneously in geological time, creatures began to evolve elaborate defences against these new visionaries of the sea. By *c.* 530 million years ago, a plethora of creatures from a number of different animal phyla started to appear in the fossil record as if they had arrived from nowhere – the famous so-called 'Cambrian Explosion'. These were creatures that had been gradually evolving for millions of years, but now survival meant having effective defences against all-seeing trilobites.

The construction of hard outer shells (which happen to fossilize easily) was one strategy pioneered by brachiopods (lamp-shells) and soon followed by the bi-valvular molluscs (ancestors of today's clams, mussels, oysters and scallops).Other strategies included establishing defensive barriers. Fossils of many creatures from the Cambrian era feature such defences, including the velvet worm Hallucigenia (see page 72), and bristle worms Canadia and Wiwaxia, oval-shaped creatures covered by overlapping shields and long, outward-projecting swords.

Other worm lineages evolved instinctively to bury themselves into the seabed or between cracks in the rocks. Some became so well camouflaged (such as the orange-coloured velvet worm called Asheaia) that even those equipped with eyes could not distinguish them from sponges rooted to the floor. Successful species, once similar in appearance, were those whose offspring had an advantageous shape, colour and form. An explosion of variety reached its peak.

The supremacy of trilobites did not last for ever. They too developed defences, initially to protect against being attacked by each other (hence their own evolution of hard outer skeletons and body armour) but later to protect them from other arthropods that had developed vision, chiefly the fearsome Anomalocaris, with its two bulging eyes and menacing pair of grasping tentacles. Growing as large as one metre long, this creature became one of the most powerful carnivores in the later Cambrian seas.

During the next evolutionary period (the Ordovician), creatures across other animal groups independently began to evolve visual senses, including annelids (bristle worms), cnidarians (box jellyfish), molluscs (such as the squid and octopus) and chordates, the phylum that eventually evolved into all vertebrate animals including sharks, fish, amphibians, reptiles, birds and mammals. So powerful a technology was the trick of sight that today over 95 per cent of all animal species have eyes.

The need to survive in a world where some had vision led to the development of countless new species with dramatically different outward forms and instinctive behaviours. It also led to the construction of hard outer bodies, giving humans an invaluable fossil record with which to puzzle over the ancient past.

Velvet Worm

PHYLUM: ONYCHOPHORA
SPECIES: PERIPATUS SP.
RANK: 56

Baby steps that became prehistory's biggest leaps.

NEIL ARMSTRONG made the most famous ever 'leap' for mankind when, on 21 July 1969, he stepped out of America's Apollo 11 lunar capsule to become the first man to walk on the surface of the Moon. An even greater achievement, however, was made by the much smaller steps of another creature which, in a bid to escape the pressures of living in the increasingly hostile seas, tiptoed ashore about 480 million years ago.

Life in the seas at the beginning of the Cambrian Period 542 million years ago became a great deal more brutal. Creatures that thrived were those that successfully adapted to fend off far-sighted predators such as the war-like trilobites (see page 70) or the metre-long carnivorous menace Anomalocaris. Some of these adaptations appeared so strange to scientists that for a while it seemed as if whole new families of now extinct creatures suddenly had evolved out of nowhere. Now, however, regardless of their bizarre looks, these creatures are thought to be, as Darwin predicted, ancient common ancestors of animal families that still populate the world today.

Among the most bizarre-looking, and for a while most misunderstood life-form, was the marine caterpillar-like creature called Hallucigenia. On the underside of its long, worm-like segmented body were what were thought to be a series of long, stiff legs, while on its topside was a brush of soft upward-pointing tentacles that groped in the water for food. Nothing like this weird creature seems to have existed before or since. That, at least, was what everyone thought until Swedish dentist-turned-palaeontologist Lars Ramskold discovered that experts had been looking at its fossils upside down! When turned the right way up the long, stiff spines could be seen to be layers of heavily armoured protection against attack; meanwhile the tentacles were instantly transformed into soft, stubby legs that allowed this creature to manoeuvre across the ocean floor out of harm's way. So effective were this animal's defences that worms with legs lived on, despite the intense competition of the Cambrian seas, gradually evolving into creatures that began to experiment with a new life on land.

Just as Armstrong's leap into the annals of human history was made possible by the Moon, so too were the first land creatures dependent on lunar forces to help them take their small, but critical, steps on to dry land. Tides that relentlessly pulled the ancient seas back and forth from the ocean's coasts created a zone that allowed small, stubby-legged worms slowly to adapt to life beyond the seas. About 480 million years ago, descendants of the heavily

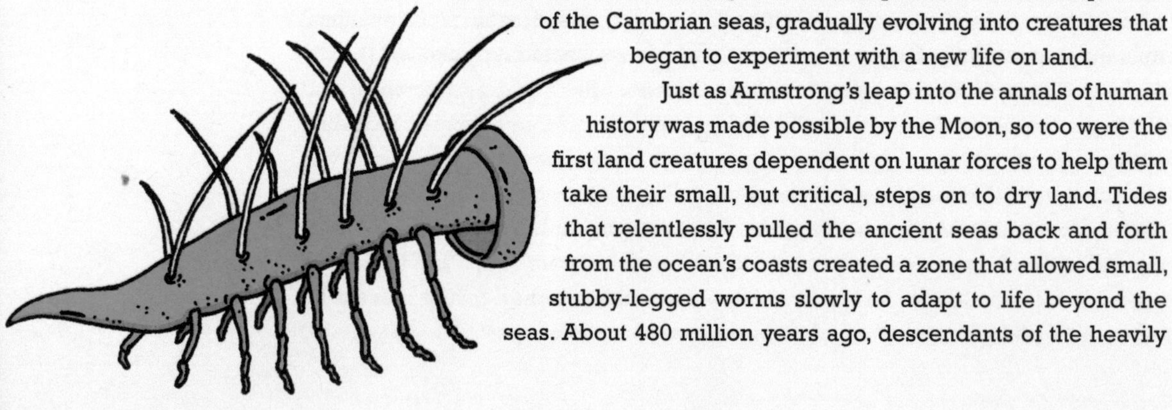

The Hallucigenia, ancestor of the velvet worm.

defended Hallucigenia were in the process of evolving into the ancestors of today's terrestrial velvet worms, adapting to life on land, which was then an almost predator-free world, allowing them to abandon their heavy body armour.

Extreme changes of environment, such as the move out of the water and on to dry land, have often caused, over generations, some of the most dramatic evolutionary transformations in the history of life on Earth. These successful pioneers of the land developed small tubes (trachea) to let oxygen in the air pass deep inside their worm-like bodies. A system for males to pass sperm directly into the bodies of females ensured sexual reproductive success in an environment where there was no surrounding water to act as a freely available transport medium for fertilization. With their bodies divided into segments, each equipped with a pair of well-adapted legs, some of these creatures are thought to have transitioned into millipedes, centipedes and eventually, through the merging of various segments, into the familiar tripartite body plan of insects, with their endless variety. Genomic fusion and hybridization, thanks to successful chance matings between distantly related species, is how some experts account for the velvet-worm-like larval caterpillar phases of today's butterflies, wasps and scorpion flies.

Today's brightly coloured velvet worms are regarded as living fossils, almost exactly resembling their 400-million-year-old ancestors. Species such as *Peripatopsis alba* (white cave velvet worm) and *Peripatoides indigo* are shy creatures that live in warm, humid caves, forests and grasslands, mostly in the southern hemisphere. They conform almost exactly to the fossils of those creatures that first pioneered the original passage on to land, hundreds of millions of years ago. Once their ancestors had established their land-based niche some adapted into species that suited other terrestrial habitats. Others continued to live apparently without need for further evolution, preserving their genetic shape, form and function thanks to the preservative power of sexual reproduction (see page 47).

Today's velvet worms are almost identical to some of the first creatures to venture out of the seas more than 400 million years ago.

More than 150 velvet worm species have been recorded to date, divided into two main families, of which the most ancient is called Peripatopsidae. Today many specimens are on the verge of extinction due to the destruction of rainforests and the draining of wetlands caused by the growth of human conurbations. Only time will tell whether the descendants of those ancient pioneers who first wriggled on to land can re-ignite their appetite for evolution, in order to save their ancient lineage.

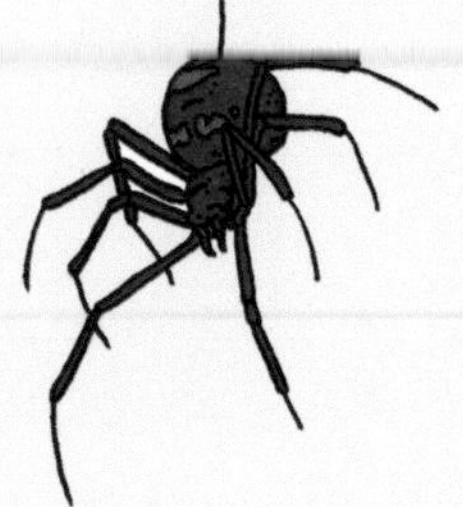

Sea Scorpion

CLASS: EURYPTERIDA
SPECIES: JAEKELOPTERUS RHENANIAE
RANK: 78

On the origins of an ancient instinct that became crucial to the survival of many different species.

A DEEP-SEATED, instinctive fear of spiders is one of the most common of all human phobias. Like the development of physical forms, such as bones, shells or teeth, behavioural instincts have evolutionary roots reaching deep into the geological past. Charles Darwin devoted an entire chapter in his *On the Origin of Species* to the question of how certain inherited instincts – like those that cause a bee to build perfect hexagonal cells or a cuckoo to lay its eggs in other birds' nests – have variously improved the chances of a species surviving in the struggle for existence. Such instincts, said Darwin, have been refined and passed on through countless generations.

The instinct in some species to flee from danger probably reaches all the way back to when the oceans were populated by extremely scary relatives of today's arachnids (spiders) called sea scorpions. That these were the largest arthropods ever to have lived is now beyond dispute following the discovery in November 2007 of a huge sea scorpion claw from rocks in a quarry near Prüm, in the mountainous Schneifel region of Germany. By *c.* 400 million years ago, sea scorpions had taken over as the 'king-pins' of the marine ecosystem following the extinction of creatures such as the Cambrian Anomalocaris. The enormous flesh-ripping claw, found by German palaeontologist Markus Poschmann, belonged to a species of scorpion called *Jaekelopterus rhenaniae*. This creature grew larger than a human – to more than 2.5 metres in length.

Sea scorpions (also known as eurypterids) fed off fish, life's original vertebrates, some of which were the ancestors of land animals, such as amphibians, reptiles and mammals, including man (see On Fish that Came Ashore, page 106). Eurypterid fossils have been found all over the world, indicating that not only were these prehistory's largest arthropod species but that they were also among the most successful. Sea scorpions spent their lives in shallow coastal areas, hunting bony fish, trilobites (see page 70) and other animals using stealth. Two pairs of eyes (similar in design to those found in today's spiders) helped these creatures burrow under the muddy sea floor where they could lie in wait, keeping at least one eye out for any tasty morsels that happened to pass by. If near enough, the scorpion would launch itself at high speed, snatching at the unsuspecting prey, ripping it to shreds with its vicious claws. Its victims usually succumbed through complete surprise.

But not always, which is where the benefit of instinctive flight comes in. Those fish with the quickest reactions survived such fearsome predators, passing on their escapology skills to future generations. Thus a highly effective instinct for flight away from such sea monsters was triggered by the slightest sign of attack. Could a link between today's human instinctive fear of arachnids (spiders) originate with sea scorpions hundreds of millions of years before people first evolved? Arachnophobia is common enough: almost 80 per cent of twenty-five to thirty-four-year-olds questioned in a UK survey owned up to having some form of fear of spiders.

Increased levels of oxygen in the atmosphere, as a result of the land being colonized by plants and trees (see page 80) may have been one reason why scorpions of this era reached such massive proportions. Another reason may have been the lack of other large predators in the seas, allowing these scorpions to increase in size to the maximum limit that the physics of the arthropod design would permit.

Thanks to the evolutionary success of the instinct of flight, vertebrate bony fish eventually bit back in the form of their own larger species, such as placoderms – the first jawed fishes – and sharks (see page 78). These creatures successfully began to populate the seas from about 390 million years ago, which coincides with when the largest sea scorpions start to disappear from the fossil record. Eurypterids finally became extinct in the Great Permian Extinction 252 million years ago (see page 65) when as many as 96 per cent of all marine species were wiped out as a result of cataclysmic environmental changes. Only their closest arthropod relatives, the arachnids, survived this traumatic period, eventually becoming one of the most numerous arthropod groups on Earth (see page 400).

Unlike symbiosis, viral infection or selective pressures, instincts are forces for conservation and preservation, not adaptation and change. Just think: so conservative are the inherited instincts of humans that most of us will baulk at the sight of a harmless spider yet mindlessly risk picking up a loaded gun in a shooting range or wielding a sharp-bladed knife in the kitchen.

It is impossible to know exactly when today's human fear of spiders originated but prehistoric sea scorpions were life's most powerful triggers for the evolution of instinctive fear in many ancient marine species and were therefore architects in the evolution of escape by fright. This powerful inherited instinct was, and still is, essential to the success and survival of many of life's most resilient die-hards.

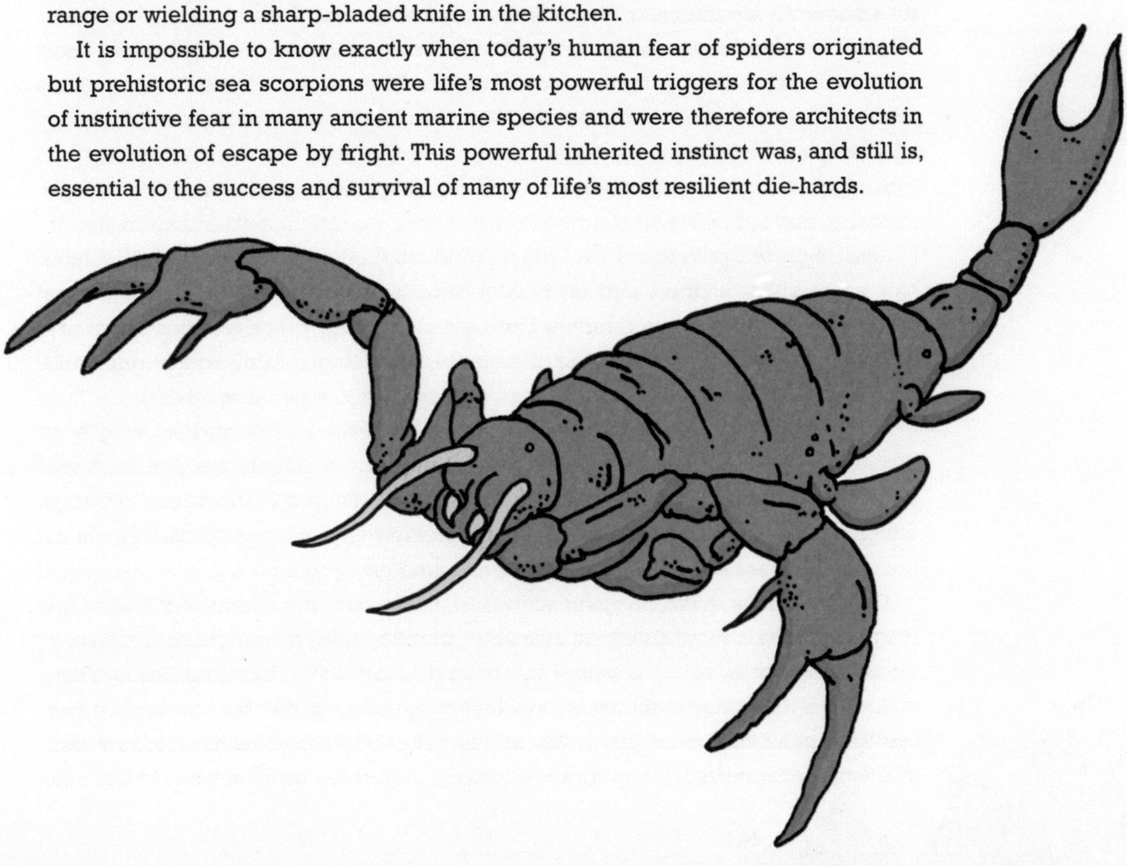

Sea Squirt

FAMILY: BOTRYLLIDAE
SPECIES: BOTRYLLUS SCHLOSSERI
RANK: 12

Ancient sea creatures that help regulate the planet's climate and could hold the keys to improved human health.

MOLECULAR BIOLOGISTS are often regarded as today's best bet for extending individual human lifespans. Recent research into countering the natural ageing process has been focused on finding ways to re-grow or repair vital human organs – such as the heart, lungs or liver – that may be diseased or suffer from old-age. Such a process of 're-vitalization' could go a long way to extending individual human lives.

A small but highly significant marine species whose family first evolved in the late Cambrian Period (*c.* 500 million years ago), and whose descendants, it is thought, eventually led to the evolution of man himself, may hold the key to how this miraculous process could one day be accomplished.

Tunicates (more commonly known as sea squirts) are small, jelly-like animals that usually glue themselves to the seabed in the luxurious environment of coral reefs (see page 66). They filter water through gill-like slits in their bodies, searching for microscopic food in the form of plankton (mostly algae and bacteria). Beneath the apparently simple body structure of these translucent, cucumber-like creatures lies a more complex system in which their vital organs (throat, heart, digestive tract, neural and reproductive systems) are protected by a layer of special connective tissue called mesenchyme, the same as that found in the embryos of humans – their latter-day descendants.

Sea squirt larvae are significant in evolutionary history as the first creatures known to have developed a notochord – a flexible, rod-shaped structure that in human embryos grows into a backbone (vertebrae). All vertebrates (which include fish, amphibians, reptiles and mammals) have backbones essential for activities such as swimming, walking, running and flying. Starting with the ancestors of sea squirts the chance change that split animal body designs into invertebrates and vertebrates began to reveal its true significance (see Hox genes, page 61).

Most sea squirts use their flexible notochord only in the earliest stages of development when they swim – a bit like tadpoles – to a new location far away from their parents to reduce competition for food. They then take root, growing either alone or into a new colony, abandoning their tails, along with their precious notochords, to lead a sedentary lifestyle attached firmly to a reef.

One tunicate species, *Botryllus schlosseri*, has caught the attention of biologists because it is somehow able to spawn new colonies directly from its blood vessels rather than from its sexual organs – a process called vascular budding. Researchers at Stanford University, California, have found that a complete new tunicate colony can be formed from scratch via this process after successive generations of 'trial and error', especially if these creatures' sexual organs are removed. Efforts are now under way to decipher the genetic code of this species to find out how to switch this asexual regenerative process on or off, chiefly to see if it could be activated in humans. Such technology could allow people to 're-grow' damaged organs such as kidneys, a liver or even limbs. The California team's optimism stems at least partly from the fact that humans are distant descendants of sea squirts and share many genetic traits.

Another family of sea squirts, the aptly named Larvacea, never attach themselves to the sea floor. Instead they stay embryo-like all their lives, using their notochord-powered tails as they swim freely in the oceans, foraging for food. It has recently been discovered that these creatures may play a significant part in the overall health of the deep-ocean ecosystem.

One example is *Bathochordaeus*, a widely distributed five-centimetre-long sea squirt that builds what are known as underwater 'houses', up to one metre wide, made out of mucus. The sea squirts use their notochord tails to beat water through these membrane-like structures to capture tiny particles of food. Once their mucus construction gets clogged up, they then abandon it, and swim off to build a new one – a process that occurs about once every four hours. The old mucus house then descends like a parachute, providing much needed nutrients to creatures that live on the sea floor.

Such house-building habits are now thought to have a quantifiable impact on the global climate. Researchers at Monterey Bay Aquarium Research Institute (MBARI), California, who have been measuring the effect of Larvacea on the oceans, believe substantial quantities of carbon dioxide are removed from the atmosphere as a result of this parachuting process, which pumps dead organic matter (which is partly made out of dissolved carbon originating from the air) to the sea floor. Their research is currently helping oceanographers and climatologists puzzle out what impact a rise in global sea temperatures may have on this natural biological process.

Despite being relatively inconspicuous and little known, sea squirts have certainly had a major impact on the planet, life and people in the deep evolutionary past as well as on deep-sea life in the present. As for the future, their potential lies in unlocking the secrets of how it may one day be possible to re-grow human tissue. Gerald Weissmann, editor of the Federation of American Societies for Experimental Biology magazine, believes that research into the reproductive quirks of sea squirts is: *'a landmark in regenerative medicine ... The biological equivalent to turning a sow's ear into a silk purse and back again.'*

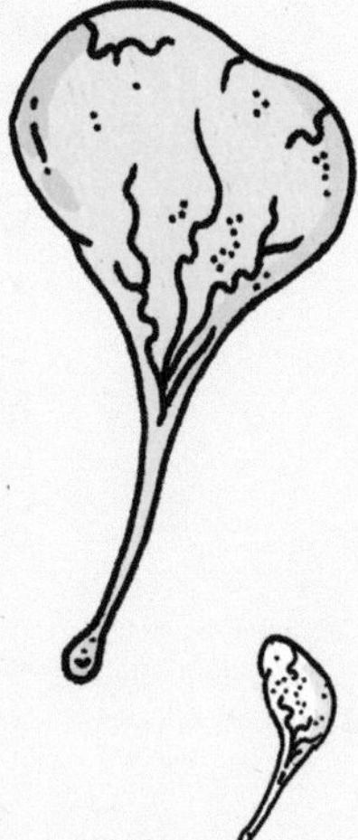

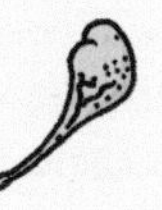

Shark

CLASS: CHONDRICHTHYES
SPECIES: CARCHARODON MEGALODON
RANK: 38

Ancient survivors that pioneered the art of adaptive immunity.

CREATURES that rank as one of life's greatest success stories do not automatically become the focus of human admiration and respect. Several of nature's most brilliant, robust and enduring designs are, in the eyes of humans, abhorrent organisms ranging from pests such as ticks (see page 384) and mosquitoes (see page 134) to monstrous, brutal killers.

Sharks are fish with jaws that evolved about 420 million years ago. Mostly they eat other fish or microscopic plankton and have no specific appetite for man. Their survival from their first forms (such as Stethacanthus) has remained almost unchanged for 400 million years. The key to their success is a combination of evolutionary innovation and an ongoing series of experiments and adaptations that has made these creatures among nature's most robust.

The evolutionary transformation of five-centimetre free-swimming, sea-squirt-like marine creatures into massive ten-metre-long sharks took place over about one hundred million years. No one knows for sure where and how jaws first appeared. They may have arisen from members of bony-headed fish called Ostracoderms. Placoderms were the first true-jawed fish, some species growing to enormous sizes owing to their ability to devour just about anything and everything in their path. One type, Dunkleosteus, had a body ten metres long and a bite more powerful than *Tyrannosaurus rex* (see page 125).

But placoderm supremacy was curtailed by a series of mass extinctions *c.* 360 million years ago and an evolutionary sister group took over as top marine predator. These were the sharks, creatures that have maintained their position at the head of the food chain ever since.

Three factors contributed to the sharks' long-term success. Firstly, sharks are among the most diverse animals when it comes to mastering the art of sexual reproduction. Some sharks reproduce by internal fertilization (copulating), others are oviparous, laying eggs with leathery cases commonly called mermaid's purses. Yet others lay eggs which females keep safely inside their bodies until they hatch. Another group behaves more like mammals (see page 157), producing live young that feed on a special type of milk secreted inside the uterus. A final group indulges in the gruesome practice of oophagy where the largest youngsters hatch inside the mother shark only to eat their smaller siblings. As a result, only the strongest offspring survive. Such reproductive versatility is good insurance against mass extinction.

Gaping jaws of the extinct giant shark *Carcharodon megalodon.*

A second success factor is size. Legendary species include the sixteen-metre-long, forty-eight tonne *Carcharodon megalodon*, a giant whale-killing shark that first appeared about 16 million years ago. Its mouth could open as wide as two metres, revealing a set of gargantuan teeth, some of which were over twenty-two centimetres long. Environmental changes finally brought about the death of this monster of the deep. As the ice ages increased their grip, about two million years ago, fatty whales migrated to live in the icy polar seas, leading to the demise of the *megalodon* through lack of food. However, other shark species remain successful today – over 400 million years after their ancestral origins – thanks to a third innovation: the extraordinarily powerful system of adaptive immunity.

No one has yet compiled a comprehensive history for the origins of adaptive immunity or properly explained why this complex system, in which specialist immune cells remember previous infections in order to ward off repeat attacks, occurs only in vertebrate animals. One theory is that about 400 million years ago repeated swarms of various viral infections colonized the ancestors of today's sharks. Many viruses (especially persistent retroviruses, see page 14) infect their hosts without causing them harm. Indeed they rely on the health of their hosts as viral replication vehicles. It is therefore in the survival interests of these infections to try to *prevent* other viruses from invading the same host organism. Over time, it is thought, tactics used by certain viruses to forestall infection by others may have helped establish the adaptive immune system that all vertebrates employ today.

Sharks display substantially greater levels of immunity from viral infection than other groups of living species – including plants, insects, bony fish, reptiles and mammals (including humans). Indeed, shark liver oil is today marketed as an 'immune system booster'. Norwegian fishermen in the eighteenth century regularly applied oil extracted from the livers of deep-sea sharks to open wounds as a healing agent. Meanwhile, Swedish researchers in the 1960s confirmed that the same immune-boosting compounds found in sharks are also abundant in human breast milk.

Mean, cruel and brutal are some ways of describing these highly successful creatures of the deep. Innovative, versatile and adaptive are others. In terms of impact, sharks have survived all five major prehistoric mass extinctions, although how well they will survive the human-induced sixth remains to be seen.

On Pioneers of the Land

How mutual collaboration between trailblazing species helped life colonize the Earth's barren landscape.

EVERY DAY I take our dog, Flossie, for a walk. Although the route sometimes differs, the experience for her must be fairly similar each time. It is grassy where we live, providing soft padding for her feet, and we always start off by passing under some rather majestic trees, mostly oaks and ashes. In the summer she charges through fields of maize and wheat. In the winter she sniffs inquisitively at the ground, ploughed up into thick strips of earth, rich with worms.

What, I sometimes wonder, would it have been like to follow exactly the same route about 450 million years ago? While life back then had already exploded into myriad forms in the seas, there was precious little evidence of its presence on land. What's more, thanks to the continuous shifting of the Earth's tectonic plates, southern England would have been located somewhere near the equator.

Back then there was no grass and there were no trees. In fact nothing green at all – just a rather barren lifeless landscape of thin, slippery mud. At any moment a steaming hot geyser might explode in front of us, shooting jets of scorching water from deep beneath the ground. About the only consolation for Flossie might have been a few weedy bushes, about fifty centimetres high, but none would have borne any flowers or leaves. This would not have been a very rewarding walk.

But wind our clocks forward another one hundred million years and we would witness a huge transformation. Suddenly we are in a forest! Giant thirty-metre-high trees with trunks like enormous telegraph poles stud the earth, which looks and feels so much richer and darker and far more crumbly to the touch. There is plenty for Flossie to sniff at here. Fallen logs, smothered with bright green mosses, carpet the ground. Tiny insects busily feed on the decaying wood. Mushrooms are popping out of the ground to spread their spores, beetles crawl up tree trunks and a spider's web glistens from a nearby branch. No dangerous creatures here, or so you would think. But, look! Flossie! What's that perched high up on a tree?

The sight of an enormous dragonfly – at least a metre wide – would be enough, I am sure, to send us both scuttling back to our time machine. This landscape, although completely transformed and a lot more familiar, is still not quite the world we are used to on our regular country walks.

Land-life ahoy!

Forests first emerged in that interval between about 450 and 350 million years ago, transforming the barren Earth into a rich terrestrial ecosystem, parts of which still cling on in the countryside, tropics and tundras of today.

Since the genesis of all land-life was ultimately dependent on these forests for food, water and shelter, it stands to reason that whichever living things were responsible for making this incredible leap from the water to the land must have had a major impact on life on Earth. Of greatest initial importance were the non-flowering plants, responsible for stocking the Earth's land with a rich store of food, the organic larder from which later land creatures could feast.

Plants evolved from algae, aquatic Eukaryotic protoctists that could transform sunlight into food (see page 54). Some turned multicellular, becoming what we now call seaweed. But, as with any dramatic change in living circumstances, successfully adapting to a life on land necessitated huge physiological changes.

The process that firmly established plant-life on land was indeed what Charles Darwin called *'modification by descent'* or *'natural selection'* – those species with variations that best equipped them to survive living and reproducing in a world surrounded by air not water were those that thrived, passing on their characteristics to future generations. But the cause of such variation had, at least initially, as much to do with mutual collaboration as cut-throat competition.

In the same way that corals recruited red algae to make food for them (see pages 66–7), so another group of once aquatic organisms entered into a series of long-lasting relationships without which life may never have made it on to land at all.

An artist's impression of a forest in the Carboniferous Period, *c.* 300 million years ago, dominated by trees such as *Lepidodendron*.

From fungal beginnings ...

Classified as an entirely separate kingdom from plants or animals, fungi have cell walls made of chitin, the same substance that forms the hard outer covering of insects. Fungi feature networks of branching thread-like hairs, called hyphae, which grow outwards to form a mycelium. Most fungi are saprobes, feeding on dead organic matter. Enzymes secreted at the end of each hair digest dead matter into a nutritious juice, which is then reabsorbed to feed the growing fungus. To reproduce, the organism usually grows a fruiting body (such as a mushroom) from which spores are released. These are carried by the wind to new places where they settle, germinate and grow.

Primitive aquatic fungi first arrived on land as spores propelled by the wind. Molecular evidence suggests the ancient split in Eukaryotic life between plants, animals and fungi happened in the oceans about one billion years ago when only microscopic life-forms existed.

Lichens were probably the first successful types of vegetative life on land. They comprise closely intermingled fungi and algae which are so totally dependent on each other that experts usually refer to them as a single, combined species. Few better examples exist in nature of how once separate organisms can develop successful survival strategies through a merger of equals.

Lichens are thought to have colonized rocks on the seashore about 500 million years ago, although, since they do not easily fossilize, firm evidence is scant. What's certain is that today's lichens are highly efficient at living away from water thanks to the food supplied by algae through photosynthesis. Neither organism can survive without the other. The fungus provides protection for the algae, the algae provides food for the fungus.

Fungi struck up a second mutual relationship that also proved essential for life on land, this time about 460 million years ago. Mosses are plants that learned to grow on the edge of rivers, lakes and swamps. They are strictly limited to a life in damp, watery environments because they have no roots to help them reach water and nutrients deep down in the ground. Their method of sexual reproduction is entirely dependent on the water too. Just as in animals, fertilization occurs through flagellated sperm-like cells that have to swim through the water from one parent to another.

The business of how flat, damp clumps of water-dependent moss transformed into forests full of tall graceful trees that could grow far inland is largely down to this second mutual bargain struck up by fungi called *mycorrhiza*.

Only relatively recently has it been established that most trees are connected by intricate networks of branching underground fungal hyphae. Indeed, over 95 per cent of all plant families rely heavily on subterranean fungi to help them gather the nutrients and water they need for effective growth. The fungi attach themselves to the roots of plants, passing them supplies of water and mineral nutrients. In return, they receive carbohydrates (starch and sugars) produced by leaves at the top of a plant or tree. Mycorrhiza massively increases the surface area available for plants to absorb essential nutrients away from the water's edge.

The enormous impact of lichenous and mycorrhizal partnerships is evident from 410-million-year-old fossils collected near the Scottish village of Rhynie. They reveal a now-extinct species of primitive plant called **Rhyniophytes** (see page 88), which became petrified (turned to stone) by hot geysers of silicate water. These fossils clearly show that a fungus acted in lieu of true roots in plants such as Aglaophyton, whose descendants live on in today's 1,200 species of club mosses.

Other plant fossils found at Rhynie, such as *Rhynia gwynne-vaughanii*, show how a clear transition was taking place from primitive herb-like plants into modern vascular forms in which tubes (called xylem and phloem) conduct water and nutrients from primitive roots up the stem of the plant like a wick in a candle. Lignin, a tightly packed mesh of cellulose cells that makes wood, was another essential adaptation that allowed these plants to stay upright even in times of drought, eventually provoking an evolutionary race for light and space.

Darwinian natural selection saw to it that, over time, thriving plant species were those that grew taller to avoid being overshadowed by their neighbours. As a consequence, by about 360 million years ago, much of the Earth's surface was covered by thick forests of trees, such as **Lepidodendron** (see page 90) that grew up to thirty metres high.

None of this could have happened were it not for fungi such as **Prototaxites** (see page 86). Fossils show that, between 420 and 370 million years ago, this giant fungus was by far the tallest living organism on land, growing up to one metre wide and eight metres tall. It was so huge its fossils are often mistaken for tree trunks. No other species better illustrates the extraordinary evolutionary contribution that fungi made to the successful colonization of inland areas by plants and trees.

While fungi provide some of the best examples of mutual symbiosis (lichens and mycorrhiza) they can also be nature's deadliest parasites. Some plants live in perpetual danger of being digested alive by parasitic fungi. Rusts, smuts, dry rot and mildew are examples of fungal species that feed on every type of plant from ornamental garden flowers to staple crops such as wheat, barley, oats and rye (see ergot, page 304). Artificial fungicides are often the only way to keep these scourges under control (see Bordeaux Mixture, page 53).

The world's ecosystems would have collapsed aeons ago without fungi because they excel in turning once-living material, such as wood, into raw nutrients that fertilize new life in the soil. Species such as oyster mushrooms (*Pleurotos ostreatus*) are such efficient recyclers that they are even being used today to clean up oil spillages, turning man-made pollution into food.

Defensive agents produced by fungi to ward off bacteria, plants, insects and vertebrate animals have also had their own profound effect on human history. By fighting bacterial disease, antibiotics such as penicillin (see page 314) have extended life expectancy and increased populations. Some fungi have even stimulated new cultural traditions and religions. Yeasts are used to brew alcohol (see page 293) and ergot to induce 'alternative' mental states (see page 304). Well into the nineteenth century, authorities in the Christian Church preached that mushrooms were creations of the devil. That, according to some experts, has had its own impact, too, putting research into fungi (mycology) at least a century behind the study of other life-forms.

... to a world of leaves

Comprehensive colonization of inland areas by plants and trees also required the development of efficient ways of turning sunlight into food. Primitive plants used small green photosynthetic scales called *microphylls* that jutted out of stems and trunks that caught the light (see *Lepidodendron*, page 90). Ferns are plants that took this a stage further by growing true leaves called *macrophylls* that radiate outwards like a fan, substantially increasing their exposure to the sun.

Ferns are the most diverse group of non-flowering plants living today, which illustrates their long ancestry. They range from ancient trees (such as archaeopteris) to small aquatic varieties such as the mosquito fern, **Azolla** (see page 93), which had its own significant impact on the Earth's past climate.

Deadly fungus: this sporulating species of rust attacks a wheat plant, hijacking its seed-producing systems.

But these plants were still unable to colonize the driest areas. Their microscopic reproductive spores are so small and so unprotected that they can successfully germinate only in moist, hospitable environments. A better reproductive system was required before plants could thrive in the hottest inland zones.

Life takes seed

Seeds are reproductive cells stocked with supplies of food and packed in a tough weatherproof case (testa). The plant families that pioneered their development were the cycads (such as the sago palm) and conifers (see page 95). These two groups came to dominate dry inland areas from about 300 million years ago, owing to the fact that they were no longer tied to the water's edge in order to reproduce.

Cones are seeds produced by conifers such as firs, spruces, cedars, pines and redwoods. The absolute success of these ancient colonizers of the land is evident from the vast boreal forests still present in the world today – despite the human obsession with deforestation. Civilizations have relied on their soft timber for construction for thousands of years. What's more, the fossilized remains of Carboniferous Period trees in the form of coal deposits makes it highly unlikely that the modern human way of life could have ever taken root without trees like these.

Resin, a thick, sticky liquid produced by conifer trees to ward off invasive insects and fungal infection, has left for posterity its own permanent record of other types of land-life. The earliest creatures found trapped in fossilized amber include bacteria, insects, fungi, spiders and plants – nature's other terrestrial pioneers.

Terrestrial transformation

Colonization of the land by plants and trees, which began about 400 million years ago, had a profound impact on many other species. Some benefited from increasing levels of atmospheric oxygen, which are reckoned to have risen as high as 35 per cent (today's level is about 21 per cent) as a consequence of the Earth's new blanket of photosynthesizing plants and trees. As a result, **dragonflies** (see page 103), typically no more than five centimetres long today, grew to a monstrous metre or more in length.

Other creatures in the sea suffered as a result of the planet's newly developed terrestrial ecosystem. Deep root systems that tall trees needed to keep them upright broke down the ground near riverbanks. As a result large quantities of loose soil leached into the seas, starving marine life of oxygen. This triggered a mass extinction of species, especially corals, stromatolites and deep-sea dwellers about 360 million years ago (see placoderm, page 78).

Millipedes, centipedes and annelid **earthworms** (see page 97) escaped the anoxic seas by adapting to breathe air. Some walked on legs, led by the semi-aquatic ancestors of today's velvet worms (see page 72). This led to an unparalleled variety of life-forms on land. Over time, the segmented bodies of these millipede-like pioneers became compressed into what we now call 'insects' with their familiar tripartite body design. Some, such as the dragonflies, took to the skies, evolving wings from ancient body-temperature regulating flaps. Others, such as cockroaches and grasshoppers, thrived on the rich vegetation of the world's first forests. Beetles evolved into a group of creatures with more variation than any other known to science. Between five million and eight million different species exist, of which only about 350,000 have been formally described (see **dung beetle**, page 100).

Invertebrate creatures like these began to fill up every available terrestrial ecological niche. Some lived as herbivores, others became scavengers or dung eaters, or parasites and predators, dominating life on and in the earth, from the coldest polar regions to the driest dusty deserts.

Up, up and away!

Despite the abnormally huge size of prehistoric dragonflies, creatures with invertebrate body designs, such as insects, are generally confined to a relatively small stature. Characteristics of small animals are their more numerous populations and shorter generation times. Tinier creatures tend to

produce more variants than larger creatures because their evolutionary clocks tick faster. It means that in a given space of time, animals like elephants (or humans) are less likely to evolve as quickly or radiate into so many different forms as smaller creatures such as insects.

All of which suggests that when life gets tough (such as during mass extinction events), genes craving immortality are best served in small creatures such as insects rather than more vulnerable (and less various) large mammals. That may account for why over 97 per cent of all surviving animals today are small invertebrates. In times of stress, size matters to the immortally minded.

Such powers of variation also explain why insects were the first creatures to take to the air. Temperature regulation, not flight, was probably the original use for flaps called cuticles on the sides of ancient insects. The knack lives on in butterflies, which increase their body temperature by spreading out their wings in the full heat of the sun. Gradually these cuticles became stretched and distorted in ways that helped ancient insects take to the skies.

Instead of crawling all the way back down a tree trunk, life's original insect flyers are thought to have experimented with jumping down from trees using these flaps to glide safely to the ground. Those with more muscles in the sides of their bodies thrived, establishing the selective pressures that made self-propelled flight a reality.

Another possibility is that primitive insects were chased into the air by other terrestrial predators. Those that flapped fast enough survived. In this scenario, species with large cuticles passed on their characteristics to subsequent generations until competent flight was properly established.

The web of life (and death)

But what predators could have driven insects to experiment with such extreme tactics for escape? Fast-running scorpions, with pincers poised, were constantly lunging out to grab at prey. They, too, had moved from the sea to the land, successfully establishing air-breathing lungs in place of gills. Their fang-like chelicerae regularly made mincemeat of invertebrate victims – usually served up with an extra dose of venom to make sure of the kill. Such habits may well have provided the original incentives for their potential prey to reach for the skies.

The scorpion's arachnid relatives never needed to take off in hot pursuit, thanks to their own ingenious adaptation. The silk of a spider's web is the strongest yet most elastic naturally occurring material in the world, just what was required to adapt to a habitat where their prey had taken to the skies. For hundreds of million of years careful, deliberate, brilliant weavers have patiently spun silken traps making a match even for the fastest flying ace (see spiders, page 400).

Such was evolution's original terrestrial Garden of Eden – an almost complete ecosystem on which humans entirely depend on for their survival today. It was created between 450 and 350 million years ago when nature's many forces of evolutionary change converged on to the land.

So compelling was this new environment, now bursting with life, that even bony fish – the distant ancestors of man – couldn't resist the temptation to explore what it might be like to live successfully beyond the seas ...

Prototaxites

FAMILY: (UNKNOWN)
SPECIES: PROTOTAXITES LOGANII
RANK: 52

A fungal giant that paved the way for life on land.

A MUSHROOM that's as high as a house sounds like something out of *Alice in Wonderland*. But no. This giant fungus dominated the Earth's early terrestrial skyline for at least seventy million years, eventually becoming extinct about 350 million years ago. Its fossils still stud the deserts of Saudi Arabia, jutting out from dry rocks like the stumps of trees.

For a long while biologists believed this ancient form of land-life was indeed a tree – an ancestral conifer, perhaps? At least that was the view of John Dawson, a Canadian naturalist, who was the first to describe these log-like fossils back in 1857. He named them *Prototaxites* (meaning first yew) because of his understandable assumption that such large structures could only be the remains of ancient woodland.

Other scientists were not so sure. Some said it was a giant stinging algae. The mystery seemed to be resolved in 2001 when, after extensive research by Francis Heuber from the Museum of Natural History in Washington, it was suggested that these were fossils of a giant fungus. Even more recently, in 2007, researchers at the University of Chicago put the matter beyond doubt. They have discovered an uneven ratio between different types of carbon (isotopes) in the fossils, indicating that this was an organism that fed on organic matter (a heterotroph) and so was not a photosynthesizing plant (an autotroph).

Confirmation that *Prototaxites* were fungi underpins the huge impact these organisms must have had in the critical transition of life from sea to land. Even now it is hard to imagine a fungus so vast that when it wanted to reproduce, its mushroom-like fruit punched up skywards taller than a two-storey building.

In the Devonian world, 420 million years ago, plants had yet to evolve their own proper roots. It was only through the symbiotic relationship with fungi like *Prototaxites* that they could survive on land at all. Hairs at the bottom of plant stems connected with the branching hyphae of *Prototaxites*, which then delivered supplies of underground nutrients and water in return for food.

No estimates have been made as to how far underground a species like *Prototaxites* may have spread, but it is fair to assume that the hairs of each organism could have extended for many miles, providing underwater plumbing services to countless dependent plants. The impact on the Earth itself must have been equally profound. Feeding on subterranean bacteria, its network of digestive hairs (mycelium) would have broken up the hard ground, transforming inhospitable rock into crumbly soil filled with nutrients vital for other life-forms.

The creation of such soils meant that over time, plants were able to reach underground unaided to find their own supplies of water and minerals, anchoring themselves more firmly with deeper roots and so reducing their dependency on fungi. However, the fact that most plants today still rely on some kind of fungal association to survive is indicative of the huge ecological contribution that was made by pioneers like *Prototaxites*.

Many extinct species have played an important role in the history of evolution. Some contributions were so powerful that, like points on a railway track, they sent evolution down an entirely new path. Ironically, such paths often prove more advantageous for other species, sometimes driving the evolutionary protagonist into extinction (see also *Lystrosaurus*, page 118).

That's what happened to *Prototaxites*. Once tree forests began to cover the land, these mighty mushrooms simply couldn't compete. Small invertebrate arthropods, such as worms, spiders, mites and millipedes, that lived on the increasingly fertile soils, now found these giants all too good to eat. As the forests grew, space for twelve-metre-high fungal fruits was at a premium. Only those fungi that adapted in size and developed new defences survived. These are the fungi we are familiar with today.

Exactly which group of fungi living today are the direct descendants of *Prototaxites* is not yet established, but the pioneering impact of these early land giants has few, if any, parallels.

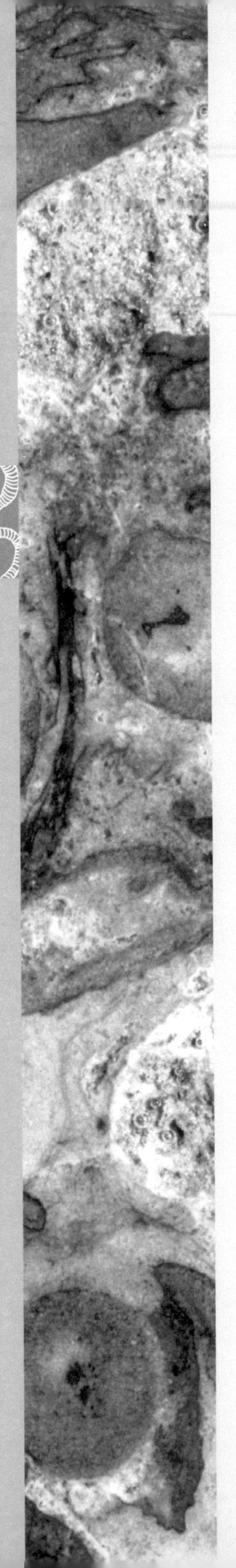

Rhyniophytes

FAMILY: RHYNIACEAE
SPECIES: RHYNIA GWYNNE-VAUGHANII
RANK: 83

Extinct weeds that were vital staging posts in the transformation of soggy mosses into tall graceful trees.

FANCY A WORLD without weeds? While the prospect may sound attractive to golfers and gardeners (see grass, page 358), it is only thanks to the curiosity of an early-twentieth-century Scottish doctor and amateur geologist that we now know what an extraordinary role ancient weedy plants played in the story of life on Earth.

One morning, in the spring of 1912, Dr William Mackie spotted some unusual-looking rocks in a dry-stone wall surrounding fields near his home just north of Rhynie, a small village in Aberdeenshire, Scotland. Being a curious sort, he took them home to study them in more detail under his microscope. He was amazed to discover they contained an intricate cross-section of beautifully preserved fossils, showing highly detailed systems of what looked like plant stems and cells.

Nearly 400 million years ago, Rhynie was nothing like Scotland is today. Located somewhere near the equator, this was a geothermic spa-land steaming with boiling pools of bubbling mud. In this hot, sweltering place, giant geysers would frequently spout huge fountains of scorching water laced with silicon from deep beneath the earth. When this silicate water landed on nearby vegetation and plants it killed them instantly and quickly cooled, turning their petrified remains into perfect fossils of stone.

The fossils of Rhynie have since shown scientists exactly how primitive plants began one of history's most remarkable transformations: from floating seaweeds to woody shrubs and ultimately into enormous trees that could colonize the land many thousands of kilometres away from the water's edge.

The plants' 'trick' was to become 'vascular'. Mosses are land plants that have no roots, leaves or stems and can survive only in damp habitats close to water. Leafy plants and trees today use a design for their survival derived from these early pioneers at Rhynie.

For plants (as for animals) life on land is dramatically different from living in the seas. For a start, land-life has to grapple with the force of gravity, which has far less effect in water. Learning to grow upright – against the force of gravity – was no simple adaptation. The plants at Rhynie show the first attempts at a solution with the development of a dead cellular structure called lignin in which layer upon layer of tightly packed cells are stacked and interwoven to make rigid stems (see apoptosis, page 48). These structures are what ultimately led to the construction of that sturdy constituent of shrubs and trees – wood.

Lignin has a host of other essential functions, too. Even in dry conditions, it supports plants so they can stay upright as close to the sun as possible. Lignin also toughens stems to support tubes (xylem and phloem) that can transport nutrients and water up and down the plant. Vascular systems stop plants drying out and allow them to grow tall and support leaves – essential prerequisites for the eventual evolution of most of the world's trees.

Living on land had other hazards such as the sun's ultraviolet radiation that can easily destroy the delicate structures within a living cell's DNA molecules (and can lead to skin cancer in humans). Early plants, like Rhyniophytes, developed ultraviolet-resistant spores that protected their offspring from the sun's radiation even if germination was delayed by a lack of sufficient water. Such compounds have been common to all land plants since.

If Rhyniophytes were solely responsible for the development of all the technology needed to make trees, then their position in our top one hundred ranking would have been a great deal higher. But Rhyniophytes mostly used scales on their stems, leaving it to their descendants, the Euphyllophytes (meaning 'good leaf plants', such as ferns, see page 83) to experiment with the concept of hanging green solar panels, in the form of leaves, from their stems to better catch the sunlight.

Rhyniophytes were also limited by having roots that grew only along the ground, rather than burrowing underneath, restricting their growth to a maximum height of about fifty centimetres. Underground roots that supply plants with water and nutrients and prevent trees from toppling over came with the Lycophytes, the world's first true trees (see page 91).

Rhyniophytes were seedless weeds. Their spores had to land on wet ground for successful germination, a limitation that became more significant as the Earth's continents converged into a single landmass, Pangaea, *c.* 300 million years ago and dried out.

Only seven species of early vascular plants have been identified at Rhynie. Other similar types (such as Cooksonia) have been discovered in other places, suggesting that the evolutionary changes seen at Rhynie *c.* 400 million years ago were also happening in parallel all over the world.

Ultimately, the significance of these early vascular plants was in starting the long, but crucial, evolutionary process of adapting to life on land from ground-hugging mosses into tall, majestic trees – a process that took about forty million years. A world without trees would have severely limited the potential for all land-life, including the evolution of primates such as monkeys and humans. Also, the atmosphere would never have benefited from the enormous capacity of the rainforests to absorb carbon dioxide, a process which has helped to keep the Earth cool and make it better suited for life on land.

It was also largely due to the emergence of early vascular plants that oxygen levels rose high enough to entice larger forms of sea life to leave the water and to evolve air-breathing lungs (see lobe-finned fish, page 112). Indeed, apart from worms and insects, most animal life would have remained limited to the seas were it not for the food and oxygen provided by these early land-dwellers.

Rhyniophytes were not, however, one of nature's long-term survivors. Their lack of deep roots, leaves and seeds led to their ultimate extinction from the record of life on land about 250 million years ago. It's only thanks to the unique snapshot into the evolutionary processes captured by Mackie at Rhynie that the significance of these earliest weedy plants can be appreciated at all.

Cross-section through a piece of Rhynie chert, sedimentary rock containing fossils showing tubular plumbing in the stems of primitive vascular plants.

Lepidodendron

FAMILY: LEPIDODENDRACEAE
SPECIES: LEPIDODENDRON SP.
RANK: 11

Ancient, gigantic, fast-growing trees that fuelled the industrial world.

GO TO GLASGOW and there, in the south-west corner of the Scottish city's Victoria Park, is a small, insignificant-looking Victorian museum. Inside, however, the building opens out to reveal one of Britain's most precious and surprising geological treasures – a group of eleven massive tree stumps – each at least a metre wide – with roots that bulge and buckle across the excavated ground. These are naturally occurring, fossilized sandstone casts from a once vast, 330-million-year-old tropical forest that stretched all the way from Norway in northern Europe to Pennsylvania in the United States of America.

The trees from which these stumps were forged were once the most prolific on Earth. They are club mosses, of which only a few species survive today. The biggest and most dominant of all were long extinct *Lepidodendron* trees that grew up to two metres wide and a staggering *fifty metres high* ...

The *Lepidodendron* was like no tree alive today. One peculiarity was that it grew extraordinarily fast – reaching its enormous height within just ten to fifteen years. For most of its life it bolted upwards like a giant telegraph pole – without branches or true leaves. Green photosynthesizing scales grew on the outside of its trunk, which branched outwards at the bottom forming networks of gnarled roots to provide structural stability and support.

Only towards the end of its life did this curious tree sprout a canopy of branches, and then they appeared only at the very top of its trunk. Clusters of small green photosynthesizing blades hung from this late-growing canopy, and, unlike today's trees, these land pioneers reproduced only once in a lifetime. Thousands of small, light cones filled with spores fell from the tips of its branches like a swansong before the tree died. Mostly these landed close by, although some would have been blown farther away by the wind, littering the forest floor with seedlings that would begin the cycle of rapid growth all over again.

At least half of all the trees in the forests that sprung up during the Carboniferous Period (360–299 million years ago) are thought to have been club mosses (Lycophytes), of which by far the most numerous was the *Lepidodendron*. Since they produced no branches or leaves until the very end of their lives, these were trees that could grow extremely close together without suffering from a lack of light. The result was that the forests they formed were the most dense of all time. It is thought that up to *2,000 trees* could grow in a single hectare of soil. Fast growth, enormous height and short lifespans meant these ancient forest floors were rapidly choked with layer upon layer of dead wood.

Over time, the enormous belt of woodland that thrived in the balmy tropical climate across Pangaea was piled high with so much dead wood that no amount of bacteria, insects or fungi could hope to rot the lot. As rapid Carboniferous Period climate changes brought frequent flooding and rising and falling sea levels, so the wood sank into the swampy forest floor, each layer of trees being buried deeper and deeper under the ground until ... their remains were discovered just a few thousands years ago by humans.

Not that Bronze Age Britons (*c.* 3000–2000 BC) had the slightest inkling that this magic black rock originally came from enormous trees that inhabited 300-million-year-old forests. Only in the relatively recent past, when countries like Britain were in the throes of a chronic shortage of wood supplies, did people first start to understand that the fuel they were burning came from the fossilized remains of the Earth's ancient forests.

Coal is the remains of ancient trees (ranging in age from fifty million to 350 million years old) that got buried in water, bogs and sediment, which were starved of oxygen causing the natural process of rotting to stop. Over enormous tracts of geological time, these non-rotted hulks of wood were transformed by pressure, heat and chemical decomposition into the black stuff we call coal. Carbon from the air, used by these trees to build their woody trunks, was therefore buried underground,

Fossilized stumps of giant *Lepidodendron* trees, unearthed in Glasgow and now preserved in a Victorian-era museum.

gradually reducing levels of carbon dioxide in the atmosphere. Over time, this helped to cool the climate, a contributing factor in the numerous ice ages that have occurred over the last three million years.

The impact of the world's coal deposits on the planet, life and people has been truly transformational. The deployment of coal in Britain during the eighteenth century to power industrial machines (chiefly steam engines) solved an energy crisis that had come about from chopping down too many trees to make ships and houses, and to clear fields and farmland. With the success of high-pressure steam, pioneered by Cornishman Richard Trevithick in 1801, came humanity's first ever form of fully independent power, freeing this species from the limitations of alternatives such as animals, water and wind. Soon everyone wanted to enjoy the benefits of industrialization. Within a generation coal was being mined in huge quantities in all inhabited continents of the world in a frenzy to feed a surge of mechanical modernization.

Without coal there would have been no industrial revolution – at least not when it occurred, in the early eighteenth century. Although oil is now the world's fossil fuel of choice (since the invention of the internal combustion engine *c.* 1880) as much as 40 per cent of the world's electricity is still powered by coal today. Also, coal (in its processed form, coke) is essential for the extraction of iron from iron ore. Everything made from iron and its derivative steel that has ultimately come to shape the modern world – from knives and forks to skyscrapers – exists thanks to these deep deposits of what were once the Earth's most ancient trees.

The best coal for human use comes from the oldest deposits because this material tends to burn more efficiently, depositing less ash and producing lower levels of sulphur (a toxic gas). The oldest deposits are those from the Carboniferous Period, when *Lepidodendron* were the predominant trees.

Despite their great success on land, *Lepidodendron* eventually became extinct because of competition from other better-adapted types of trees. Seeds, broad leaves and the recruitment of other creatures (such as insects) to aid their sexual reproduction were just some of the innovations made by other species that eventually led to the *Lepidodendron*'s demise. Yet few plant species have had quite such a profound impact. If carbon capture techniques, currently being trialled in Germany and the US, are successful, 'clean' coal looks likely to play an increasingly important part in the world's medium-term energy mix, because there is still so much of the fiery black stuff available to be mined from deep underground. Even with current averages running at approximately six billion tonnes of coal extracted out of the ground by humans each year, known reserves will last at least another 300 years. Quite how much more lies buried down deep is anyone's guess.

Azolla

FAMILY: AZOLLACEAE
SPECIES: AZOLLA PINNATA
RANK: 60

A prolific fern that caused the Earth's climate to flip from a greenhouse to an icebox.

FEW SPECIES in our survey of those that have had the greatest impact on the planet, life and people have as bizarre a claim to supremacy as the small, water-dependent fern called *Azolla*. Its influence pivots on a dramatic period in the history of the world's climate.

Ferns were the first plants to develop true leaves, called megaphylls. These broad, green solar panels provide a large surface area for photosynthesis, giving this family a significant survival advantage over more 'primitive' plants, such as Lycophytes, that used microphylls – scales and thin hair-like structures on their stems, trunks and branches (such as *Lepidodendron*, see page 90).

About 20,000 species of ferns exist today, ranging from common woodland bracken to the more exotic *Cyathea*, a genus of tree-ferns that grow up to twenty-four metres high in the forests of Australia and New Zealand. Other, less conspicuous species still hug the water, harking back to their evolutionary origins. Ferns, unlike conifers and flowering plants, lack flowers or seeds. Instead, like fungi, they produce spores that rely on moisture for successful germination. Fertilization also requires a fern's sperm-like reproductive cells to swim through water. The story of how one genus of tiny floating water ferns, called *Azolla*, managed to have quite such a big impact on all other life begins about forty-nine million years ago.

At that time in the Earth's history, about sixteen million years after the mass extinction event that wiped out the dinosaurs (see page 159), the Arctic polar region, familiar to us today as a continent-sized ice pack, was a balmy semi-tropical swamp. Alligators and snakes slithered through the stagnant water, and giant mosquitoes grew as large as a human head. Research from an Arctic expedition has revealed that for about one million years, the Earth's temperature rose dramatically, owing to abnormally high quantities of carbon dioxide in the air. What caused levels of this greenhouse gas to rise to as much as 3,500 parts per million (compared to 330 p.p.m. today) is still unknown. A sudden increase in volcanic activity perhaps? Or maybe the mysterious release of millions of tonnes of methane (another greenhouse gas) that is normally trapped as frozen deposits on the sea floor?

Whatever its cause, the effect was dramatic. Samples (cores) taken from deep down under the seabed show how average temperatures at the North Pole soared to about twenty-three degrees centigrade. Because of the way the Earth's continents were arranged at that time, the Arctic sea was almost completely isolated from the rest of the world's oceans and so had few or no deep-water currents. River water poured into the northern seas following an increase in rainfall, which formed a layer of low-density fresh water on top of the high-density salty sea.

Azolla, the water fern, grows like crazy in the right conditions and may have switched the Earth's climate from greenhouse to icebox.

Conditions for the water fern *Azolla* to make its mark could not have been better. Like legumes (see *Rhizobia*, page 40) but unlike most other plants, *Azolla* had established a symbiotic partnership with nitrogen-fixing organisms, in this case cyanobacteria. These bacteria live inside the leaf cells of the fern, turning nitrogen from the air into a nitrate solution that the plant can use for growth. In return, the fern feeds the bacteria sugars derived from its photosynthesis. So powerful is this partnership that *Azolla* is able to grow extraordinarily fast – it's been dubbed a 'super-plant' that can more than double its mass every two to three days.

About forty-nine million years ago was a great time to be alive for *Azolla* plants. With abundant supplies of carbon dioxide in the air, an internal source of nitrates, a favourable climate and plenty of nutrients in the water, nothing could stop this fern from runaway reproduction.

And for about 800,000 years nothing did.

Arctic cores show that a massive four-million-square-kilometre swamp of *Azolla* grew right over the top of the world and as far south as Britain. In the process, it sucked up so much atmospheric carbon dioxide that levels fell steeply to 650 parts per million, reducing air and sea temperatures and flipping the world from a greenhouse to an icebox. Average temperatures have been cooling ever since, providing a better environment for the globe-trotting habits of mammals, the spread of grasslands and conditions that eventually encouraged apes to come down from the trees and experiment with walking on two feet (see *Australopithecus*, page 176).

The climatic contributions of this group of ferns are set to continue well into the future. As polar ice melts from the injection of ever-increasing levels of carbon dioxide into the atmosphere by modern man, geologists are now discovering what may have happened to the remains of that enormous prehistoric swamp. When the ferns died, they sank to the salty depths of the sea where a lack of oxygen prevented their proper organic decomposition. As layers of sediment piled on top, this dead organic matter is thought to have become compressed and crushed over millions of years into what may be vast fields of yet-to-be-discovered crude oil and natural gas.

As the polar ice melts, exploration of the Arctic seabed and drilling for cores has become big business. Russian submarines, Norwegian ice-ships and the Canadian military are all preparing to make new territorial claims in anticipation of a dash for the huge reservoir of oil trapped at the roof of the world.

If future generations of humans do find it and then pump it out and burn it, releasing the trapped carbon into the atmosphere, maybe these small ferns will be needed again? Not that their efforts are likely to provide a quick-fix solution to the current problems of global warming. Judging from their previous efforts, mopping up is a process that takes the best part of one million years.

Norway Spruce

FAMILY: PINACEAE
SPECIES: PICEA ABIES
RANK: 22

A highly celebrated species that boasts today's oldest living individual.

WHILE CHRISTMAS commemorates the birth of Jesus Christ more than 2,000 years ago, the origin of the Norway spruce, pulled out of the ground, taken indoors and decorated as part of this annual festival, is a great deal older. Researchers from Umeå University in eastern Sweden recently discovered a single specimen of Norway spruce (*Picea abies*) in the Swedish mountains that they say has been alive for more than 9,500 years.

This venerable tree has been nicknamed Old Tjikko and may be the oldest-known living organism in the world. The key to its longevity is simple. Whenever the main trunk dies (typically about every 600 years), its ancient roots simply sprout a new one in its place. No need for fertilization, no need to spread seeds.

Old Tjikko belongs to a group of trees called conifers, which evolved after the club mosses (see page 90) but long before flowering plants (see page 129). Between 300 million and one hundred million years ago, these trees were the predominant forms of plant-life on land. They out-competed other trees, such as the telegraph-pole like *Lepidodendron* (see page 90), with a range of survival skills, of which Old Tjikko's cloning trick is but one. With the rapid rise of flowering trees and plants, particularly in the last hundred million years, conifers have finally been beaten into retreat. Except when deliberately cultivated by humans their domain is usually restricted to the world's least hospitable climates – or to soils that are so thin and sandy that few other types of tree can effectively compete.

A range of evolutionary innovations accounts for the conifers' historic success. Their use of seeds instead of spores allowed these trees to spread inland, far away from the water's edge. They developed branches that grow outwards all the way up their trunks, and not just from the top, allowing them to absorb greater quantities of light, especially important for survival in higher latitudes where the Sun stays lower in the skies. Perhaps the biggest impact of conifer trees is down to their ingenious method of self-defence – resin. This thick, sticky liquid slowly oozes down the sides of the tree's branches and trunk, plugging up holes bored by insects. As it solidifies, it protects the tree against fungal and bacterial infection.

Amber fossils (formed from resin deposits) date back 345 million years, although those containing the remains of ancient insects are common only from about 150 million years ago. Today's most prolific amber deposits are found in the Baltic, a region where conifer trees continue to dominate the landscape despite the region's cool climate and the impact of various ice ages in the last three million years. Resin has had its own special influence on human culture. Its sticky, viscous qualities are perfect for use as a natural glue and varnish for woodworking, incense for religious ceremonies, and for rosin, a material which helps 'sharpen' a string player's bow.

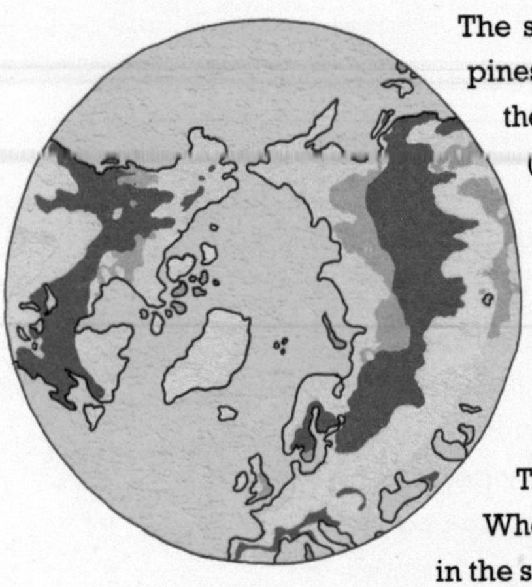

The success of conifers is manifest from their wide variety, ranging from pines, firs and spruces to cypresses, junipers and redwoods. They include the tallest (*Sequoia redwood*), oldest (*Norwegian spruce*) and broadest (*Ahuehuete cypress*) living species in the world.

In their heyday, conifers were as ubiquitous in the tropics as they are in today's northern latitudes – as evidenced in coal strata laid down over hundreds of millions of years. The flattened shape of the juniper is living proof of a tree that has adapted to environments where light shines directly from above, as it does in the tropics, maximizing the area available for leaves to gather light.

The most hallowed conifer in New Zealand is the ancient kauri tree. When its acidic leaves fall to the ground, they decompose other nutrients in the soil (such as nitrogen and phosphorous) sterilizing the earth so that the kauri is free to thrive without competition from other plants.

A ring of coniferous Boreal forest (shaded dark green) still dominates in northern latitudes.

But, like all trees, the kauri has not been immune to the impact of man. Maori settlers chopped down vast forests of kauri to make houses, boats and carvings – they even used its resin as a form of chewing gum. The process accelerated dramatically after the 1820s when European settlers started hacking them down to make masts for their sailing ships. In the last 200 years there has been so much deforestation that only about 4 per cent of New Zealand's kauri forests still stand.

Other conifers include the world's most massive trees. The giant redwoods of North America have developed immunity to what, for other trees, would be the greatest threat of all – fire. Redwoods, like some pines in Mexico, have become so used to forest conflagrations that they cannot properly regenerate without them. Their seeds spring loose from their cones only in the extreme heat of a fire. Such events clear the forests, providing sufficient light for new redwood seedlings successfully to germinate. They also destroy harmful predators such as carpenter ants. Unfortunately, efforts by modern humans to prevent forest fires have been just as damaging to these species as logging has to the kauri of New Zealand. Native American traditions encouraged forests fires as part of their culture's cyclical relationship with the forest. Invading Europeans were never so ecologically astute.

For more than 200 million years conifers eclipsed all other trees in the forest – club mosses, ferns, cycads – even pushing some of them into extinction (for example the *Lepidodendron*). Their success stemmed from a combination of good design, unfussy habits and a range of innovative systems for self-preservation.

Without man, however, the long-term survival of the Norway spruce and other conifer species (and many of its cousins – the pines, firs, larches, cedars and hemlocks) looks a lot less secure. Cloning may have worked brilliantly for Old Tjikko, but it is not a strategy that bodes well for future generations. Diversity is the surest protection against climate change or viral infection (see page 47). The lack of species variety amongst the conifers of northern latitudes today is a symptom of this once great family's gradual demise. To find out why, skip on to the rise of flowering plants and trees (see page 129) ...

Earthworm

FAMILY: LUMBRICIDAE
SPECIES: LUMBRICUS TERRESTRIS
RANK: 1

Subterranean wrigglers whose constant burrowing fertilizes the soil.

ACCORDING TO CHARLES DARWIN, no living thing has had such a profound impact on history as has the earthworm. So fascinated was he by these humble creatures that he devoted an entire book, published in 1881, to the formation of soil (then called 'vegetable mould') through the action of worms:

> *'The plough is one of the most ancient and most valuable of man's inventions; but long before it existed, the land was in fact regularly ploughed, and still continues to be thus ploughed, by earthworms. It may be doubted whether there are many other animals which have played so important a part in the history of the world, as have these lowly organized creatures.'*

Few people have ever spent as much time analysing the behaviour of earthworms as Darwin. Towards the end of his life, playing with worms became something of an obsession. He blew various musical instruments at them to see if they would react to sound, concluding that: *'worms do not possess any sense of hearing. They took not the least notice of the shrill notes from a metal whistle, which was repeatedly sounded near them; nor did they of the deepest and loudest tones of a bassoon.'*

He put them on his grand-piano to see how they would respond to the vibrations of its hammers hitting the strings:

> *'When the pots containing two worms which had remained quite indifferent to the sound of the piano were placed on this instrument, and the note C in the bass clef was struck, both instantly retreated into their burrows. After a time they emerged, and when G above the line in the treble clef was struck they again retreated.'*

Finally, he wanted to find out if these small, slimy creatures could think intelligently for themselves. His experiment involved cutting out 303 elongated paper triangles and coating them with fat to stop them shrivelling up from the morning dew. After sprinkling the leaf-like triangles on the grass outside his living-room window, Darwin withdrew overnight to see how the worms responded. In the morning he recorded that 62 per cent of the time the worms tackled the triangle's apex, the easiest way to manoeuvre the paper shape cleanly, before dragging them into their burrows. *'We may therefore infer,'* he wrote, *'that worms are able by some means to judge which is the best end by which to draw triangles of paper into their burrows.'*

This cartoon was published in the British magazine *Punch* in 1881 – the same year Darwin brought out his book on earthworms.

Whether these experiments tell us more about Darwin than about worms is a moot point. They absolutely reveal a man with an insatiable passion for creative tinkering. His life was blessed with no financial worries (Darwin's family was well off and his wife came from the wealthy Wedgwood ceramics family) affording him the luxury of unlimited time at his disposal. He used it to write his books, correspond with other naturalists around the world and design intricate (and sometimes rather eccentric) experiments to probe the inner workings of nature through the observation of species. Bees, beetles, barnacles and pigeons fell under his critical gaze but, as his last book on vegetable mould demonstrates, no creatures fascinated him quite so much as nature's perennial ploughs.

Earthworms are annelids, a phylum of creatures whose evolutionary past stretches back at least to the Cambrian Explosion *c.* 530 million years ago when the trilobites first developed sight and marine creatures evolved bones and shells. *Burgessochaeta* is an ancestral, twenty-segmented sea worm whose fossils were found by Charles Walcott in the Burgess Shale (see page 60).

Descendants of these marine creatures came ashore at the time of the first invertebrate invasions of the land, *c.* 450 million years ago, making their living in damp soils broken up by bacteria, fungi and the roots of colonizing plants. These earthworms have been ploughing up the earth, ventilating the soil and nourishing terrestrial ecosystems with their excrement ever since. Five mass extinctions have occurred over the last 500 million years, some of which devastated up to 96 per cent of all marine species and 70 per cent of all land species, but none of them ever touched these creatures. Slice a worm in half and it re-grows as if nothing happened. Divide one half and the same thing happens. One worm even survived forty such butcherings, all in the name of science.

The effects of worms on human history are as profound as they are unwritten. French scientist-cum-poet Andre Voisin was one of the few experts who properly highlighted the role of worms in the birth of ancient human civlizations. Were it not for their continuous regeneration of soils around damp river valleys such as the Nile, Indus and Euphrates, early agricultural societies in Egypt, India and Mesopotamia could never have succeeded in building humanity's first large-scale urban communities. Even the Egyptian pyramids, said Voisin, were built thanks to the nourishment of the soil by earthworms. It was only because of their hard work that farmers could take time off from tilling the soil themselves to work as a labour force for their pharaoh's ambitious building projects.

Throughout human history earthworms have unintentionally but undeniably triggered the rise of civilizations. Wherever earthworms plough, people thrive. When worms perish, societies collapse. Infertile soils led to the demise of the people of ancient Sumeria. Rising levels of salt as a result of irrigating the land with sea water killed off the worms around the mouth of the Euphrates river and the soil turned sour. By 2000 BC their civilization was so weak from lack of food that they fell easy prey to Assyrian invaders from the north.

It might be easy to think that worms matter little today, replaced by artificial fertilizers and pesticides that guarantee soil fertility anywhere and everywhere they are spread. But no. Once again it was largely thanks to the earthworm that the unsustainable nature of using such methods was originally exposed.

Rachel Carson was a teacher and environmental campaigner of the 1950s and 1960s. She is famous for warning that Americans might one day wake up to discover they could no longer here the birds singing in the trees. The reason, she said, was that artificial pesticides, such as DDT, were poisoning the soil. While robust earthworms are able to tolerate such toxins, for those creatures that eat worms it was a different story. As few as eleven worms that had ingested DDT are enough to poison a robin – either killing it or making it sterile. Since robins regularly eat up to twelve worms an hour the use of DDT put their populations, along with those of similar birds, at risk of annihilation.

The prospect was terrifying enough to have DDT banned in the USA by popular demand (although it is still used in some African countries, such as Kenya, to support the global cut-flower trade – see rose, page 341). Carson's book *Silent Spring* (first published in 1962) inspired the founding of the US Environmental Protection Agency in 1970. Therefore, the robust digestive system of the earthworms is, in a curious way, inextricably linked to the birth of the modern environmental movement.

But which worm species has had most impact? There are some 15,000 species of segmented worms in the annelid phylum, including leeches, and marine polychaetes – as well as earthworms. They range from the now rare but enormous purple-headed Giant Gippsland (*Megascolides australis*), a native Australian worm that grows up to three metres long, to the extremely common red wrigglers (*Eisenia foetida*), vermicultural alchemists that turn kitchen vegetable scraps into rich garden compost.

Lumbricus terrestris, the European earthworm, is now probably the most prolific and invasive species in the world. Its success is largely thanks to the spread of Europeans from *c.* 1600 onwards. Immigrant farmers inadvertently brought these earthworms, sometimes called 'night-crawlers', to the Americas in everything from the soil in their potted plants and their horses' hooves, to the treads of their boots and the wheels of their wagons. Today, there is hardly a region of North America where Europe's earthworms have not made a home for themselves. There they continue to plough, ventilate and fertilize the soil to the general benefit of life in and on the Earth.

But their presence is not always welcome. Artificially introduced invasive species almost always have a darker story to tell (see also eucalyptus, page 264, and rabbit, page 330). Native American redwood forests that were once free of worms are now being invaded by this European species. It is relentlessly munching its way through carpets of fallen leaves – a vital habitat for native American insects, amphibians and ground-dwelling birds. Without leaf litter on the soil's surface, the seedlings on which the future of these forests depends will not germinate. Within a generation or two, some experts fear that those forests not already mined by man may instead be undermined by the worms he has introduced.

Darwin's fascination with these ubiquitous, noiseless digesters of the Earth was well placed, confirming their rank as our Number One species (see table on page 393). Whatever they lack in glamour, colour or a sense of adventure (most worms move only about fifty metres in their four-year lifetime), they make up for in their constant ploughing, harrowing, fertilizing and recycling of that most precious of all the planet's assets – the living earth itself.

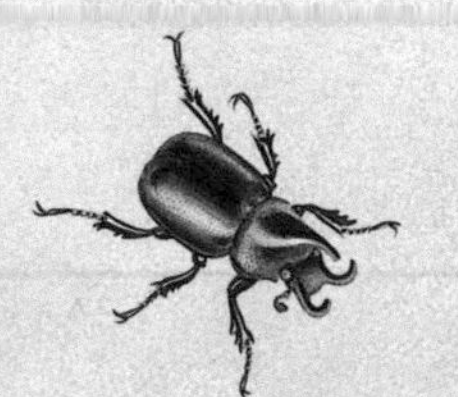

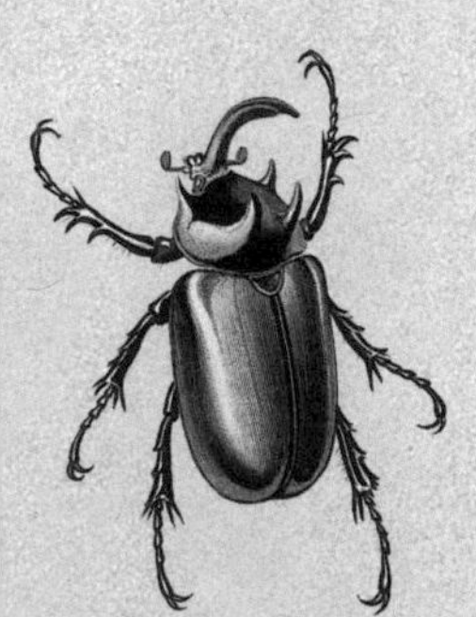

Dung Beetle

FAMILY: SCARABAEIDAE
SPECIES: SCARABAEUS SACER
RANK: 68

A venerated recycler and worthy representative of the most diverse group of animals on Earth.

THERE IS A STORY about the distinguished British-born Indian geneticist J. B. S. Haldane (1892–1964) who once found himself in the company of theologians. He was asked what one could conclude as to the nature of the Creator from a study of His Creation. Haldane is said to have answered: *'An inordinate fondness for beetles ...'*

Beetles come in a greater variety of shapes and forms than any other order of creatures. They comprise more than *a quarter of all known animal species*. At least 350,000 types have been described to date but between three million and eight million species are thought to exist in total. These are nature's most varied group of multicellular creatures and have endlessly fascinated man.

Why so many species? Beetles are not the first insects to have evolved, although fossils show their origins date back at least 280 million years. Beetles are also generally small (ranging from 0.25 millimetres, *Nanosella fungi*, to twenty centimetres, *Titanus giganteus*), inclining their populations to be numerous and their generation times short. Both are factors that increase the statistical likelihood of genetic mutations and, therefore, of spawning into so many different species (see pages 84–5).

But these are factors common to most small, terrestrial insects. What makes the beetle order of animals (Coleoptera) so extraordinarily varied?

The answer may lie in a branch of evolution that has only recently come under scrutiny – hidden perhaps because it runs in a contrary direction to the competitive state of nature emphasized by Darwin in his *On the Origin of Species*. Co-evolution is best described as an intimate evolutionary dance between different types of living things in which a spiral of mutually beneficial adaptations triggers a diversification into numerous slightly different species.

Beetles were the original pollinators of plants and trees, more than 100 million years before the emergence of birds, bees or moths (see pages 122, 152 and 260). Beetles pioneered a mutually beneficial relationship with the first seed-bearing trees, the cycads (see page 400), which evolved about 280 million years ago. Until then, the sexual reproduction of plants and trees was a mostly hit-and-miss affair. Wind power was the only viable alternative transport medium for plants on land. Aquatic plants used water currents (as green algae and mosses still do today) to carry male genes (pollen) to female eggs. But in a world far away from water, wind was all there was. By 280 million years ago, a new mechanism became essential as the world's landmasses converged into the one large super-continent of Pangaea, where rainfall patterns shifted and inland areas became drier than ever.

An inordinate variety of beetles attracted the attention of Victorian engravers and coleopterists.

Successful plants and trees were those that produced enormous clouds of pollen – the more pollen, the greater the chance that the wind would blow at least one or two particles towards their intended targets. But for any plant, producing so much pollen was expensive and wasteful. All power to those species that found an alternative means of transportation for their precious male genes.

Enter the beetle. Even the earliest fossils show how well these creatures adapted to hot, arid environments. Folding wing designs allowed them to squeeze into narrow places, inside the cracks in bark or under rocks, where larger fixed-wing predators could not penetrate (see dragonfly, page 104). This knack of retreating into tight corners served these creatures well in the harsh inland environments of Pangaea. When the ground froze, beetles retired into deep crevices or sheltered under logs. Their forewings (called elytra) that protect the abdomen and hind wings (used for flight), also reduced water loss, allowing them to live in some of the driest places on Earth.

Cycads were trees that developed their own innovative response to the challenges of life on a dry super-continent (a few species survive today). Producing seeds allowed their offspring to germinate without the need for constant moist, damp conditions, as required by spores. Seeds contain a store of food, packed around the embryo and surrounded by a tough outer coat designed to survive the heat. A second innovation was the ancient cycads' production of scents that attracted beetles to feast off their pollen and seeds. As the beetles roamed around neighbouring trees, they spread the cycads' pollen – a far more efficient and targeted form of transport for the trees' male genes than having to rely on their pollen blowing about willy-nilly in the wind (see also honey bee, page 152).

As cycad species spread across the globe so did subtle changes in the aromas they produced, each species evolving its own special scent. As clusters of similarly inclined beetles found themselves attracted to the same types of trees, their gene pools became localized – or isolated – in the process. Whenever gene pools become more limited (either through competitive selection, geographical isolation or, as in this case, the allure of a particular fragrance), over time, each community evolves into its own separate species as successful characteristics become predominant.

So much for the effects of co-evolution in generating the conditions necessary for the extraordinary variety of beetles (and ancient cycads). How humans have become so familiar with such a wide variety of species is largely thanks to the efforts of European beetle hunters. One Russian collection, begun by Tsar Peter the Great (ruled 1682–1725), runs to more than *six million individual specimens*, each one separately housed, mounted, listed and labelled. The fascination for beetles in Darwin's day among British Victorian naturalists was just as pronounced. On one occasion, Darwin was out hunting for beetles and was holding a specimen in each hand when he saw another, even rarer example. His eminently practical response was to put one of the beetles he was holding into his mouth for safekeeping so he could retrieve the third. In the process he made yet another scientific discovery – in confined spaces some beetles eject a noxious poison ...

How to pick out a single representative from the 350,000 or so known species living today presents another interesting challenge. One candidate, venerated since the beginnings of recorded human history, is the scarab or dung beetle, a creature that feasts off the excrement of larger animals. The scarab's preference is for the dung of herbivores such as elephants, camels, cows and sheep. So enthusiastic are these creatures for their share of the cake, that in some parts of the world thousands of beetles converge within moments of a fresh helping landing on the ground. Some Australian dung beetles (a genus called *Onthophagus*) have even developed a knack of clinging to the hairs around a wallaby's anus, just to make sure they are first in line when it comes to fresh deliveries.

Scarab beetles knead dung into spherical balls that they roll across the ground to a suitable place for burial, providing a rich store of food. Female scarabs lay their eggs inside these balls, giving their young larvae a source of protection as well as nutrition. Ever since large terrestrial herbivores began to make their mark, starting in the age of dinosaurs (see page 109), the ecological contribution of these assiduous recyclers has been immense. Even today it is estimated that the US cattle industry alone saves $380 million a year thanks to the burying of livestock faeces by scarabs and other insects.

The ancient Egyptians venerated one species, *Scarabaeus sacer*, as a living god on Earth because of its dung-rolling antics, recorded in many scarab-shaped amulets (funereal broaches). Pictures of scarabs were also painted or etched on to the walls of ancient Egyptians' tombs.

It was the same scarab that inspired one of Greek storyteller Aesop's most famous fables. *The Eagle and the Scarab Beetle* is a tale in which a small, wily scarab takes its revenge on a mighty eagle that had devoured its friend, the hare. So angry was the scarab that it relentlessly searched for the eagle's nests, destroying the bird's store of eggs whenever it found one. So concerned was the eagle about the future of its offspring that it sought the protection of Zeus, king of the gods, who promised to look after the eagle's eggs in the safety of his lap.

But the small scarab was not so easily put off. After it had dropped an especially large pellet of dung on Zeus's head, the king of the gods jumped up in disgust. The eagle's eggs fell from his lap and smashed to the ground ...

Scarab beetles were venerated in ancient Egypt, as shown in this cartouche, which signifies the Pharaoh Amenhotep II (reigned *c.* 1427–1400 BC).

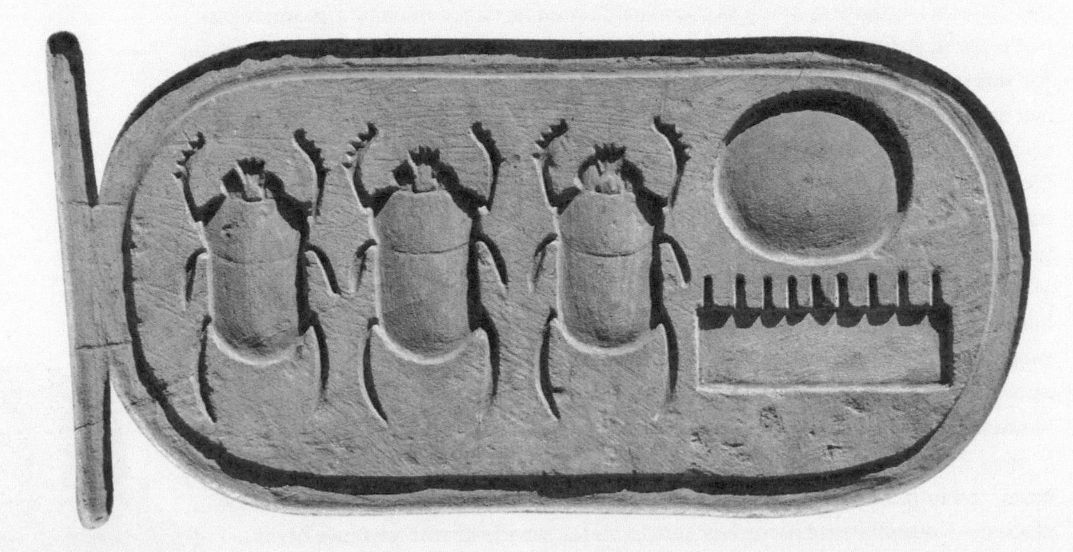

Dragonfly

ORDER: PROTODONATA
SPECIES: MEGANEUROPSIS PERMIANA
RANK: 50

A fearsome fixed-winged insect that once dominated life on land from the skies.

AGGRESSION is an instinct most often found in species at the apex of the 'pyramid of life' – a natural arrangement of creatures that occurs in all ecosystems. Light from the sun is absorbed by plants and turned into sugars. Herbivores (creatures who feed on plants) stand to gain only about 10 per cent of the sun's trapped energy, since the rest is used by the plants for their own growth and reproduction. Further up the pyramid, the number of species and the size of populations gets smaller and competition for survival increases. Carnivores (creatures that feed on other animals) stand to gain only about 10 per cent of the energy trapped in herbivores. Therefore, in their bid to survive, these species have to be the most aggressive of all.

Many of the most spectacular examples of the survival of the strongest over the weakest (one of Darwin's recurrent themes) come from species at the top of this pyramid. Over the course of evolutionary history, the fortunes of the 'super-species' at the top have risen and fallen like stocks and shares today, sometimes descending into a niche or occasionally going bust and sinking into extinction.

Among the land's earliest animal king-pins were the dragonflies. Today, these extraordinarily beautiful creatures represent only a faint echo of the absolute predominance their ancestors once commanded. But instincts die hard. Deep within the behavioural routines of today's dragonflies lie some of the most spectacularly aggressive habits in nature.

The aggression (and success) of ancient dragonflies arose partly from the fact that these insects were among the first to take to the skies. High levels of oxygen in the Carboniferous Period's atmosphere and the increasing height of plants as they morphed into the Earth's first trees are likely to have assisted insects' first flight. It is not too difficult to imagine how it may have happened, even if no fossil evidence has yet been found of insect 'proto-wings'. Flaps originally designed to regulate body temperature could have gradually adapted into wings and some daredevils probably experimented with gliding from tall plants (or trees) to the ground rather than crawling all the way back down. Over time, muscles developed to assist the gliding process, which was made easier by the dense atmosphere that contained greater quantities of atmospheric oxygen, itself derived from the forestation of vast tracts of formerly barren land. Higher oxygen levels also help to explain why the ancestors of today's dragonflies were able to grow quite so large – powering them with the extra energy needed for growth and flight.

Griffinflies, an extinct order related to today's dragonflies, grew as large as eagles. Swooping down from the skies with a wingspan of more than half a metre, these creatures preyed on unsuspecting terrestrial animals. Compound eyes, perhaps descended from those of the trilobites (see page 70), reached their zenith in flies like these. Typically the lenses featured more than 30,000 individual facets, providing a near 360-degree view of the world.

Such huge flies could either indulge in a small, juicy four-legged vertebrate (such as a frog) or alternatively devour less powerful insects. A major impact of their intense predation was the development of the folding wing in other insects (a characteristic of the group Polyneoptera – see dung beetle, page 101). The design, thought to have evolved as a defence against dragonflies, allowed vulnerable insects to creep under logs or into crevices and cracks in rocks to escape becoming a fixed-wing predator's prey. It is therefore thanks to dragonflies and their ilk that the highly successful escapologist instincts and skills of house-flies, beetles, wasps, bees, ants and other flying, folding-wing insects ultimately evolved – a process that began from the later Carboniferous Period onwards.

For as long as fifty million years (*c.* 320–270 million years ago) the skies were ruled by giant dragonflies (such as *Meganeuropsis permiana*) and griffinflies (*Meganeura moni*), the largest insects ever known. But, ultimately, being big came at a price, making them more susceptible to sudden environmental change. Declining oxygen levels probably contributed to the demise of these enormous creatures. Insects do not have lungs. Instead they rely on oxygen diffusing through their bodies via small tubes called trachea. Large flyers therefore depend on high concentrations of oxygen so the life-giving gas can properly permeate throughout their body. When oxygen levels fall, the maximum possible size for flying insects falls too.

Today's smaller dragonflies thrive nonetheless. They express their powerful inherited instincts for survival in at least two remarkable ways. First, a dragonfly's genius at acrobatic flight usually means it can use its near 360-degree vision to catch its hapless victim while still on-the-wing. Spine-like studs jut out from its legs forming a cage-like grip from which there is no hope of escape. Some flies take their meals back to their roosts, others prefer to consume their catch fresh while still aloft.

A second highly aggressive survival strategy is sexual. Charles Darwin wrote about 'sexual selection' as a force of variation between genders of the same species in his *On the Origin of Species*. The sexual antics of dragonflies offers an extreme example of sexual competition between males (for other examples, see *Archaeopteryx*, page 124, and sperm whale, page 166).

Having identified a compatible mate (colour and flight patterns act as species hallmarks), a male dragonfly will grasp a female by the neck in mid-flight and hold her there for as long as its takes to transfer his sperm successfully. The two of them will form what is known as a 'copulating wheel', the male not releasing his firm grip until the female yields, even if it takes several hours. Then, in a unique evolutionary adaptation, the male's genitals are designed to scoop out any sperm already inside the female that may have been placed there during a previous liaison. The dominant male will then attempt to hold the female hostage until she has laid her eggs, a final bid to ensure other males of the same species do not repeat the trick he performed. In such a manner do genes from the strongest, most dominant individuals get passed on, via inheritance, to future generations. Such tactics are nature's surest way to cultivate a predatorial master-race.

Carnivorous species are aggressive because in order to survive they have to *fight* or *kill* to get sufficient energy to live. That's a feature of life at the top of the food pyramid. Selective breeding in favour of aggression, as practised by the dragonfly, is one way to help ensure fitness for such a way of life persists.

Dragonflies were the land's top predators until reptiles, dinosaurs, and eventually birds knocked them off their perch. The survival of their kind today demonstrates an impressive biological resilience, despite their top predator status belonging to such a long-lost age. Superlative skills in selective breeding, passed down from their ancestors who ruled in the ancient Carboniferous forests, continue to give these creatures a good chance of surviving whatever ecological hazards the modern world throws in their way.

Then there is the insect's-eye view. Were it not for the enormous dragonflies of the Carboniferous Period, the hugely successful folding-wing design of subsequent species might never have evolved: no wasps, bees, beetles or house-flies – perhaps, even no flowers or fruit, since these are plants that mostly rely on folding-wing flyers for transporting their precious pollen and seeds. Just think what a different place the world might have been were it not for dragonflies and griffinflies – these highly aggressive flying aces of the skies.

Checkmate: the mating rituals of dragonflies are among the most aggressive sexual antics in nature.

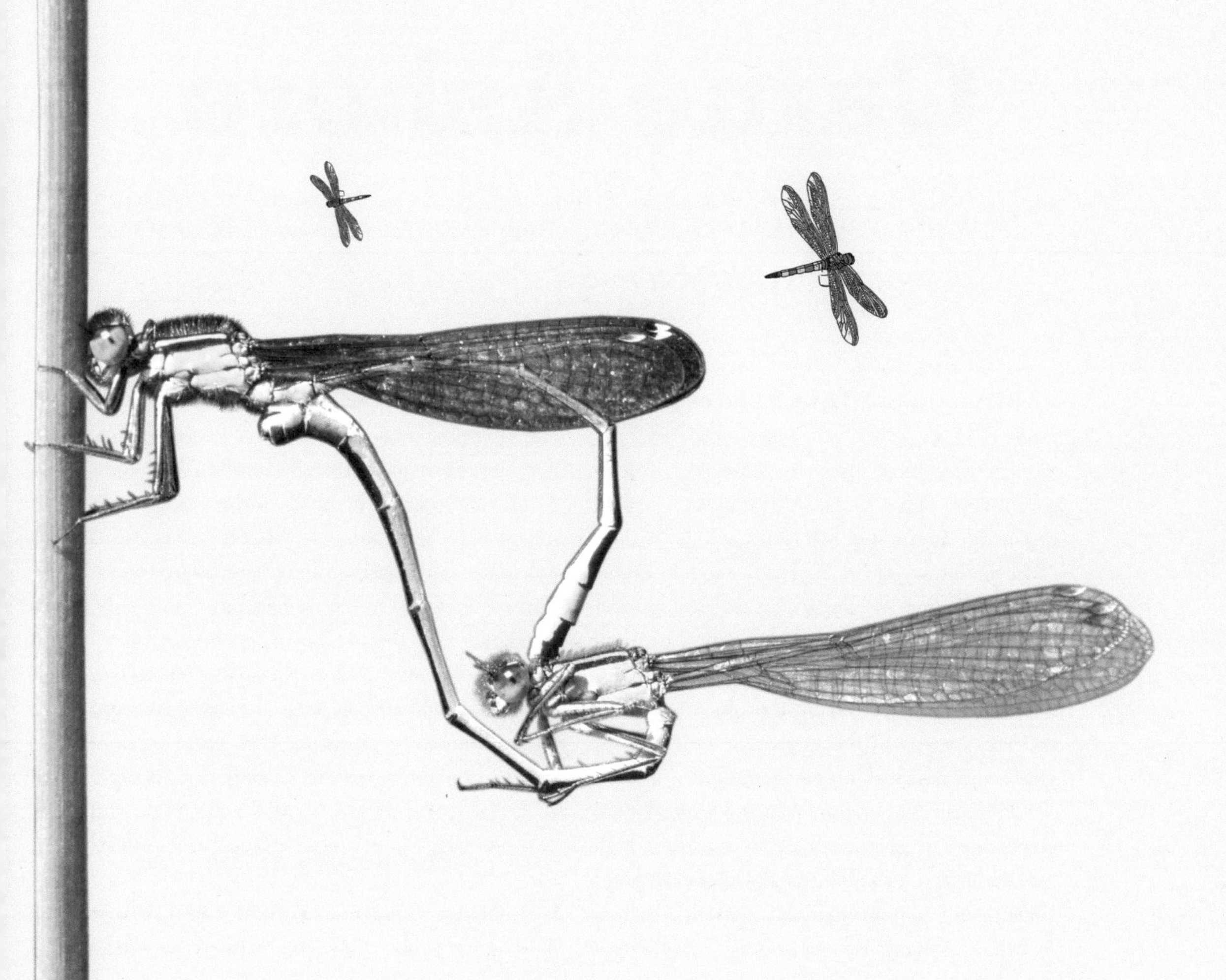

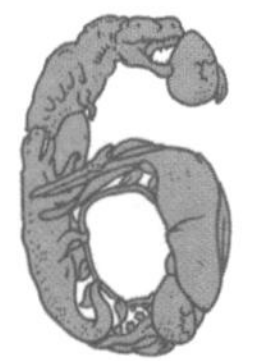

On Fish that Came Ashore

How the descendants of bony fish clambered ashore and were jerry-wrenched into a diverse range of forms.

HAROLD MACMILLAN, Prime Minister from 1957 to 1963, was once asked what he thought was most likely to blow the British government off course. Macmillan replied succinctly: *'Events, dear boy, events ...'*

In the history of life on Earth, random events have had a dramatic impact on the evolution of species. In the period from 350 million to sixty-five million years ago, just as the first Carboniferous forests were properly taking root, luck was often the prime mover in the survival or extinction of species that ventured out on to the land.

On at least two occasions catastrophic geological events threw the evolutionary pack of cards so high into the air that no amount of good body design or extreme survival skills could account for the success or failure of one kind over another. On the first occasion, about 252 million years ago, so extreme were environmental conditions on Earth that up to 96 per cent of all marine species and 70 per cent of all land species perished. Then, about 187 million years later, another mass extinction event took place. Once again, some of the strongest and most robust forms of life on Earth were suddenly wiped out.

Such dramatic interventions by non-living forces profoundly affected the course of evolutionary history, sometimes thrusting life's more modest species to the top of nature's Champions League.

The jigsaw of life

Anaximander (*c.* 610–546 BC) was a Greek philosopher who came up with some truly extraordinary ideas. After an intensive study of some fossils that he had found in rocks near his home in Miletus (in today's south-west Turkey), Anaximander came to the radical conclusion that humans were distantly related to fish.

Little direct evidence remains of this great man's thinking because most of his works have since been lost. But later Greek historians (writing in Roman times) have left tantalizing clues. According to Plutarch (*c.* AD 46–120):

'Anaximander taught that the first animals were begotten in moisture ... But as they grew older they came out on to dry land.'

Another Roman writer, Censorinus, said that according to Anaximander either fish, or animals very like fish:

'... sprang from heated water and earth, and that human foetuses grew in these animals to a state of puberty, so that at length when they burst, men and women capable of nourishing themselves proceeded from them.'

Ever since Charles Darwin first articulated his theory that all living creatures have common ancestors and that each species is simply a modification of a previous form, the search has been on to re-order the living world into its logical evolutionary sequence.

In recent years brigades of traditional palaeontologists (fossil hunters) have been joined by gaggles of geneticists, evolutionary biologists and embryologists, all focused on piecing together this evolutionary puzzle.

The result is an astonishing vindication of the general idea behind what ancient Greek philosopher Anaximander was hypothesizing more than 2,500 years ago. Humans, along with all vertebrate land animals – amphibians, reptiles, birds and mammals – technically belong to a group of sea creatures called the bony fish (Osteichthyes), although even the most broad-minded experts still baulk at the consequences for traditional forms of scientific classification (see page 112).

The tale of how bony fish in the seas morphed into vertebrate land animals – from frogs to dinosaurs and birds to man – is only now fully beginning to reveal itself thanks to advances in modern science.

Terrestrial adaptations

The story began some 400 million years ago with a class of fish called Sarcopterygii that swarmed in the shallow Devonian seas. Whether they were hounded out of the waters by fearsome predators such as placoderms (see page 78) or sea scorpions (see page 74) or whether they opportunistically sought to explore terrestrial pastures new, now rich in vegetation and invertebrate life, is still a subject of scientific debate. Primitive air-breathing lungs and fins that helped some fish bounce themselves along the shallow bottoms of the shores and river estuaries developed in now extinct groups – although their descendants survive today in the form of the **lobe-finned fish** (see page 112) and lung-fish.

Although these creatures came ashore to feed (or escape becoming food), they still primarily lived in the water. Many retained gills for breathing through water, but also developed air-breathing lungs, adapted from cavities behind their heads called swim-bladders that originally provided their ancestors with buoyancy.

For bony fish to adapt completely to living on the land rather than in the sea, a number of further anatomical modifications soon proved necessary.

First, their bodies had to get used to experiencing gravity. Astronauts wishing to practise space walks traditionally take to underwater tanks for their training because, thanks to the buoyancy of water, exercising below the surface simulates a low-gravity environment.

To cope with the lack of buoyancy in air, the process of natural selection favoured terrestrial creatures with strong backbones. Structures like these were best placed to support the powerful muscles needed for moving about effectively on land. Inherited traits, amplified by the passing of

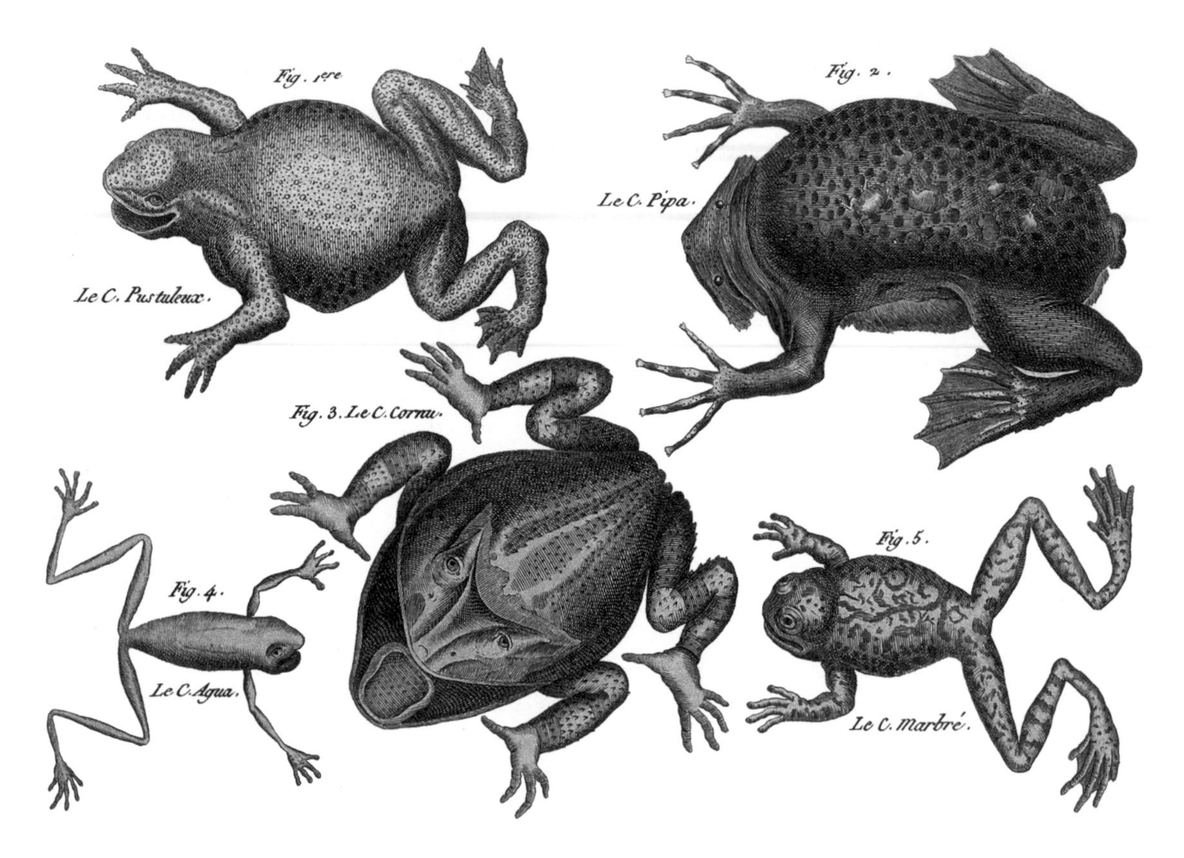

Toads like these belong to the pioneer class of 'semi-terrestrial fish' – the amphibians.

generations, gave way to other useful features. Fossils of ***Tiktaalik*** (see page 114), a half-fish, half-crocodile amphibious creature, which emerged from the seas about 375 million years ago, represent the first land-based fish to feature a neck, allowing it to move its head from side to side to look for food or detect danger. The four bony fins of its lobe-fish ancestors had now strengthened to become proto-legs that lifted *Tiktaalik*'s body off the ground, allowing it to crawl across uneven terrain. Even its back was arched for extra support, being balanced by a tail that grew rearwards as a rudimentary counterweight.

Amphibians went on to became the king predators of vertebrate life on land – dividing their time between land (for feeding) and the sea (for laying their eggs). *Amphi* in Greek means 'two' and *bios* means 'life'. Today's frogs, toads, salamanders and newts, although ancient and highly successful survivors of those first terrestrial vertebrates, are but a pale echo of the enormous variety of amphibians that once roamed the shores. At over two metres long, *Eryops* was perhaps the most fearsome amphibian that ever lived, the king carnivore of its age (*c.* 275 million years ago). Indeed all bony fish that came on to the land were initially carnivorous, feeding off insects or each other, since the concept of a herbivorous, vegetarian-only diet was not one inherited from vertebrate ancestors that lived in the seas.

Sturdy backbones, strong limbs and the first successful attempts to hear sound in air were other modifications pioneered by these amphibious descendants of bony fish. Jaw bones that originated in marine species became modified in amphibious creatures such as *Eryops*, receding into their skulls to form a bone that could amplify vibrations from the air, passing them on as sound signals to the brain. In later mammalian species, three such jaw bones would converge to become the bones of the middle ear (malleus, incus and stapes). This greatly improved their ability to hear higher-frequency sound waves travelling through the air. Repurposing old bones, rather than inventing news ones, is one of nature's favourite tricks.

By about 310 million years ago events were unfolding fast – at least on a geological timescale.

A severe threat to the long-term survival of amphibious creatures began to present itself with the unceasing movements of the Earth's continental plates. While sea creatures remained mostly unaffected by the gradual movement of giant rafts of continental crust, for those on land intercontinental collisions grew increasingly threatening ...

The impact of Pangaea

Between 300 and 250 million years ago, the Earth's continents converged into a single giant super-continent, Pangaea, dramatically reducing the quantity of shoreline. For creatures living near to the coast that had to return to the ocean to breed, competition intensified because forests rich in vegetation that were too far away from the water's edge were simply off limits.

An opportunity now presented itself for those species that could get round this limitation by building their own transportable 'oceans' from which their young could emerge, fully able to support themselves on dry land. The consequence of the random collision of the Earth's continental plates into one giant super-continent therefore spawned the evolution of creatures that laid the world's first shelled eggs.

Crack open a chicken's egg today and the link between reptiles (birds are reptilian offshoots, see page 122) and their fishy ancestry is clear to see. A chicken embryo grows in its own protected sea with the outer shell allowing air in but not letting precious water out.

Amniotes, the group to which all creatures that have eggs with embryonic membranes belong (including reptiles, birds and mammals), were much better equipped for coping with Pangaea's vast inland areas, much of it now situated thousands of miles from the seas. *Hylonomus* is the most ancient reptile discovered to date (estimated at 310 million years old) and therefore the genus currently credited with developing the first amniotic egg. The presence of five fingers and five toes at the end of each limb is another hallmark of amniotes (although sometimes modified, as in hoofed mammals). These were highly effective adaptations for spreading body weight in a gravity-dominated world.

A tough, waterproof skin provided the finishing touch for amniotic reptiles. With an abundance of vegetation available in inland areas, the first vertebrate non-meat-eating, herbivorous species began to emerge, such as *Edaphosaurus* (see page 400). The tell-tale sign of this momentous switch in terrestrial eating habits came from the flattened shape of these creatures' teeth, as revealed by their fossils.

The most successful reptiles were those that were adapted in ways that let them hunt at the least competitive times of day, such as the hugely successful sail-back pelycosaurs (see ***Dimetrodon***, page 116). These creatures became the dominant land carnivores for about forty million years.

Others developed adaptations that helped them avoid being eaten, such as becoming small (to hide out of harm's way), hunting at night (to avoid being seen in daylight) and developing a more varied, flexible diet (becoming omnivorous). Such changes gradually led to the development of a new branch of land-loving bony fish – the mammals – represented at its earliest stage by cynodonts such as *Thrinaxodon*, the 'dog-toothed cat'.

When the world was thrown into chaos by the Great Permian Extinction, 252 million years ago, fate looked kindly on a creature called ***Lystrosaurus*** (see page 118). This strange beast, which looked rather like a hippopotamus crossed with a pig, was one of the few land vertebrates to survive the massive trauma. Up to 70 per cent of all land species perished, including the – until then – enormously successful *Dimetrodon*.

Dinosaurs!

It was from a surviving reptile group called the *Sauropsida* that these most famous of all prehistoric creatures came. Dinosaurs emerged to take full advantage of the ecological vacancies afforded by prehistory's most devastating extinction event.

They dominated the land between 235 and sixty-five million years ago. By contrast, the Great Apes, the group to which humanity belongs, have lasted about fifteen million years to date. Humans, the fourth Great Ape branch, emerged only in the last 2.5 million years.

Dinosaurs were special because many of them learned to walk on two feet rather than on all fours. The secret was a small but significant adaptation – a ball and socket joint – that connected the leg bone (femur) to the hips (pelvis). This arrangement allowed their hind legs to drop directly under their bodies, supporting the full weight of their upper bodies like pillars. Such a 'fully improved stance' gave these creatures the strength and mobility to dominate all other forms of life on land. The same arrangement later helped bipedal apes (hominids) rise to their current position of predominance (see page 176). It is also a feature of all birds, the only surviving branch of the dinosaurs left, which first took to the skies about 150 million years ago thanks to the adventures of **Archaeopteryx** (see page 122).

Current evidence suggests that the first dinosaurs were theropods. These were bipedal, mostly carnivorous reptiles that radiated across Pangaea, growing in size from the relatively small *Eoraptor* to the positively gigantic *Allosaurus*, a species that hunted in packs and fed on hulking great herbivores like *Diplodocus*.

The presence of such fearsome carnivorous predators probably triggered land-life's second airborne sortie. Like the dragonflies more than one hundred million years before, pterodactyls were the first of three separate amniote groups to take to the skies, a venture that probably followed an initial attempt to flee from danger by climbing up trees. Pterosaurs initially thrived as crow-sized creatures, with their fourth fingers reaching out into struts that supported web-like wings. But some species (such as **Quetzalcoatlus**, see page 120) grew into giants with a wingspan stretching more than fifteen metres, making them by far the biggest flying creatures of all time – more than a third wider than a World War II Spitfire aircraft.

Other reptiles escaped predation by reviving ancient sea-borne instincts. Preceding similar actions in mammalian dolphins, whales and sea cows by about 200 million years, ichthyosaurs re-evolved their land-adapted bodies back into species that became, once again, perfectly well adapted for swimming in the seas (see page 400). As their history shows, when environmental circumstances change, nature readapts, even if it means turning full circle.

Some species became gigantic, such as *Shonisaurus sikanniensis*, which at twenty-one metres grew into the largest marine reptile that ever lived. Meanwhile, fossils of their relatives, *Ophthalmosaurus*, show a spectacular innovation which, like flight, later evolved independently in both placental mammals (see page 161) and sharks (see page 78) – the ability to give birth to live young.

Of all the dinosaur groups, the theropods lasted longest. The **Tyrannosaurus** (see page 125) was one of its last representatives, dying off in a great mass extinction 65.5 million years ago, when a random event blew evolution on to a different course. This time it is thought that a massive meteorite strike killed off all the dinosaurs, leaving just one class of theropod offshoot – the birds.

This dramatic intervention into the unfolding process of evolution thrust species otherwise at the margins of terrestrial society into the limelight.

Crocodiles and turtles emerged alongside the dinosaurs as sister reptile groups, although their modern descendants date from about one hundred million years ago. Both survived the mass extinction that wiped out the dinosaurs. Quite why or how remains a mystery. Meanwhile, the mammal descendants of *Thrinaxodon* and *Lystrosaurus* clung on by staying small, squirrel-like and nocturnal. The survival kits they developed in the form of breast milk, fur and giving birth to live young eventually proved their worth after that ten-kilometre-wide meteorite struck the Earth 65.5 million years ago, travelling at a speed of about 70,000 miles per hour (see pages 157–8).

With the death of the dinosaurs, the glory days of reptilian creatures were over. Today's 6,500 species are a faint remnant of more than 500,000 species of reptile that are thought to have lived since their initial rise 300 million years ago.

Still sounds fishy?

If you're still not convinced that amphibians, reptiles, birds and mammals (including humans) are best described as terrestrial species of ancient fish then consider what all families of vertebrate land animals share in common with *Tiktaalik*, that half-fish, half-crocodile-like creature that waddled on to land about 375 million years ago: four limbs, a neck, two eyes, a backbone, one mouth with teeth, nostrils, a jaw bone, a nasal cavity. Indeed the body plans of all land vertebrates share the same basic design: a top and bottom, a front and back, a left and a right, skulls, brains, backbones and limbs.

Triassic theropod: this *Eoraptor* was one of the earliest dinosaurs, a fast mover with sharp teeth and thin legs.

The key difference between species of bony fish that still live in the sea and vertebrates that now live on land are the ways in which a single set of basic components has been adapted over time to accommodate the gravitational challenges of living on land and coping with environmental changes wrought by random events and the struggle to survive.

There are many links to our aquatic past, as first proposed by Anaximander. Land animals, including humans, are born in a fluid, watery environment – be it the amniotic sac of a human baby or translucent albumen of a chick's egg. Male spermatozoa are curiously bacteria-like, propelling themselves with flagella-like tails in a sea of fluid, and vertebrate lungs rely on dissolving oxygen in the air through hundreds of millions of microscopic sacs called alveoli – each one lined with fluid.

What Charles Darwin suspected about the common descent of all species is most clearly illustrated by the transformation of vertebrate life from the sea to land. The more detailed our knowledge of evolution becomes, the more clearly the process of descent by modification can be seen. Even if not traditionally classed as such, today's land animals are, in fact, part of one big group that emerged from the seas about 400 million years ago – we are terrestrial fish. However, it seems that the success of one species over another has at least as much to do with events (dear boy) as with competition – as much about the luck of the draw as about specific excellence in corporeal design.

Lobe-finned Fish

CLASS: SARCOPTERYGII
SPECIES: PANDERICHTHYS RHOMBOIDES
RANK: 51

A semi-aquatic fish that began the process of adapting to life on land.

NOTHING IN SCIENCE creates more fuss, controversy, angst and antagonism than its constant preoccupation with determining which species belong to what part of life's family tree. Long before Darwin published his *On the Origin of Species*, humans were using their powers of mental reasoning to order the natural world into groups. Ancient Greek philosopher-cum-scientist Aristotle wrote several influential books on the natural world and for him observation was the only method available for creating order out of biological chaos. More recently, Carolus Linnaeus (1707–78), a Swedish naturalist, founded the modern science of classification (called taxonomy), dividing creatures into groups based entirely on their observable similarities and differences.

The gradual acceptance of Darwin's thesis – that all living creatures are descended from common ancestors – has dramatically pulled the rug from under traditional scientific taxonomy. Today, little else causes as much confusion and debate (and often fury) in science than the preoccupation with biological reclassification, ever since evolutionary descent rather than outward form became the accepted new order. Willi Hennig (1913–76), a German biologist, first formalized the quest to reveal life's true family tree and originated today's system of putting species into groupings or branches, called 'clades', based on evolutionary descent.

One of the biggest problems is that as more data comes to light from sequencing the genomes of different species, the more it seems that evolution works convergently – that is, when ecological opportunities arise, unrelated organisms often end up evolving in similar ways.

One of the most bizarre consequences is that it is now incontrovertibly clear that humans belong to a clade of creatures called Osteichthyes – the bony fish. In fact, in cladistic terms, all land creatures – from birds, cats and cows to elephants, mice and humans – are technically divisions within the world of bony fish (see page 111). So, if land animals are descended from bony fish, whichever species were the first to make it possible for vertebrates to clamber ashore must have been of enormous evolutionary significance.

Air-breathing apparatus, prototype arms, legs, fingers and toes – all vital equipment for moving about in a gravity-dominated environment – originally evolved from the four lobed fins of a class of fish called the Sarcopterygii, common in the Devonian seas.

The full potential for exploring a new life beyond the seas reached its zenith with lobe-finned fish such as *Panderichthys rhomboides*, now long extinct, which lived about 380 million years ago. Fossil evidence shows that this species pioneered opportunities for feeding beyond the water's edge. Because of the low levels of oxygen in the shallow waters near the coasts at that time, the most successful of these fish were those that could absorb small quantities of oxygen by gasping in air through their mouths. At the same time, pushing a long, fish-like body along the shallow sea floor using rear fins made perfect sense for a creature that cruised along the coast, its eyes emerging from beneath the waves – hunting and prowling for fresh food – half fish, half crocodile. *Panderichthys* fossils clearly show how bones in its rear (pelvic) fins helped it rise out of the water. By pushing these fins against the ground it could then drag its body along like a modern, walking catfish.

Several elements critical to the success of animal life on land came together in this environment, about 375 million years ago. Bones such as the humerus, radius and ulna can be traced back to lobe-finned fish, as can several joints that have since become limbs, wrists and ankles in land animals. Tetrapods, terrestrial four-legged descendants of lobe-finned fish, soon began to populate the seashores. Lines of immature seal-like creatures called *Ichthyostegas* sheltered from aquatic predators by seeking the comparative safety and protection of the mostly untrodden land.

Lobe-fins were, for a long while, among the most dominant fish in the seas. One huge genus, *Hyneria*, was the king fish-killer of its era. This fast-swimming two-tonne, four-metre-long carnivore roamed close to the shore, using its powerful front fins as legs to haul itself along the sea floor, searching for fish stranded by the tide.

Of course, a bony spinal column, air-breathing lungs and four jointed limbs aren't the only requirements for success on land. Sophisticated senses such as hearing and even the capacity for making sounds – the ancient origins of birdsong and music – are now thought to have first evolved in bony fish.

Researchers at Cornell University, Seattle, have discovered that certain male midshipmen fish make humming sounds by resonating their swim-bladders at a frequency of about 100 hertz. Females of the same species are sensitive to the same frequency, which helps them locate the males. Analysis of brain patterns in these fish suggests to these researchers that the ability to create and hear sounds originated from a form of mate selection that evolved in early bony fish as long as 400 million years ago. Singing and listening skills like these were later passed down through the lobe-fish line to four-legged tetrapods, from which all vertebrate animals, including birds and mammals (and humans), are descended.

There is little doubt that all vertebrate animals (including humans) are, essentially, relatively recent forms of mutant bony fish. Chief among our lobe-finned fishy predecessors is *Panderichthys rhomboides,* long since extinct, but with a highly impressive evolutionary CV – the first known tentative vertebrate explorer of a brave new terrestrial world.

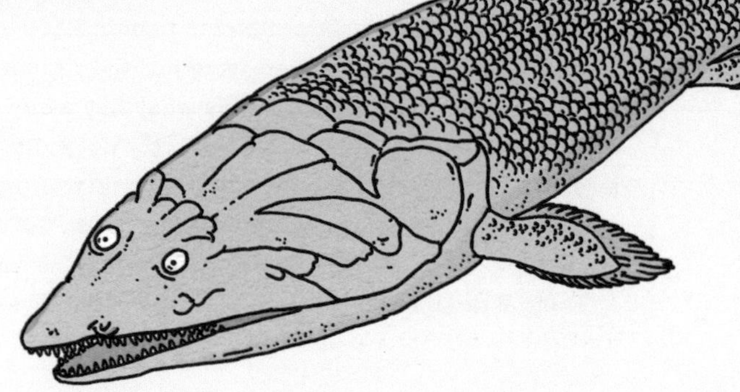

Tiktaalik

CLASS: TETRAPODOMORPHA
SPECIES: TIKTAALIK ROSEAE
RANK: 88

Caught in the act: the flat-headed half fish, half land animal that boasted the world's first neck.

EVERY SO OFTEN the discovery of an entirely new species fills in a conspicuous gap in the jigsaw puzzle of life. Fossils provide vital but often patchy, random and incomplete pictures of the countless life-forms that have ever lived. Some are known only by the remains of their teeth. Others can be reassembled into complete fossilized skeletons enabling artists to show how they would have looked when alive (see the *Tyrannosaurus rex* reconstruction on pages 126–7).

The most prized discoveries are known to science as 'transition fossils'. These are the imprints of creatures that show the process of evolution during a clear transition from one group of living things to another. Such creatures provide snapshots of the evolutionary process at work. One of the most famous examples is *Archaeopteryx* – half reptile, half bird (see page 122).

Experts who discover a transition fossil tend to get rather excited – which is exactly what happened to two American fossil-hunters, Neil Schubin and Steve Gatsey, who went out camping on one of the most remote islands on Earth in the summer of 2004.

The pair had already spent six long years on a quest to find the missing evolutionary link between bony fish and four-legged land creatures, called tetrapods. The challenge they set themselves was to find a species that could clearly be identified as a genuine half-fish, half-tetrapod pioneer. If Darwin's theory that all creatures are descended from common ancestors was correct and if, as other fossil finds indicate, all life began in the seas, then somewhere in the rocks there should be fossils that show the transition from a purely sea species to one that could thrive on land.

The sun never sets in the summer at Ellesmere Island, in northern Canada. But in the winter, it is dark twenty-four hours a day. Here there is no shelter from the elements – no trees, no houses, nothing but a vast expanse of desolate, barren terrain where winds regularly whip up to more than fifty miles per hour.

Yet somewhere, in an arc stretching about 1,500 kilometres wide, were rocks that Schubin and Gatsey reckoned were just the right age for their transition fossils to be found there. They knew roughly what they were looking for. The buried remains of the creature they sought was probably not more than about a metre long and at least 370 million years old. All they had to do was dig. After spending four summers over the previous six years on the same quest, the summer of 2004 was, by mutual agreement, their last-ditch attempt. The pressure was on.

Then, at the beginning of July, Schubin struck fossil hunter's gold:

'I cracked the ice and saw something I will never forget: a patch of scales unlike anything else we had yet seen in the quarry. This patch led to another blob covered by ice. It looked like a set of jaws. They were, however, unlike the jaws of any fish I had ever seen. They looked as if they might have been connected to a flat head ...'

One day later, and after some rather delicate hammering, the unmistakable snout of a second mysterious beast could be seen poking out of the rocks. Gatsey spent the rest of the summer recovering the fossilized creature, bit by bit, so it could be taken back to the laboratory for analysis.

The creature has been called *Tiktaalik*, the Inuit word for large, freshwater fish. It had scales on its back like a fish, but, like a land animal, it also featured a flat head and neck. Most amazingly, its fins were composed of bones that correspond to those found in all land animals. These were the upper arm, forearm and parts of the hand with their corresponding joints that in later creatures became shoulders, elbows and wrists.

The conclusion that the pair drew from their finds was incontrovertible. Here was a genuine 375-million-year-old transition species showing the evolutionary link between lobe-finned fish and four-legged, land-vertebrate tetrapods.

So what of the impact of a creature such as this? No one can be sure if the adaptations seen in *Tiktaalik* are unique, since who knows what other secrets lie hidden in the icy rocks? What's more, other creatures such as *Eusthenopteron* and *Acanthostega* have also been discovered that show evolutionary transitions away from the seas and on to the land. But *Tiktaalik is* special because it is the oldest creature found so far whose fins clearly show the arrangement of bones that in later land animals would became wrists, palms and fingers. *Tiktaalik* was a creature that could do the world's first push up – or, in the words of Schubin, it was able to '*drop and give us twenty*'.

Any time you bend your wrist back and forth, or open and close your hand, you are using joints that first appeared in the fins of the half-fish, half-amphibious creature called *Tiktaalik*. That's its place, and therefore its significance pretty much speaks for itself.

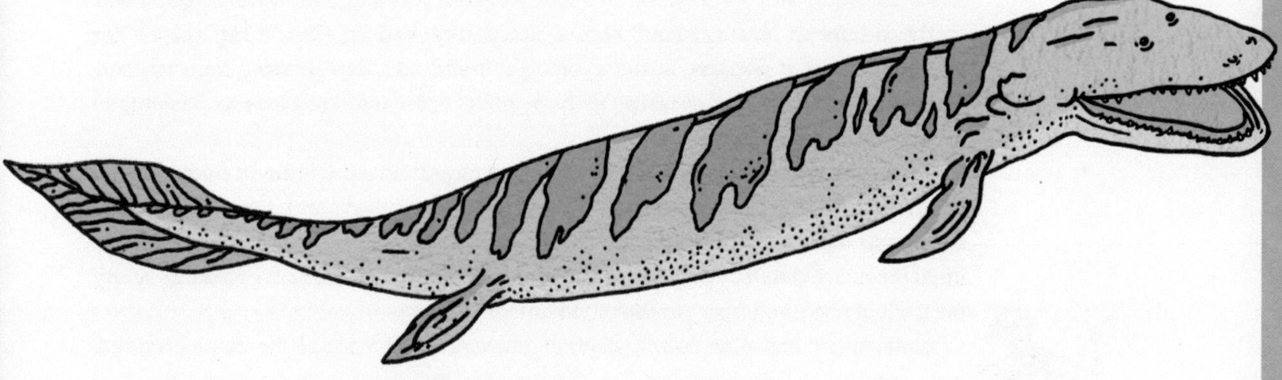

Dimetrodon

ORDER: PELYCOSAURIA
SPECIES: DIMETRODON GRANDIS
RANK: 82

The sail-back pelycosaur that found a way of keeping itself at just the right temperature.

JERRY MACDONALD is a detective who has spent the last forty years exclusively investigating footprints. Old ones. Very old ones. In fact, the prints that MacDonald specializes in are typically about 270 million years old.

MacDonald's scene of investigation is also fairly eccentric. High up in the mountains of New Mexico he has removed more than one hundred tonnes of rock, revealing a series of ancient footprints that tell their own fascinating story of lives once lived. MacDonald then takes them away for storage in the nearby town of Las Cruces, where they are stacked, catalogued and sometimes sent off to laboratories or museums for further analysis.

This is 'Pelycosaur Heaven' – the name of MacDonald's fossil trackway quarry in the mountains of New Mexico. Its name comes from its record of footprints belonging to a group of ancient creatures, the most successful land animals of their age.

Dimetrodon, a member of the pelycosaurs, grew up to three metres long, making it the largest reptile of its day. First appearing about 286 million years ago, this genus was one of the prehistoric world's longest survivors. For more than thirty million years, *Dimetrodon* was a top predator on land. Unlike amphibians, its waterproof skin held moisture inside its body, allowing it to migrate far inland and, if necessary, to survive in hot, arid conditions. And like all reptiles, *Dimetrodon* laid its eggs in watertight shells, freeing itself from the highly competitive environment near the coasts where amphibious creatures were forced to breed.

But *Dimetrodon* is most famous for the dramatic-looking sail on its back. This ingenious adaptation worked like a radiator, allowing its body to heat up quickly in the morning sun. Packed full of blood vessels, this sail meant *Dimetrodon* was ready to hunt at least an hour before non-sailbacked reptiles, a big reason for its long period of success. In the evening it could turn its sail away from the sun, allowing heat to escape, ensuring its body always remained as close as possible to its operational optimum.

These traits are also common in mammals, such as humans. Our body automatically keeps to the same temperature, within a few degrees, night and day, regardless of conditions outside. Reptiles do not. Their behaviour is limited by their need to warm up in the sun. *Dimetrodon*, therefore, represents the first significant milestone along the path of evolution from reptiles to mammals.

Dimetrodon has also left a distinct impression thanks to its teeth. Indeed, *Dimetrodon* literally means 'two teeth'. These were the first reptiles known to chew their prey, mashing it up into pieces before swallowing (unlike other reptiles that gulped down great chunks of food whole). Different types of teeth of varying sizes made this possible. Some were adapted for tearing up flesh, others for chewing it. This gave these carnivorous creatures an advantage because it takes less time to digest and draw energy from pre-processed, chewed-up food. As with the ability to regulate body temperature, their varied teeth and eating habits were features that later expressed themselves in *Dimetrodon*'s mammal-like descendants.

MacDonald's footprint investigations leave no doubt as to the success of these impressive beasts, which emerged a good fifty million years before the first dinosaurs. While fossils often reveal what an animal may have looked like, they do not generally say so much about how it behaved. Footprints, however, offer a snapshot of what animals were actually doing at a given instant in time. They are long, unbroken trails of prehistoric walking, darting, swerving and pouncing.

Dimetrodon tracks from the mountains of New Mexico have been found submerged by mud and silt along what was once the shoreline of a vast inland sea. They show that invertebrate animals and small shellfish were frequently exposed by the receding tide, leaving large predators, like the *Dimetrodon*, to feed on the pickings.

After thirty to forty million years of uninterrupted success, the *Dimetrodon* suddenly died out during the Great Permian Extinction (see page 118). It did not disappear through bad design, however, or even because of the rise of a rival species. These creatures were swept away by powerful geological forces, leaving a void at the apex of the pyramid of life, without which a new group of reptilian dinosaurs might never have arisen to take their place ...

Sail-back pelycosaurs ranged in size from one to three metres long.

Lystrosaurus

ORDER: THERAPSIDA
SPECIES: LYSTROSAURUS OVICEPS
RANK: 66

The barrel-chested herbivore that survived when all around it perished.

I FEEL a sense of impending doom as we approach the story of *Lystrosaurus*. It isn't helped by the front-page headline in this morning's newspaper: 'The Methane Time-Bomb' tells the story of Dr Örjan Gustaffason, leader of a marine research project studying methane emissions from beneath the Arctic sea. His team from the University of Sweden has discovered a series of 'chimneys' on the sea floor from which large undissolved bubbles of methane gas are steadily discharging.

In some places, says the report, concentrations of methane are escaping from the sea floor at more than one hundred times background levels.

'The conventional thought,' Gustaffason is quoted as saying, *'is that the permafrost "lid" on the sediments of the Siberian shelf can hold the massive reservoirs of shallow methane deposits in place ... The growing evidence is that this lid is starting to get perforated and thus leak methane ...'*

What seems to be happening beneath the seas today is frighteningly similar to what is thought to have happened much earlier in Earth's history. About 252 million years ago vast quantities of this gas started bubbling up to the ocean surface. It was released from huge frozen methane deposits (called hydrates) under the sea floor that had been accumulating over millions of years through the bacterial decay of organic matter. The slow grinding together of the Earth's tectonic plates to form the super-continent Pangaea may have caused a large enough increase in global temperatures (thanks to carbon dioxide spewing out of volcanic vents) to have provided the trigger.

The eruption of giant bubbles of methane was bad news for life on Earth, leading to the largest mass extinction of all time. Up to 96 per cent of all sea creatures and as many as 70 per cent of all land species perished in the Great Permian Extinction. Nothing as appalling has happened to life on Earth since or before – not even the extinction that killed off the dinosaurs 65.5. million years ago.

Not only is methane about twenty times more potent as a greenhouse gas than carbon dioxide, it also reacts with atmospheric oxygen to make carbon dioxide and water. About 252 million years ago this chemical reaction caused global levels of atmospheric oxygen to plummet from 30 per cent to just 12 per cent over the course of about 20,000 years – the merest fraction of geological time.

Suddenly having to cope with an atmospheric oxygen content of 12 per cent, compared with 30 per cent before the Great Permian Extinction, is equivalent to living at the top of Mount Everest today. Large reptiles that depended on meat for food never had a chance, which spelled disaster for the thirty-million-year reign of *Dimetrodon* (see page 116). Giant griffinflies the size of birds (see page 103) were also at a hopeless disadvantage, dependent as they were on high concentrations of atmospheric oxygen to power their over-sized bodies.

Plants suffered, too, especially those in forests. One of the tell-tale signs of death by suffocation is the absence of coal deposits. The 'coal gap', dated to between about 250 and 243 million years ago, is explained by a decline in oxygen levels that forced peat-forming plants into extinction. It wasn't until several million years later, when new species of plants developed better tolerance to lower oxygen levels, that the coal-making process could begin afresh. Only those creatures with lucky adaptations that made them less susceptible to low levels of oxygen, and those that didn't depend on a supply of fresh meat, could thrive.

Enter *Lystrosaurus*. Judging by the fossil record, for at least two million years after the Great Permian Extinction, this metre-long herbivorous creature – half-hippopotamus, half-pig in appearance – represented about half of all vertebrate animal life on land. So numerous were they that their remains have been found in the rocks of every continent, once again confirming that 250 million years ago all the world's continents were indeed joined together into the super-giant Pangaea.

The success of *Lystrosaurus* may have been down to its barrel-like chest that had an especially large breathing cavity, putting it in a good position to survive the plummeting oxygen levels. It used its two horn-like canine teeth to burrow into the ground. Perhaps its subterranean habits explain this creature's large lung capacity, a happy adaptation for survival in stale, low-oxygen underground air.

Lystrosaurus gained a survival advantage in being able to regulate its body temperature (see also *Dimetrodon*, page 116). Being vegetarian was good news, too. With the extinction of so many other species, there was just enough plant food left for herbivores like *Lystrosaurus* to survive.

The consequences of *Lystrosaurus*'s survival through the Great Permian Extinction were immense because, without it, the line of warmblooded vertebrates that eventually evolved into mammals would probably have died out 252 million years ago, along with *Dimetrodon* and all the rest.

Yet, following this most traumatic of evolutionary episodes, the baton of supremacy among land creatures passed to a different group, one that led to the dinosaurs. The descendants of *Lystrosaurus* retreated into the margins of terrestrial society, becoming smaller and furrier, and limiting themselves to hunting mostly at night. Then events once again took evolution in a new direction with another extinction event, which unfolded just as dramatically 65.5 million years ago (see pages 159–60).

Quetzalcoatlus

ORDER: PTEROSAURIA
SPECIES: QUETZALCOATLUS NORTHROPI
RANK: 49

The world's largest flying creature and the first global circumnavigator.

DREAMS OF HUMAN FLIGHT reach far back into history to the myths of ancient Greece. The youth Icarus is said to have donned the wings of his inventor uncle Daedalus only to fly too close to the Sun, causing his wings to melt and sending him plunging into the sea. More than two thousand years after that story was first told, a real man, an Italian artist called Leonardo da Vinci, sketched practical designs for a set of artificial wings. He imagined they could be strapped to the arms of a human who, after jumping off a cliff, would be able to fly like a bird. But no volunteers stepped forward to try it out.

But during the twentieth century flight for humans became an everyday, humdrum affair. With the rise of fossil-fuel-powered, propeller-driven and jet-engined craft, flying lost much of its mystique. At any one time today, as many as 250,000 people are up in the air, criss-crossing the globe in pressurized aeronautical tubes.

This human achievement was preceded by that of other vertebrates that took to the skies aided by nothing more than the power of evolution some 230 million years ago. What's more, these animals mastered the power of flight a good seventy-five million years *before* even the world's most primitive birds could be heard singing their first morning chorus (see *Archaeopteryx*, page 122).

Pterosaurs were reptiles that evolved alongside the dinosaurs, soon after the Great Permian Extinction 252 million years ago. Their bodies were adapted for flight by the growth of an extraordinarily long and strong fourth finger, which protruded like a strut to support the front edge of a giant webbed wing. Exactly which species first took to the skies and why is unknown. Fossils of small crow-sized creatures called *Sharovipteryx* with flaps of skin stretched between their fore and hind limbs have been found in Kazakhstan. Reptiles like these were well adapted for escaping predation by leaping off hills, stones, cliffs and ledges. Over time, their success probably led to the first species of pterodactyls, which could actually fly and not just glide.

Exactly how pterodactyls were capable of powering themselves into the air, and not just gracefully gliding through it, was demonstrated by American inventor Paul MacCready. In the 1980s this engineer rose to the challenge of building a half-size working model of the largest flying reptile that ever lived.

Quetzalcoatlus northropi was a genuine giant. Fossils found in Texas suggest its wingspan stretched to as much as twelve metres – larger than many small aircraft today. This awesome creature, which lived between eighty-four and sixty-five million years ago, was a masterpiece of aeronautical engineering, reaching what experts believe are the maximum limits possible for any living creature – it was the Howard Hughes *Spruce Goose* flying boat of its day.

Quetzalcoatlus: the jumbo jet of the Cretaceous Period with a wingspan of up to twelve metres.

And, as MacCready's successful simulations demonstrated, this behemoth really did fly. The half-size model, launched from a trolley in the Mojave Desert, California, on 27 January 1986, made six successful sorties using battery-powered electric motors instead of organically-fed muscles. With wings flapping, looking just like a Cretaceous pterosaur, the model spiralled, dipped, and swooped through turns as sharp as 270 degrees, re-creating the graceful winged magic of a long-lost age.

Pterodactyls were the first creatures to have had global oversight of the world, capable of travelling from pole to pole. For the first time in the history of life, the entire planet could become a single creature's habitat. Large species, like *Quetzalcoatlus northropi*, could fly for thousands of miles at a stretch, connecting disparate lands as the single continent of Pangaea continued to break apart to form the discrete continents we know today.

Who knows what opportunistic species hitched a ride on this global transport system? What seeds, spores, bacteria or viruses were spread by this new form of long-haul flight? Birds would later reshape the world by spreading all manner of organisms, from influenza to fruit. Pterodactyls most likely did something similar, but many millions of years before.

Pterodactyls never made it through the catastrophic extinction that struck 65.5 million years ago, despite their enormous success over the previous 170 million years. This trauma killed off not only the dinosaurs but eight out of nine orders of birds (see page 124). Now only one order survives, providing life's single link with the era of vertebrate flight pioneered by pterodactyls many aeons ago.

Archaeopteryx

CLASS: AVES
SPECIES: ARCHAEOPTERYX LITHOGRAPHICA
RANK: 41

A boastful show-off that spawned new flights of evolutionary fancy.

IN HIS LATER YEARS Charles Darwin became increasingly fascinated by sex. He wasn't interested in the salacious kind beloved of today's tabloid newspapers and lifestyle magazines. His focus was far more scientific. It was trained on the extraordinary differences in appearance and behaviour between males and females of the same species. Why are male gorillas so much larger than females? Why are male peacocks so colourful compared with female peahens? Why do batteries of male birds of paradise with elaborate plumage perform eccentric dancing rituals in front of crowds of fussy females?

Darwin believed all these variations could be explained by a second type of natural selection pressure. Sexual selection was not, he said, a struggle for existence, but a competition *'between the males for possession of the females'*.

Birds are creatures that have often followed this trend, as in the case of peacocks and birds of paradise. Females of these species are attracted to males that show off. They wish to mate only with the most brightly coloured or best performing males since they instinctively reckon the chances of their children mating will be greater if they inherit the best 'showing-off' characteristics. In this way the selection pressure for genes responsible for brightly coloured feathers – or elaborate dancing rituals – in male birds becomes ever more pronounced over time, as has happened in some bird species (for example, peacocks, parrots and drakes).

The curious natural history of sexual discrimination is now thought to have provoked one of the most significant developments for life on Earth in the last 150 million years – the flight of the world's first bird (for another example of the effects of sexual selection, see *Homo erectus*, page 181).

Archaeopteryx is one of the most famous fossil discoveries of all time. The impression in rock of a half-reptile, half-bird creature caused a sensation when the fossil was dug out of a limestone quarry in Germany in 1861. Just two years after Darwin published his *On the Origin of Species*, here was a fossil that showed a creature in an evolutionary transformation. Like *Tiktaalik* (see page 114), *Archaeopteryx* was (and still is) one of the most stunning transition fossils ever discovered.

Its skeleton was unmistakably reptilian but it had the feathers of a bird, even though they jutted out from each vertebra along the length of its tail, and not in a clump from the final vertebra, as in modern birds. Its discovery turned up the heat in the nineteenth-century debate between the Creationists and the Darwinists.

Evolutionary sceptic Andreas Wagner, a German zoologist, was under no illusions about how Darwinians would interpret this find:

'Darwin and his adherents will probably employ this new discovery as an exceedingly welcome occurrence for the justification of their strange views on the transformation of animals ... '

Today theories as to the origin of modern birds and the place of *Archaeopteryx* have themselves evolved. Traditionally it has always been thought that the definition of a bird was a creature (usually able to fly) that is covered in feathers. But the amazing story of 150-million-year-old *Archaeopteryx*, the world's ancestral bird, has been complicated by the discovery in China in 1998 of (non-flying) dinosaur fossils showing the unmistakable imprints of feathers. If dinosaurs had feathers then perhaps birds are actually a branch of the dinosaurs?

Most experts are now agreed that birds do represent the one and only surviving group of dinosaurs. They are descended from theropods (see page 125), fossils of which have recently been discovered also showing feathery features.

So, if feathers evolved *before* birds, what was their original purpose? Chinese dinosaur fossils show that theropods with feathers were not flying creatures because they had no wings, leaving experts to suppose that feathers in dinosaurs evolved first as a means of insulation, and were later adapted – as is so often the case in nature – by *Archaeopteryx* and its descendants for flight.

But now another theory is gaining ground. It goes back to Darwin's acute observation of the major differences in looks and behaviour between the sexes in some species. Feathers are outgrowths from scales, a common feature of reptilian creatures. Some reptiles are renowned for their elaborate courtship rituals and displays, for example, the male komodo dragon.

A cast from an *Archaeopteryx* fossil found in Germany in 1877 – was this the first feathered flyer?

Such dynamics may have led unwittingly to feathered flight in birds. Somewhere along the dinosaur-to-bird evolutionary highway, flashy scales that refract the light, producing pretty colours, proved irresistible to females. This could have caused a runaway evolutionary spiral that resulted in elaborate plumage as seen in *Archaeopteryx*. Once endowed with such a covering, the benefits of extra insulation and later the re-orientation of feathers for the aerodynamics necessary for flight were adaptations no more remarkable than the jaw bones of fish receding in the skull to become bones for amplifying vibrations in the air allowing terrestrial animals to hear (see page 108).

But what caused the feathers of *Archaeopteryx* to become adapted into wings that supported flight? Its habitat was surrounded by lagoons and sparse vegetation no more than three metres high, suggesting a ground-up origin of flight. Like its theropod relatives, ancestors of *Archaeopteryx* probably ran along the ground, occasionally leaping into the air to catch insects to eat. As they did so, the ones that flapped their feathers were increasingly successful, until eventually those with the largest, most aerodynamic arrangements spawned wings that supported take-off.

Archaeopteryx is the oldest-known relic of that branch of dinosaurs that survived the mass extinction 65.5 million years ago and went on to fill our skies with the songs of thrushes and skylarks we are familiar with today. The impact of bird flight, which co-existed with the soaring pterodactyls for about seventy-five million years, has been one of the most powerful forces of evolutionary change. Unlike pterodactyls, which ate fish and insects, many birds are herbivores. Such a flexible diet helped their long-term survival, enabling them to radiate into myriad modern forms.

Seeds wrapped in fruit, originally designed as large packets of fast food for roaming dinosaurs, evolved into smaller bribes (such as berries) that could be carried in the belly of a creature the size of a dove and deposited in its excrement many miles away from the plant's parental origin. The overpowering success of angiosperms (flowering and fruiting plants and trees) over other types of trees (such as conifers, cycads) in the last 150 million years is living proof of the incalculable effect birds have had on the shaping of the modern world (see page 129). The story began with *Archaeopteryx*, the first known flying avian, whose appetite for sexual display seems to have led to an outcrop of feathers that were subsequently adapted for the purpose of better survival – jumping into the air – eventually turning them into the lords and ladies of the skies.

Tyrannosaurus

ORDER: THEROPODA
SPECIES: TYRANNOSAURUS REX
RANK: 34

An iconic terrorist that forced other creatures to find new forms of survival.

WHAT'S THAT coming over the hill – is it a monster? From the myths of King Gilgamesh, the fifth Sumerian ruler, to the stories of ancient Greece, monsters, such as the giant one-eyed Cyclops and the multi-headed Hydra, have been a constant reminder of the vulnerability of man and his artifice against the forces of nature.

Imagine the surprise of Western naturalists and scientists when incontrovertible evidence began to emerge in the early nineteenth century that enormous monsters *really did* once roam and indeed rule the Earth. Fiction became fact. And what of the possibility that humans, following on from Darwin's deductions regarding evolution, were themselves descended from creatures such as these? To most people living then, the very idea was, well ... monstrous.

Today evidence for the existence of now-extinct giant lizards called dinosaurs is universally accepted. Fossilized bones have often been exquisitely preserved in the rocks, allowing experts to piece back together the precise shape and form of these creatures that once ruled the world. Although extinct for more than sixty-five million years (apart from the birds, see page 123), dinosaurs had a profound impact on the planet, life and people that lives on today.

Some trees turned the threat of being eaten into an opportunity. They bribed these roaming herbivorous beasts such as *Diplodocus* (living from approximately 154 million to 144 million years ago) by wrapping their seeds in a juicy meal – a packaging system we call fruit. At an enormous forty-two kilograms, the mature, buttock-shaped sea coconut (the fruit of a coco de mer tree) evolved into a form that was attractive to peckish dinosaurs. These creatures would then spread the seeds in their dung. A few of these ancient palms still grow in Madagascar, Malaysia and New Guinea today. Owing to the demise of dinosaurs, however, these species are long past their prime.

Meanwhile, other trees established elaborate techniques for defence, such as the maze of branches of the monkey puzzle tree. A group of plants called monocots even learned to grow 'upside down', keeping their buds hidden beneath the ground to prevent grazing dinosaurs from eating the newest, freshest growth first. These species evolved into grasses and palms that have also transformed the world (such as wheat, page 199; and rice, page 230).

The most successful dinosaur group of all was the theropods. Not only were these among the first dinosaurs to have emerged, about 230 million years ago (*Eoraptor*) but they are the only surviving branch of dinosaurs to exist today. Birds, with their common theropod characteristics of wishbones and three-toed feet, are this group's living descendants. The rest disappeared in a giant mass extinction 65.5 million

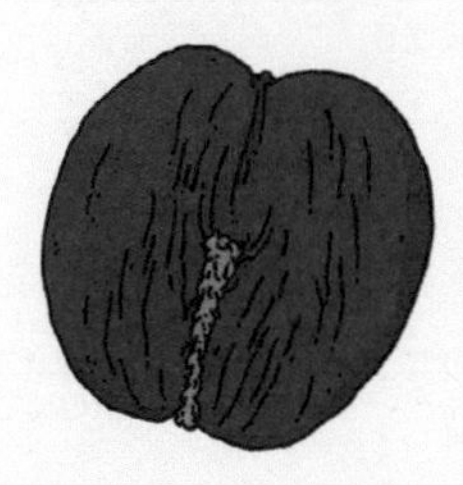

years ago – the same catastrophe that finished off all the other dinosaurs, as well as the pterosaurs (see page 121).

The most fearful theropod, and the most famous dinosaur fossil, was *Tyrannosaurus rex* – a gruesome creature that lived between seventy-five million and sixty-five million years ago in North America and Canada. Other similar creatures, such as *Tarbosaurus*, roamed the forests of East Asia, while *Allosaurus* and *Gigantosaurus* lived in South America. Collectively, these dinosaurs would have created a climate of fear among other vertebrate animals during the Jurassic and Cretaceous Periods (206 million to 65.5 million years ago).

Tyrannosaurus rex was at the apex of the prehistoric terrestrial ecosystem before being wiped out by a giant meteorite.

Anyone living close to Chicago, in the US, can get a good first-hand impression of the sheer size and power of creatures like these by taking a trip to the city's Field Museum. Here the largest and most complete fossilized *T. rex* skeleton is to be found. She is named Sue, after Sue Hendrickson who discovered her in 1990 near the Cheyenne River Indian Reservation in western South Dakota. Sue clearly shows why the *Tyrannosaurus* was such a fearsome creature, with bone-crushing jaws more than 1.2 metres long and one metre wide. Its individual teeth were curved and serrated, many of them longer than a human hand, allowing the beast to crush its victims to death almost instantaneously, swallowing up to seventy kilograms of meat in one go.

Key to these creatures' success was a fully improved stance that meant their entire weight was taken on their hind legs, allowing them to balance effectively on two feet and increasing their manoeuvrability despite their enormous body size (twelve metres long).

The devastating event that killed off all the dinosaurs 65.5 million years ago, including *T. rex*, was probably the chance impact of a ten-kilometre-wide meteorite. This cataclysm plunged the Earth's climate and atmosphere into a new dark age where luck, not evolutionary fitness, became the main criterion separating survival from extinction. Large terrestrial reptiles were most at risk – especially those that fed on flesh. As populations plummeted, food became scarce. If the actual impact or climate didn't kill off *T. rex*, then starvation soon finished the job.

Yet long after their demise, the impact of theropods like *Tyrannosaurus* continues to reverberate in the shape, form and function of much of the biological world today.

Descendants of *Dimetrodon* (see page 116) and *Lystrosaurus* (see page 118) thrived by staying small and hidden in the woods and trees. These were little, shrew-like mammals, which spent time protecting their young, and benefited from a covering of fur to keep their bodies warm so they could hunt at night when there was less threat of attack by theropod monsters.

Fertilized eggs that grow *inside the mother's body* were other adaptations that ultimately evolved in many mammals as a direct response to the threat posed by dinosaurs (including birds and pterosaurs) that regularly helped themselves to eggs. Mammals were oviparous – they gave birth to live young. They also had mammary (milk-producing) glands, which meant females did not have to leave their nests to find food for their young until it was safe to do so. In a world dominated by giant meat-eaters like *T. rex*, mammary glands evolved as an essential component of the mammals' Jurassic survival kit (see also pages 157–8).

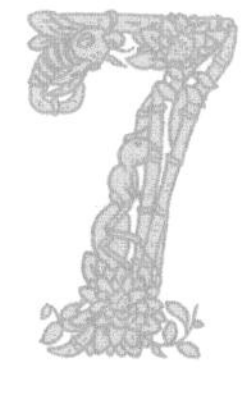

On Biodiversity

How beauty, collaboration, deception, parasitism and vice wove terrestrial life into a rich carpet of countless species.

AS HE GREW OLDER, Charles Darwin's obsessive powers of observation made it increasingly difficult for him to reconcile the idea of a benevolent Creator with the machinations of the natural world. The creatures that gave Darwin his biggest cause for religious disquiet came from a family of parasitic wasps called Ichneumonidae. Darwin wrote agonizingly about their gruesome lifestyles in a letter to his American botanist friend, Asa Gray:

> *'I cannot persuade myself that a beneficent and omnipotent God would have designedly created the Ichneumonidae with the express intention of their feeding within the living bodies of caterpillars.'*

From the time when the dinosaurs began to dominate the Earth (*c.* 230 million years ago) an increasingly rich and interconnected pattern of lifestyles began to emerge within and between the kingdoms of plants and insects that had, by the time of the dinosaurs' demise 65.5 million years ago, literally changed the face of the Earth.

The wasps that so unsettled Darwin were typical of many insect groups that evolved in the early Triassic Period (*c.* 230 million years ago). Today there are about 100,000 different species of parasitic wasp, which makes it one of the most diverse groups of living things.

Hymenoepimecis argyraphaga is a typically gruesome example. Its prey is a species of spider called *Plesiometa argyra*. The wasp temporarily paralyses the spider with venom and then injects its eggs through a tube, called an ovipositor, deep into the live spider's body. Inside, the growing wasp larva manipulates the spider's web-spinning system to make a cocoon of silk. The hapless

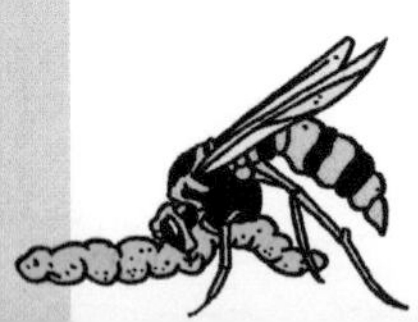

arachnid is then gradually eaten alive from the inside out.

Parasitism is a lifestyle in which one species draws nourishment from the living body of another. Most free-living animals today play host to one or more types of parasite. Some of them, such as **mosquitoes** (see page 134), **fleas** (see page 137) and **tsetse flies** (see page 140), have also had a dramatic impact on human history by carrying and spreading some of nature's most pernicious diseases.

Parasitism has acted as an evolutionary catalyst, greatly increasing the diversity of living things. Just as the most successful host organisms were those that *adapted to resist* the worst effects of parasitic attack, so successful parasites were those that *adapted to evade* the defensive systems of their chosen hosts.

'The Busy Bees and Their Cousins', by Maud Scrivener; these insects are now under threat from Colony Collapse Disorder.

Flora meets fauna

Angiosperms (flowering plants) evolved into a world of parasitic insects. Exactly when and why flowers emerged is still something of a puzzle – Darwin himself called it a most 'abominable mystery' – but the development of this radically new way of life in the plant kingdom, about 140 million years ago, marked a watershed in the evolution of all life.

Angiosperms are plants that, over the last 140 million years, have become life's most successful group by far. Today as many as 400,000 different flowering species dominate plant-life from the tropics to the Arctic, dwarfing in number the more ancient lineages of ferns, conifers and cycads (see page 96).

Such success is derived from their unsurpassed powers of persuasion. Descendants of ancient parasitic wasps were mesmerized by the irresistible allure of the world's first flowers. As a consequence, some of them – such as **honey bees** (see page 152) – turned their backs on the lifestyle habits of generations and learned to live on sugary water and protein-rich pollen instead of preying on other animals. Ultraviolet colours, invisible to the human eye, run along the sides of petals as guidelines, like landing lights on a runway, leading the insect precisely to where treasure, in the form of sugary nectar, is to be found. Along the way, pollen from the flower's male parts rubs on to the insect's legs, tummy and back, while pollen, gathered from a previous flower, is deposited on a pouting style – a sticky tube that juts up expectantly.

In this way sexual reproduction through the transfer of the pollen from one plant to another is provided by the power of an insect's flight. The trick of employing armies of mobile insects, first learned by the flowerless cycads (see page 101), was substantially improved by flowering plants thanks to the power of visual attraction and the bribe of nutritious rewards.

Pollination of flowering plants by insects triggered a dramatic co-evolutionary dance between new plant and insect species. Since the goal of any flowering plant is to have its pollen spread to another plant of the same species, it often made genetic sense to try to attract one *specific* type of insect pollinator that would *always* be attracted to that same species of flower. In this way fertilization (through the transfer of pollen from one plant to the eggs of another of the same species) could be targeted precisely in a way that random and chaotic pollination by wind power never could.

The butterfly's long, coiled proboscis evolved to feed on nectar in flowering plants.

One of the consequences of targeted fertilization was that it allowed certain flowering plants to get away with producing much less pollen. Savings in one direction allowed greater investment in others, such as in the production of nutritious rewards (in the form of nectar), larger, flashier flowers (to attract attention), and sweet-smelling scents. (See On Beauty, page 336, for the effects of angiosperms on humans.)

Flowers became highly competitive in their bid to attract different species of insects. Those that struck on successful combinations of beauty, shape, scent and reward became the most successful, with their powers of persuasion amplified via inheritance as each new generation passed. In this way, between 140 million and one hundred million years ago, vast numbers of new angiosperm flowering species emerged, co-evolving with a wide variety of insect pollinators in countless different ways.

The power of floral designs to influence the shape and design of other species was (and still is) at least as profound as the power of a virus or a parasite to infect and modify its host (see pages 43–4). The long coiled proboscis of a butterfly or moth is just one example of a specialist tool that evolved for the purpose of sucking up flower-juice from precisely placed nectar glands – delicately positioned to make sure that these thirsty visitors always brushed past pylons of plentiful pollen.

A marriage of convenience

Nature's floral pageant resulted in certain types of insects becoming, in a sense, 'married' to particular species of flower. Take the case of the carpenter bee which has co-evolved alongside a variety of pink gentian that thrives in South Africa. Three large stamens lunge outwards from the gentian's spread-out pink petals, apparently offering its protein-filled pollen to any flying passers by. But no. Only the carpenter bee knows how to unlock the plant's trove. By lowering the tone of its buzzing wings to the pitch of about middle C, the carpenter bee causes the anther to resonate at just the right frequency to release the plant's pollen-packed stamens. The bee then gathers up its haul, which it stuffs into specially adapted packs on its back, before continuing on its merry gentian jaunt.

How efficient a system is this! The mutual association between the gentian and carpenter bee has the precision of a targeted modern direct-marketing campaign. Adaptations in both species led to mutually beneficial cost-savings. The carpenter bee knows exactly how to access its exclusive store of food, while the gentian's own pollen is never wasted on insects that flirt with all and sundry.

Owing to the considerable benefits on offer for both pollen-maker and pollinator alike, the fashion for co-evolution has been another of nature's most powerful protagonists for species change. Flowering plants, like gentians, that allied themselves with a single pollinating species could count on their progeny spreading as far and as wide as their chosen pollinators could fly, knowing that success in future generations could be secured regardless of which way the wind blew.

Dangerous liaisons

Orchids are flowers that have co-evolved with all manner of highly specific flying insects (see vanilla, pages 346–7). It helps explain why, with more than 22,000 recorded species, Orchidaceae is the largest family of angiosperms alive today. Of particular interest is a species called *Angraecum sesquipedale* that grows in Madagascar. These are flowers that store their nectar in string-like spurs up to half a metre long that trail from the base of their flowers. Only one species of hawk moth (*Xanthopan morganii*) has a proboscis long enough to extract the juice, again providing a perfectly targeted pollination service like that offered to the gentian by the carpenter bee.

But such symbiotic relationships between different species do not always give both sides an equal advantage.

The low-growing mirror orchid that thrives in the western Mediterranean produces blue-metallic oval-lipped petals edged with a yellow border and long red hairs. These are flowers that have evolved specifically to *mimic* the looks of a female bee. They also produce a perfume that exactly resembles a female bee's pheromones. Therefore, when a male bee flies by, he is instantly excited by the sight and smell of the flowers, which trick him into thinking they are females of his own kind. Grasping a flower as if it were a mate, the bee begins to copulate with it. In the process, some pollen is glued to his head while other pollen deposited by a previous flower gets stuck on the orchid's well-placed style. More than one hundred different kinds of European orchid use mimicking systems like this to dupe insects into becoming pollinators.

Butterflies and moths are pollinating insects that – like orchids – have become highly skilled in the art of mimicry. In their larval forms, certain caterpillars are capable of disguising themselves as sour-tasting bird droppings to make them look unattractive to eat. They even secrete a strong-smelling form of white uric acid to complete the illusion. Adult *Heliconius* butterflies are among nature's most efficient mimics. They have evolved to look like other poisonous species even though they are safe to eat. Why bother mixing toxins yourself when a masquerade may be just as effective and much less trouble to produce than the real thing?

One of the most frequently mimicked insect families is the vespids (wasps). Yellow and black stripes in nature have become synonymous with a venomous sting, which wasps generally use for attack and defence. So effective are these markings that other non-stinging flower-visiting insects – chiefly hoverflies, moths, and some beetles – have copied vespid colour markings as a way of fending off attack. Others mimic ants, themselves descended from wasps, benefiting from their fearsome

Deceptive beauty: the mirror orchid (*Ophrys speculum*) mimics the appearance and smell of a female bee.

reputation for defending against attack with a powerful bite and formic acid spray.

For beauty, deception and mimicry read advertising, seduction and fraud. All were lifestyle strategies that began with the co-evolution of flowering plants and their insect pollinators beginning about 140 million years ago.

Matters of taste

Larger animals were also bewitched by a form of floral bribery at least as powerful as visual beauty, but this tactic was directed towards their stomachs. The fertilized seed-filled ovaries of flowering plants often swelled into juicy meals that were designed to be eaten by vertebrate creatures. Tough seeds were built to endure the highly acidic conditions in a dinosaur's stomach while waiting to be shot out far away from their parent plants in a pile of manure. In such a way flowering plants were able to spread their seeds as far and as wide as possible to avoid competing with their parents for light, water and nutrients. Wide seed dispersal also provided excellent insurance against dramatic local environmental changes that could otherwise wipe out the entire species if restricted to growing in one particular place.

Chemical warfare: the deadly nightshade plant (*Atropa belladonna*) is both lovely and lethal.

The exchange was a fair trade – in return for spreading a plant's seeds in their dung, larger birds and animals were offered a fruity shot of well-packaged fast-food.

Most fruits we eat today have been artificially selected and cultivated over hundreds or thousands of years to suit human diets and convenience agriculture (see page 336). However, there are some that still survive from the Jurassic past, such as the **durian** (see page 148), which became essential food for herbivorous dinosaurs. These are fruits that have remained true to form for millions of years.

The expertise of flowers in manufacturing chemicals was just as effective when directed towards their needs for defence. Plants cannot move. This single fact is the main reason why these life-forms have become the word's pre-eminent experts in chemical warfare. The making of poisons and toxins is a specialist science developed by some of nature's most successful species of flowering plants.

Tannins, toxic alkaloids, terpenes and oils are all chemicals produced by plants for the explicit purpose of defending themselves against attack by insects, bacteria, fungi, herbivorous animals and parasites. Thanks to their superb chemistry skills, flower orders such as the Solanales (to which the potato belongs) can survive as soft-tissued, herbaceous, fast-growing plants that have no need for external defences such as wood or bark – see potato (page 206), tobacco (page 289) and chilli pepper (page 356).

In the wild, plants like these can be deadly. Tropane alkaloids produced by flowers such as *Atropa belladonna* (deadly nightshade) are among the most lethal in all nature, often inducing convulsions, hallucinations, coma and death if eaten by animals. Having bitten one, there's little chance of a grazing animal surviving long enough to become twice shy. As a survival strategy, the power of floral chemical factories is one of nature's most robust.

Strategic design

Chemicals were not, however, the flowers' only form of defence against attack. One group, called monocots, developed an entirely new type of body design that turned a potential disadvantage into a recipe for success. Instead of fresh growth sprouting outwards from the edges of its leaves, as is the case in conifers, these species grow upwards and outwards from a central apical bud. Such a design meant the newest, most valuable tissues were always furthest away from the jaws of a hungry vertebrate, be it a herbivorous dinosaur or grazing mammal.

The most successful monocot families today are palms and grasses, both of which radiated extensively after the demise of the dinosaurs. The world's first grasses were ancestors of today's rice plants (see page 230) and **bamboo** (see page 150). Daffodils, tulips and crocuses are all monocot flowers with bulbs that have apical buds that are safely tucked underground, out of harm's way.

Many of today's most prolific flowering plants thrived with the emergence of grasslands in the last seventy million years. Daisies and sunflowers (see pages 400 and 401) belong to a family called Compositae, flowers that have evolved the remarkable technique of populating their faces with hundreds of individual tiny flowers called florets. Ingenious biological pumps ensure that passing insects are richly covered in pollen. In the event of there being a dearth of animal transporters, the many florets (some male and some female) are so tightly packed that the flowers can self-pollinate, if need be. Versatility is what gives plants like these a survival advantage.

Similar reproductive strategies were also key to the success of some of the world's most significant flowering trees such as the **oak** (see page 143), which first evolved in tropical regions about ninety million years ago. Prodigious quantities of acorns are designed to be spread by animals such as squirrels. If necessary, however, these are flowering trees that can reproduce without the need for fertilization by re-growing out of dead or damaged trunks – even after a whole forest has been destroyed by fire (see also eucalyptus, page 264).

Strength in numbers

Perhaps the most ingenious lifestyle to evolve alongside flowers and dinosaurs was that of living in a society, a technique which succeeded through sheer force of numbers. Creatures that live like this have fascinated scientists for hundreds of years. **Ants** (see page 154), honey bees (see page 152) and termites (see page 400) are noted for their highly sophisticated lifestyles. Their collective behaviour is so well developed that in some respects these insect societies behave like a single organism (witness the swarming of a bee colony).

Significant differences exist between insect societies and humanity, such as their methods of communication. Insects generally rely on chemical smells and dances, whereas humans have evolved the ability to communicate through vision, speech, sound and song. Despite this, organizational hierarchies, systems of social care, teaching, farming, soldiers and slavery – systems normally associated only with human history – originally emerged in social insects alongside the evolution of flowering plants between 150 million and one hundred million years ago.

In fact, many characteristics of human societies have prehistoric origins that can be traced back to the highly varied states of nature that became predominant with the rise of biodiversity in flowering plants, parasites and social insects. Versatility, sociability and mutualism, today regarded as positive lifestyle choices, had their origins here alongside the more negative modus vivendi of parasitism, bribery, seduction, mimicry, deception and exploitation as well as chemical attack and defence.

So clearly did Darwin associate the characteristic behaviours of mankind with the unscrupulous, amoral and apparently godless lifestyles found in nature that he knew his theories on human evolution would cause many people distress, as he openly admitted in the last part of his book *The Descent of Man*:

> *'The main conclusion arrived at in this work, namely that man is descended from some lowly organized form, will I regret to think, be highly distasteful to many . . .'*

Mosquito

FAMILY: CULICIDAE
SPECIES: ANOPHELES GAMBIAE
RANK: 14

Nature's female hypodermic injectors that carry species-changing infections.

NEVER UNDERESTIMATE the power of a parasite. Never misjudge the meanderings of a mosquito. And be especially fearful of females. They are the ones that bite.

Mosquitoes evolved at about the same time as the dinosaurs, 230 million years ago, and have probably had a bigger impact on vertebrate animals than any other living creature. Females puncture the skin of animals with nature's equivalent of a hypodermic syringe, sucking out the blood to use as a source of protein for producing eggs. Such behaviour has had far-reaching consequences for the evolution of life on land.

There is no desire in these bloodthirsty females to inflict damage or disease on the hosts that they feed on – far from it. It's just that mosquitoes are perfect vehicles for the transmission and replication of other types of parasitic life. All manner of microscopic parasites have found a way of spreading themselves among the land animal population by inserting themselves into the mosquito's saliva, thanks entirely to the power of a female mosquito's bite and flight. Males feed on nectar and honeydew.

Immediately after a female mosquito inserts her syringe into an animal, she injects a shot of saliva that contains anti-coagulants designed to prevent the host's blood from clotting too fast. It's via this saliva that other parasites – such as worms, bacteria, viruses and protozoa – are able to gain entry into the vertebrate bloodstream, spreading rapidly through the complex network of arteries and veins.

Once inside, these parasites try to compromise their host's immune system so that they can successfully breed. Their offspring either spread via the host's guts and faeces, sometimes through body liquids such as breast milk, or they may lie in wait until another mosquito draws out its own small vial of blood. The parasite then infects the saliva of this new mosquito, preparing for when she bites another host – and so the transmission of the parasite species moves successfully along.

Dr Charles Laveran was commemorated on this Algerian stamp in 1954 for his discovery of the malaria-causing parasite *Plasmodium*.

About 300 diseases that affect humans are transmitted via mosquitoes, including some of the most dramatic in history as well as many of the worst afflictions still affecting populations today. The most famous is malaria – so called because until 1894 it was thought to be caused by stale or 'bad air' (*mal'aria*) since the disease was frequently contracted in marshy, wetland areas. But when French physician Charles Laveran took blood from an infected patient and studied it under a microscope, he spotted a tiny amoeba-like creature moving around, powered by a flagellated (whip-like) tail. His deduction that this

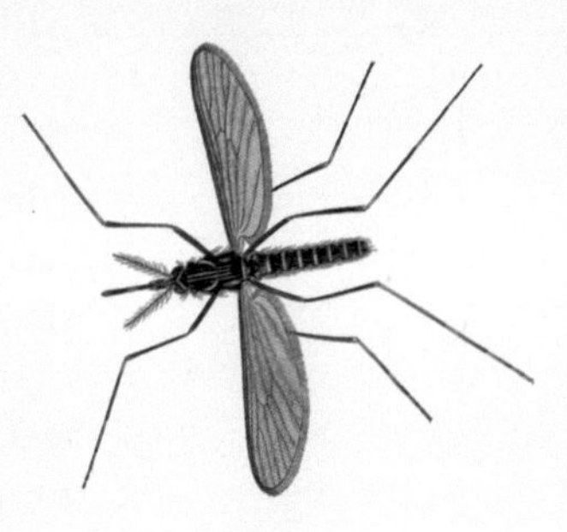

protozoan parasite, *Plasmodium*, was spread via mosquito bites earned him a Nobel Prize in 1907.

Mosquitoes: prolific transporters of species-changing infections.

Laveran's discovery was momentous. Malaria is a disease that can cause all manner of symptoms from fever and flu to coma and death. Today approximately 515 million people have malaria worldwide. Deaths from the disease total from one million to three million people a year – most of them in Africa. Historically, this disease has been one of mankind's major killers, ranking alongside smallpox (see page 24) and influenza (see page 16), although its origin is at least ten times more ancient than most other diseases, which emerged in humans alongside the domestication of farm animals *c.* 10,000 years ago (see On Agriculture, page 191).

Scientists are sure that humans have been affected by malaria for such a long time because they can trace a genetic mutation that has arisen in some people as a result of the disease. This mutation stretches back at least 5,000 generations and amply demonstrates the impact that mosquitoes, as flying vectors of disease, have had on human evolution.

Some people in Africa, which is where malaria has maintained its grip for the longest period of time, have developed genetic immunity. Their red blood cells have become flattened and rigid, a response that makes it hard for the *Plasmodium* parasite to attach itself and enter bodily organs. This is known as sickle cell trait. Although it brings immunity to disease, the genetic differentiation carries a high cost. Flattened blood cells are less efficient at carrying oxygen around the body and they cannot always squeeze through their blood system's narrowest passageways, called capillaries.

Sickle cell trait symptoms are usually mild. But some individuals inherit a more serious form of the condition called sickle cell anaemia. This greatly shortens life-spans and can severely affect people's quality of life. But its origin as a genetic response to the *Plasmodium* parasite represents a classic example of how one species (a disease-carrying mosquito) can alter the evolutionary genetics of another species (humans) by acting as a conduit between a parasite (*Plasmodium*) and its host.

Other parasites have also latched on to the powerful transmission opportunities afforded by female mosquito saliva. Forms of yellow fever (caused by a virus of the family Flaviviridae) result in symptoms ranging from headache to coughing up of blood and internal bleeding, and may also result in death. Several appalling epidemics have been recorded in human history since the fall of the Roman Empire, mostly in Africa, Europe and the Americas. Outbreaks have sometimes caused hundreds of thousands of deaths.

In 1802 Napoleon sent an army across the Atlantic to suppress a rebellion in the Caribbean island of Haiti. His forces were decimated by yellow fever, as a result of which any lingering French aspirations to invade the recently liberated mainland of America via French-controlled Louisiana were finally curtailed. Consequently Haiti – a land of enslaved African cotton labourers – became only the second European colony to gain sovereignty (after the recognition of American Independence in 1783).

The cause of yellow fever was finally discovered by the brave, if not foolhardy, actions of a number of American and Cuban medics including Jesse Lazear (1866–1900), the first man to suspect that mosquitoes carried the disease. Whether it was ambition or selflessness (or a mixture of the two) Lazear allowed himself to be bitten by infected mosquitoes in a bid to prove his theory correct. Curiosity about the disease cost him his life and he died at the age of thirty-four. By 1937 scientists had developed a safe vaccination against yellow fever (although there is still no cure), but an estimated 30,000 people still die of the disease each year in unvaccinated populations.

Wuchereria bancrofti, a type of nematode roundworm (see page 68), also employs the services of nature's most prolific parasitic transporter, the mosquito. These worms can grow as long as six to eight centimetres and live inside the human lymphatic system, especially around the groin and legs. Elephantiasis is a highly unpleasant disease caused by these worms that has been known to cause a man's scrotum to swell to the size of a football. Despite the existence of effective treatments, an estimated forty million people worldwide suffer from its effects, especially in countries such as Ethiopia.

According to recent virological research, the success of the many poxviruses that infect animals today probably resulted from their ability to hitch a ride on flying, biting mosquitoes more than 200 million years ago. Such viruses initially adapted to reptilian hosts, then to birds and, finally, following the demise of dinosaurs 65.5 million years ago, to mammals, monkeys and humans. During this time, each infected creature's immune system will also have adapted to resist persistent parasitic attack.

Charles Darwin believed that the 'struggle for existence', which underpinned his evolutionary theory, was driven by the power of overpopulation which: *'inevitably follows from the high rate at which all organic beings tend to increase ... As more individuals are produced than can possibly survive, there must in every case be a struggle for existence ...'*

The disease-carrying mosquito subverts this basic Darwinian dictum by controlling population levels, sometimes before they get large enough to compete aggressively with other species. Therefore, those best adapted to *withstand parasitic attack* become those best placed to survive, as shown by the case of human sickle cell trait. As a result, at least a few individuals in a given species will have immunity to disease. Their survival potentially gives rise to new species (see also HIV/AIDS, page 18).

Diagrams that illustrate the web of life usually put producers (such as plants) at the bottom and carnivores (such as humans) at the top. That is wrong. Biting insects such as mosquitoes – see also fleas (page 137) and tsetse flies (page 140) – are the real king-pins of wild ecosystems through their constant infecting and modifying of the most powerful species through the transfer of disease-causing parasites. In the process they have shifted and shuffled endless evolutionary strands throughout the countless generations of life.

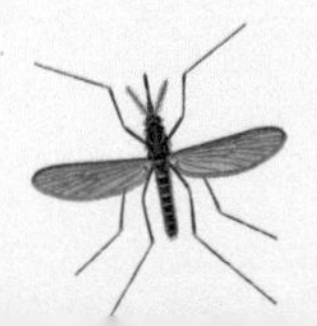

Flea

ORDER: SIPHONAPTERA
SPECIES: XENOPSYLLA CHEOPIS
RANK: 18

Spectacular jumping insects that have plagued human history.

MIRIAM ROTHSCHILD (1908–2005) was one of the twentieth century's most brilliant naturalists. Belonging as she did to the famous Jewish banking family, it is to the great benefit of science that money was never an obstacle in her life – except once. At the age of six, Miriam and her family fled from a holiday in Austro-Hungary on the eve of World War I. Her father was in such a hurry to get back to England that he forgot to take enough cash to pay for the family train ticket. He was forced to rely on the charity of a fellow Hungarian passenger who later declared: *'This is the proudest moment of my life. Never did I think that I should be asked to lend money to a Rothschild!'*

Charles Darwin's father also 'arranged' his financial affairs so that his son could focus on studying the natural world. Perhaps it is thanks to this financial independence that both Miriam Rothschild and Charles Darwin left such valuable and impressive legacies.

Rothschild's father, Nathaniel, enthused his daughter with a fascination for the natural world. Home-educated until she was seventeen, Miriam used her copious quantities of spare time in a relentless pursuit of discovery. Her father died when she was fifteen, but her uncle Walter, also a great naturalist, continued providing the young Miriam with a constant flow of inspiration. He built a museum at Tring Park in Hertfordshire – now part of the Natural History Museum. Miriam and her brother Victor were frequent visitors, admiring his collections of giant tortoises, cassowaries, wallabies and the other exotic creatures that lived there, roaming free.

Not 'Lord of the Flies', but 'Queen of the Fleas': Miriam Rothschild hard at work poring over parasites.

So inspired was Miriam by her youthful experiences of living close to nature that she became an early champion of conservation. She campaigned vigorously against all kinds of cruelty to animals and converted to being a strict (albeit eccentric) vegan, famous for wearing Moon boots in the winter, training shoes in the summer and white wellington boots in the evenings. But her most celebrated passion was the study of parasites – from flukes to cuckoos. She even found time for a certain rather rare worm *'which lives exclusively under the eyelids of the hippopotamus and feeds upon its tears'*. Among creatures with these lifestyles, her greatest fondness was saved for a species of tiny jumping insect, which is why she was often called 'Queen of the Fleas'.

Fleas were in the family. Miriam, like her father, was an avid collector. She spent thirty years cataloguing her father's world-famous collection of fleas, donated to the British Museum in 1923. She insisted on keeping her personal, live specimens sealed in cellophane bags in her bedroom, *'so I can see what they are doing and so that the children do not annoy them'*. As Rothschild explained in the foreword to her classic study on parasites, *Fleas, Flukes & Cuckoos,* these minute jumping insects have had an intimate relationship with mankind over thousands of years. Their impact scales from the simple bed bug that hides in cracks and crevices at night, and *'steals out to suck blood surreptitiously from a sleeping beauty's breast'* to being the prime vector of devastating infections that have killed hundreds of millions from pandemic plague.

Fleas are small parasitic insects that live by sucking blood from their animal hosts. Their ancestors diverged from flies about 160 million years ago, giving up the use of wings, preferring instead to leap from meal to meal using a pair of extremely powerful hind legs. Indeed, for their body size, fleas are among nature's most powerful jumpers, far out-leaping the best human Olympiads. They can cover a distance vertically and horizontally more than *200 times* their body length.

Almost inevitably highly mobile fleas that feed on other animals' blood became prime targets for less mobile life-forms that wished to hitch a ride. Fleas, like many flying parasites (in particular, mosquitoes and tsetse flies) carry a wide variety of passengers including worms and bacteria. One in particular, a meddlesome variety called *Yersinia pestis,* has proved the most lethal to humankind.

Yersinia pestis infects rodents – rats and marmots in particular. Fleas that bite these creatures pick up the bacteria, which then thrive in their guts (apparently without harm to the fleas), only to be spread via flea bites to other mammal hosts, including humans.

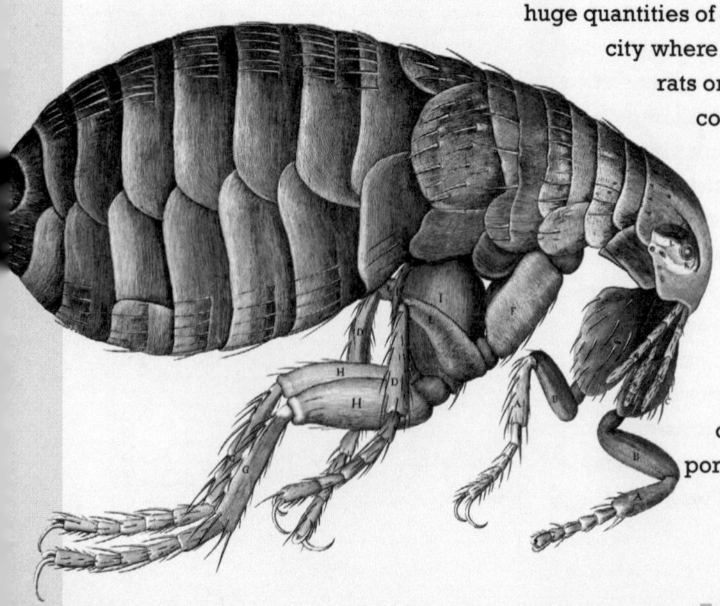

Meticulous 1660 engraving of a flea by English scientist Robert Hooke (1635–1703).

The first human pandemic caused by this bacterium, bubonic plague, occurred in the sixth century AD just as Emperor Justinian I was trying to reunite the ancient Roman Empire after its disintegration at the hands of Germanic invaders such as the Vandals, Huns and Goths. The outbreak of plague is thought to have originated in Ethiopia from where the imperial capital Constantinople imported huge quantities of grain. Plague thrived in the narrow streets of the city where fleas could easily leap off the backs of infected rats on to humans who lived in cramped, grimy, urban conditions. At its height, the pandemic is thought to have killed as many as 10,000 of the city's inhabitants *every day*.

Even this devastation didn't have as great an impact as the second outbreak – the most famous occurrence of plague in all history. The Black Death (1347–51) originated in China, and was spread to Europe by Mongol invaders. Little did Jani Beg, leader of the Mongol horde that attacked the Crimean port of Kaffa in 1346, know what his flirtation with biological warfare would unleash on the hapless continent of Europe. Carcasses of dead Mongol victims of the plague were flung over the city walls on giant slingshots called trebuchets, infecting the Italian soldiers inside. After escaping back to port in Genoa, 90 per cent of the Italian troops had died, leaving just enough survivors to pass the disease to the European mainland from where, thanks to the leaping fleas and rampaging rats, it ravaged human populations. As many as sixty million people died in China, forty million in Europe and ten million in Africa.

Huge numbers of books have been written about the impact of the Black Death on world history, coming as it did just prior to the overseas expansion of the European powers. It is beyond doubt that the decimation of populations in countries such as England, France and Italy caused a significant transfer of power away from landowners to the surviving peasants who could command higher wages for their services. Some were able to free themselves from serfdom. In less populous countries that were less affected by plague, particularly those in eastern Europe, serfdom continued unabated until well into the nineteenth century, partly accounting for the divergent political histories of the two halves of that continent.

Others have speculated that the origins of the Christian Reformation can be found in the lax standards of church discipline that followed the decimation of clergy during the Black Death; or that new forms of pronunciation in languages such as Dutch and English can be attributed to the mass migration of peasants from the countryside into the towns in their quest for new employment. One theory even proposes that it was by only the tiniest margins that the world avoided being plunged into another ice age because of a significant decrease in carbon dioxide levels and deforestation that followed the annihilation of so many people.

There is a small caveat, though, to the trilogy of fleas, rats and bacteria being to blame for triggering everything from religious schism to climate change across medieval Asia, Europe and North Africa. At least two rival theories have recently been proposed that question traditional explanations for the cause of such depopulation by flea-borne disease. One suggests that mass death was the fault of a virus, such as ebola, while another proposes anthrax (see page 36). The once cosy consensus that disease spread by minute fleas was to blame for this most turbulent episode in human history is now a cauldron of hot historical debate.

There is no argument about the third pandemic, though – one that spread through China beginning in 1855, killing more than 12 million people across Asia. Unmistakable evidence of the plague bacterium *Yersinia pestis* has been traced to the bodies of its victims.

It was on a foraging expedition to Egypt in 1903 that Nathaniel Rothschild – Miriam's father – found the precise flea that accounted for such pandemic devastation: a member of the *Xenopsylla* genus that he named *Cheopis* after the famous pyramids of the same name. Not that this is the only flea capable of transmitting disease and not that humans are always the targets of flea-borne bacterial attack. The history of flea-bite infection amongst vertebrate animals – from cattle to birds and chickens to bats – would be no less rich, had it been afforded the opportunity to have been recorded in history. Unfortunately, no such record exists.

Tsetse Fly

FAMILY: GLOSSINIDAE
SPECIES: GLOSSINA MORISTANS
RANK: 37

Bloodsuckers that helped preserve the last remnants of human hunter-gatherer lifestyles in Africa.

STUDENTS of the history of human civilizations seldom dwell long on the people and countries of Africa. Their stories are mostly ignored except for the occasional glamorous highlight, such as the glories of ancient Egypt.

Africa failed to follow a similar pattern of development to the settled civilizations on which most recorded history is based. Over the last 10,000 years, the relationship between man and nature in Africa has followed a different trajectory from that of almost any other place on Earth, partly thanks to the exploits of a rather large flying insect, known today simply as the tsetse.

Tsetse are large flat flies that are resistant to being crushed when slapped, an adaptation most probably provoked by the relentless swatting of animal tails. Like the mosquito (see page 134), these flies act as vectors for parasites. Unlike mosquitoes, though, males and females both enjoy a good bloodsucking feast. The disease they carry is one that has directly affected the ability of people to establish settled civilizations, contributing to the reason why today Africa is one of the least 'developed' parts of the world.

Tsetse are carriers of a single-celled parasite called *Trypanosoma brucei* that causes trypanosomiasis – a serious illness known as sleeping sickness in humans and nagana in cattle, horses, camels and pigs. The parasite is spread by these flies as they suck blood from one infected mammal and pass it on to another with subsequent bites. Sleeping sickness is almost always fatal unless treated. The parasite travels to the vertebrate lymphatic system and then crosses into the nervous system, eventually reaching the brain where it induces extreme lethargy and finally death.

Poised to strike: the tsetse fly spreads fatal diseases, such as sleeping sickness in humans and nagana in cattle.

In cattle, horses and pigs, the parasite impairs growth, milk production and eventually, as in humans, leads to their sickly, premature demise. Although the tsetse evolved alongside mammals between sixty-five million and fifty-five million years ago, their biggest human impact seems to have occurred in Africa over the last few thousand years.

From *c.* 8000 BC humans in some parts of the world began to experiment with breeding wild animals in an attempt to turn them into compliant domesticated species that would yield an ongoing source of meat and milk (see On Agriculture, page 191). From about 3000 BC powerful new civilizations arose in countries such as Egypt and Nubia (modern-day Sudan) based on the growing of crops and the domestication of farm animals. But due to endemic infection by parasites carried in the tsetse, the practice of living in towns and cities, supported by surrounding agriculture and farming, bypassed African people living south of the Sahara Desert for thousands of years. Instead,

Tsetse infection in Africa

How cattle rearing was thwarted throughout equatorial Africa.

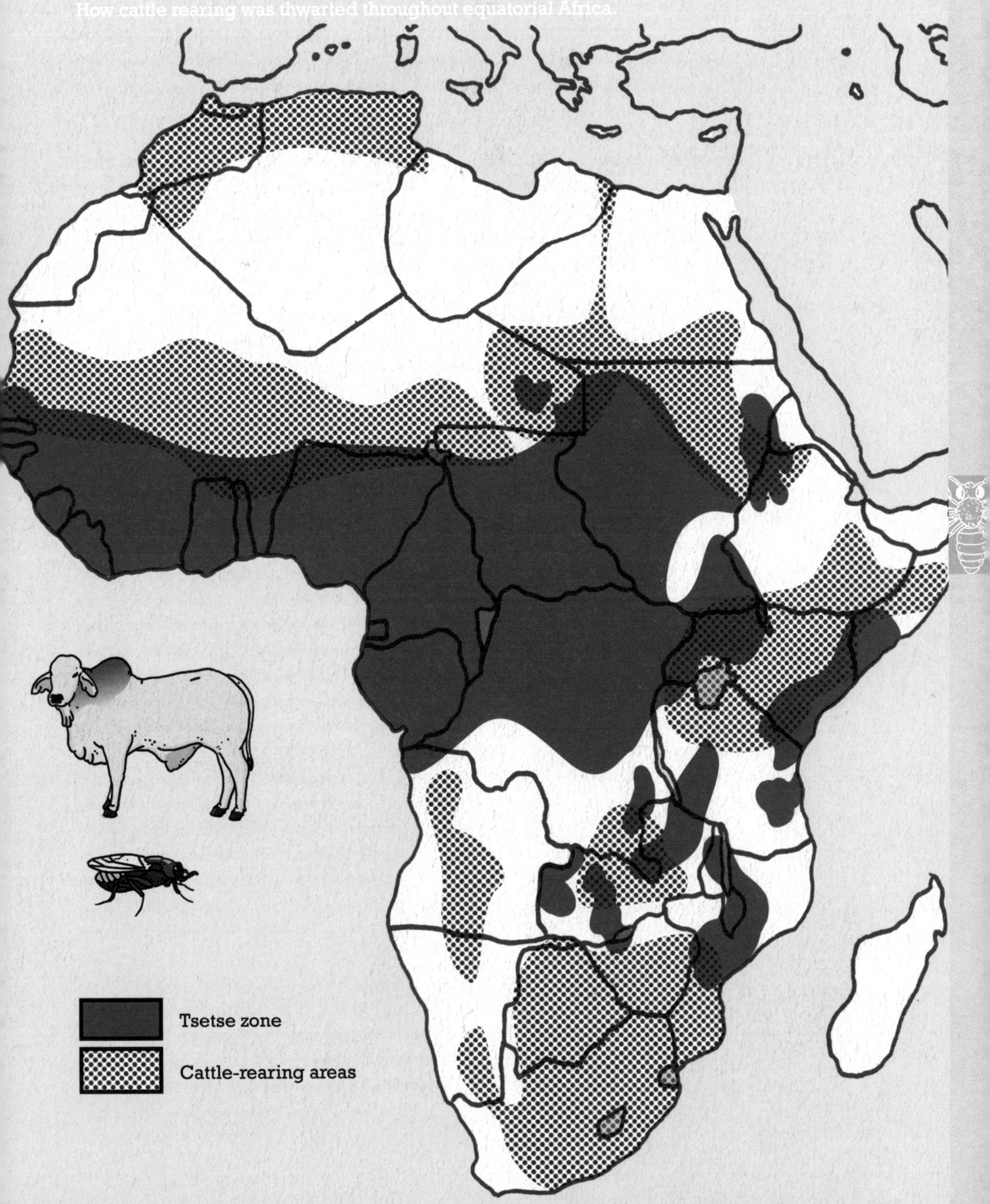

lacking domesticated farm animals, most African people clung to a hunter-gatherer lifestyle until well after the rise and fall of the Greek and Roman empires.

Domesticated cattle introduced from the Middle East to west, central and southern Africa had no natural immunity to nagana. It was only thousands of years later when West African Bantu herders finally succeeded in breeding cattle that could resist the parasites carried by the tsetse that the populations of farming people began to increase significantly.

With domesticated herds came copious supplies of meat and milk, supporting ever-larger families and enabling herding nomads to fan out across the continent. By about 500 BC, they had reached the East African coast where they mastered the art of fashioning tools of iron, further helping to establish their culture as the dominant force in Africa. With their tsetse-immune animals, Bantu people then expanded southwards for more than 2,000 miles reaching as far as the Natal in south-east Africa by AD 300. Eventually, with the addition of crop farming, they created powerful kingdoms such as Great Zimbabwe, built around the Zambezi river between AD 1200 and 1500. In the process, many of the less populous hunter-gatherer people such as the Central African Pygmies and southern African Khoisans were either driven out, enslaved, or killed.

Infection by parasites also hindered attempts by medieval and early modern North African, Muslim and European invaders to advance into the interior of Africa. War horses, for many years the world's most potent military force (see page 242), could not survive in the tsetse-infected heartlands of Central Africa where imported livestock had little or no immunity against parasites. Not until Vasco da Gama and other Portuguese explorers ventured by boat around the Cape of Good Hope in the late fifteenth century did European gunpowder introduce a new, even deadlier ingredient. It was only from about the 1870s onwards – a period known as the 'Scramble for Africa' – when steam power on sea and over land finally opened up the interior of this continent, that European invaders were liberated from the need to travel or wage war using easily infected animals. Africa's story since is a relentlessly grim tale of colonization, exploitation, armament and division from which its people still reel today.

Modern humans have waged war with partial success against the tsetse, reducing its numbers through campaigns of eradication using pesticides (such as DDT) or by irradiating males with gamma rays causing them to become infertile, thus reducing populations. Success is manifest in the gradual decline in the reported numbers of people suffering from sleeping sickness (estimated at *c.* 60,000 Africans). Epidemics still break out sporadically, such as one in Uganda that began in 2008. This latest bout is nothing on the scale of an outbreak between 1896 and 1906 that killed more than one million people.

It is impossible to know how African history would have turned out were it not for this parasite-carrying, bloodsucking fly. It is a curious twist of fate to think that without it, the last remnants of the Kalahari bushmen's ancient way of life would probably have been swept away thousands of years ago by opportunistic agriculturalists. Without the tsetse, the fate of African people today would certainly be very different indeed.

Oak

FAMILY: FAGACEAE
SPECIES: QUERCUS ALBA
RANK: 21

A highly robust flowering tree that hosts its own rich ecosystem but defies specific description.

FEW SPECIES in the plant kingdom have had as big an impact on the history of humanity as the oak. Reaching far back to the beginnings of recorded history, these flowering trees have captured people's imagination for many different reasons including religion, culture and artifice.

More specifically, these are trees that have dominated the minds of people living in the northern hemisphere – in North America and Europe in particular. The reason goes right back to the evolutionary origins of the family Fagaceae (beeches, chestnuts and oaks), which begins about one hundred million years ago.

No one knows precisely where the ancestors of oak trees first emerged. As the super-continent Pangaea started to break up 250 million years ago, it split into two enormous terrestrial crusts. Gondwana was the southern mass, which later fragmented into South America, Africa, India, Australia and Antarctica. Laurasia, the northern half, split up into Europe, Asia, Greenland and North America. It was somewhere on this northern mass that oaks evolved, becoming key flowering species that rafted across the northern hemisphere as the continents we are familiar with today gradually spread out far and wide.

Ancient oak woodlands are vibrant habitats for mosses, insects, birds and mammals.

Oak trees are the terrestrial equivalent of a tropical coral reef (see page 66). A typical oak acts as host to at least thirty species of birds, forty-five types of bugs, and more than 200 species of moths. Insects are especially fond of laying their eggs inside its leaves, which swell up with small growths called galls that provide larvae with nutrition and protection. Tiny parasitic wasps (see page 128) are particularly keen on these breeding grounds, as are the larvae of midges, moths, worms and caterpillars.

Not only is the reproductive strategy used by oaks hugely successful, it is also one of the most generous in nature. A mature oak may yield a crop of as many as 100,000 acorns (a type of seed/nut) a year. The result is a highly nutritious feast for scores of different animals ranging from pigs, deer and bears to squirrels, rodents and birds. For thousands of years wood pigeons have devoured up to 120 acorns each per day; birds have migrated across continents just to join the feast. Magpies and jays enjoy them too, and also profit by dining on the eggs of birds that nest in oaks. Owls and bats make their homes in cracks and holes that open up in the bark of old oak trees, which can survive for as long as a thousand years.

There is more subtlety to the oak's reproductive strategy than simply swamping the ground with a surfeit of nuts, hoping that some take root. Acorns are rich in fats and protein and so animals carry them off to hide away in makeshift food stores. It is important that these animal transporters do not eat *all* the acorns, since the seeds do not pass through the gut undigested. Oaks therefore lace their nutrient-rich acorns with a carefully measured dose of tannin, a chemical that can upset the digestive system when consumed in large quantities. This encourages the animals to bury their cache as a long-term supply of food, and not eat too many all at once. Recent medical research demonstrates how effective is this precise dose of tannin. Small quantities have been shown to be therapeutic to animals that feed on acorns, an effect similar to a human having a glass of red wine a day to help prevent hardening of the arteries. Large doses, however, can hinder the animal's ability to digest proteins.

Some animals die prematurely and their acorns remain uneaten. Others forget exactly where their food stores are buried. In this way at least some of the oak's crop of acorns are successfully planted by animals, usually far away from the parent tree, and left to germinate in peace.

Tannin has had its own impact on human history. Vats and barrels made of tannin-rich oak are largely responsible for giving wine and spirits their intense aromatic flavour. Similarly, tannin obtained by soaking wood from oaks in water has been used for thousands of years for preserving leather hides that are subsequently turned into everything from boots to book covers.

But the impact of the oak tree on human culture extends a great deal further. These trees, sometimes dubbed 'Lords of the Forest', have been at the core of the mythology and folklore of the people of the northern hemisphere since the beginning of recorded history.

Dodona is the most ancient oracle in Greece, far older than its more famous rival at Delphi. As long as 4,000 years ago, Neolithic farming people gathered there to consult their Mother Goddess. She spoke to them through the rustling of leaves in the oak forest all around. Herodotus, the Greek storywriter, said that a mighty beam of oak from this forest was used in the construction of Jason's famous boat *The Argonaut*, giving its crew the prophetic foresight they desperately needed to help them complete their epic voyage.

To the Norse people the oak symbolized Thor, the god of thunder, because it was thought to be struck by lightning more often than other trees. One of the most famous Norse outdoor shrines was an ancient tree, 'Thor's oak', which stood proud for hundreds of years. It was worshipped by flocks of Germanic pagans who came to Geismar, in Northern Hesse, to seek their own protection against the fickle gods of nature. The tree survived until the year 723 when, according to legend, a passing English Christian evangelist called Winfrid hacked it down to prove that Christ was the one and only true God. When the locals saw that Winfrid was not struck down by lightning for his mischievous deed it is said they converted to Christianity in droves. In such a way did this new religion come to replace ancient Norse veneration for these majestic trees and later Winfrid became Germany's patron, Saint Boniface.

Take that! Winfrid hacks down Thor's oak without being struck down by lightning.

Medieval Christendom's zeal for destroying trees put the relationship between man and oak on to an entirely different footing. From *c.* AD 900 onwards the greatest landowners were the monastic orders, such as the Cistercians, who have been described as the 'shock-troops' of deforestation during the medieval period. Clearing forests to create fields for farming that could then be offered for rent was one of the most successful routes to riches in medieval Europe. Oaks also provided perfect timber for building monastic settlements. Like Norse shrines, Christ's temples were surrounded by oak. Unlike pagan centres of worship, though, they were generally made out of dead wood and ceremonies took place indoors, not in the living forest.

By the sixteenth century, European axes were falling on the continent's once prolific oak forests with renewed vigour. Wood from oaks was the material of choice for building transatlantic ships that took European explorers anywhere and everywhere in the world. The strongly preservative qualities of oak proved ideal for making vessels that were sturdy enough to weather oceanic storms and tough enough not to rot.

Britain was once a land smothered by oak forests from shore to shore. But oak became such an essential part of the new way of life that by the seventeenth century, the country was facing a severe deficit. Only the discovery of coal as an alternative fuel saved Britain from a chronic energy shortage and other timbers from the pine forests of the Baltic and America were sourced to make up for a dearth of suitable indigenous building timber.

There is a final curiosity – a kind of postscript to the story of the highly successful oak. There are about 400 described species of oak, but it is almost impossible to identify one single species as more or less significant than another. The reason is that these are trees that defy the traditional scientific definition of what a species actually *is*. Oaks are among the 10 per cent of flowering plants that have stuck with wind power as their chosen medium for pollination. Not needing to attract insects or birds with a showy display, their flowers (called catkins) are small and discreet. Compared to the targeted approach of pollination by insects (see pages 129–30), wind power is extremely imprecise and haphazard. Oaks have overcome this potential negative by defying one of the basic rules in the human understanding of nature. They habitually breed *between* different species, creating perfectly fertile hybrids. If pollen from their own kind doesn't blow their way, pollen from a different species will usually do.

Such promiscuity is another reason for the oak's great success. But it's something that leaves scientists rather scratching their heads wondering: what's the use of a term such as *species* in a world of hybrid oaks?

Acacia

FAMILY: FABACEAE
SPECIES: ACACIA TORTILLIS
RANK: 36

An angiosperm with superlative powers of survival and symbiosis.

A TREE STOOD all alone in the African desert for hundreds of years, at least 400 kilometres away from its nearest rival. With the ever-increasing heat and dryness of the Sahara Desert, what was once a rich forest finally faded into history; all, that is, except for this single surviving specimen – the *Arbre du Ténéré* – the most isolated tree in the world.

For centuries desert travellers worshipped this umbrella thorn, believing its mystical powers would secure them safe passage across the arid terrain. As French African commander Michael Lesourd wrote in 1939:

> *'One must see the Tree to believe its existence. What is its secret? How can it still be living in spite of the multitudes of camels that trample at its sides? There is a kind of superstition, a tribal order which is always respected. Each year the azalai gather round the Tree before facing the crossing of the Ténéré. The acacia has become a living lighthouse; it is the first or the last landmark for the azalai leaving Agadez for Bilma, or returning.'*

Acacias, of which there are about 1,300 species, are among the most robust flowering plants ever to have evolved. Their range of survival strategies is so impressive that these trees will continue to populate the world despite climate change or human endeavour.

They belong to a legume family called Fabaceae. Like all legumes they can fix atmospheric nitrogen thanks to their symbiotic relationship with *Rhizobium* bacteria (see page 40). Such self-nourishment puts them high up nature's list of die-hards. It also means that their seeds are highly nutritious to eat.

These plants do not team up with bacteria alone to maximize their own chances of survival. Several acacia species, especially those found in the dry deserts of Africa, Texas and Central America, play host to vast colonies of ants (see page 154). A queen ant lays her eggs in an acacia's large spiky horns, which are designed to protect against herbivores. When her workers hatch they act as the acacia's police

The *Arbre du Ténéré* was a precious natural beacon in the arid Sahara Desert.

force, protecting the tree against other insects that might happen to try their luck by feeding on the legume's nutritious leaves or stem. In return for their guardianship, the acacia produces nectar from glands conveniently located to feed the ants. It also secretes small, orange, highly nutritious beads from the tips of its tiny leaves as additional food for ant larvae.

This is a powerful partnership. Experiments have shown that cutting off an acacia's thorns and removing the ants can cause the plant to wither and sometimes die. The ants will even patrol the perimeter of the plant, destroying the seedlings of any other plants that are germinating nearby to preserve the acacia's precious supplies of water. Branches of trees that approach from above also get short shrift. Armies of ants march across and devour the rival branch until it is no longer a threat.

Devotion, loyalty and fair exchange are hallmarks of this natural association between two unrelated species whose mutual dependence is essential to the survival of each. It is easy in this case to see why and how evolutionary adaptations in nature have occurred, as the bonds of association between such species became strengthened by their mutual success, one generation after the next.

Acacias are pioneering species – a term that applies to life-forms that are able to colonize new lands with the minimum of fuss. Their seeds will happily germinate in the driest, thinnest, most undernourished soils in the world. They dominate the arid Australian outback, and the deserts of Africa and the Americas. If conditions are simply too tough, acacia seeds have been known to lie dormant for up to sixty years, sprouting into seedlings when conditions for life are sufficient to support new growth.

The impact of acacia trees, and the umbrella thorn in particular, has been no less profound in human culture. Wood from this tree is recorded in the Jewish Torah (Christian Old Testament) as being the material that God instructed his chosen people, the Israelites, to use to construct their Tabernacle – the Ark of the Covenant. Inside this chest they are said to have placed the Ten Commandments carved by God on tablets of stone that their celebrated leader Moses brought down from Mount Sinai. Later in the Bible an acacia tree bursts into flames – the famous burning bush, sent as a sign from God to Moses.

African people have traditionally used acacias for their everyday needs from making wagon wheels, fence posts and cages, to constructing a type of string from its bark. Its nutritious pods and leaves provide fodder for animals (camels and giraffes particularly) while its sap, an edible form of gum, is now used as a stabilizer in modern, mass-produced foods (additive number E414).

The most famous acacia is still the *Arbre du Ténéré*, the tree that stood for so many hundreds of years all alone in the desert. Its trunk acted as a beacon for travellers for miles around while its roots were a testament to an unsurpassed thirst for survival, stretching downwards more than *thirty-six metres* to the water table below. But like many of life's great legends, its tragi-comic end eventually came not through its lack of resilience against nature, but through the negligence of man, in the form of a drunken Libyan truck driver, who carelessly knocked it over in 1973. Its remains now lie in state at the National Museum of Niger and, in its place, local people have erected a commemorative metal post.

Durian

FAMILY: MALVACEAE
SPECIES: DURIO ZIBETHINUS
RANK: 89

Jurassic fruits that advertise their presence for miles around and may have been the forerunners of the first flowering plants.

SOME WHO have tasted the durian fruit have said that it's like eating *'pig-shit'*, *'turpentine'* and *'onions garnished with a gym sock'* ... Others say this Indonesian delicacy is like eating *'sweet raspberry blancmange in the lavatory'*.

Albert Russell Wallace, the nineteenth-century British naturalist who independently came up with his own theory of evolution by natural selection at about the same time as Charles Darwin, was a little less scathing:

> *'A rich custard highly flavoured with almonds gives the best general idea of it, but there are occasional wafts of flavour that call to mind cream-cheese, onion-sauce, sherry-wine, and other incongruous dishes.'*

Whatever your sense of smell or your taste bud configuration, the durian fruit is certainly one of nature's most spectacular when it comes to advertising its presence. In the tropical forests of Borneo its smell can be detected from as far as half a mile away. This large, fleshy, spiky-skinned fruit grows up to thirty centimetres long and can weigh up to four kilograms – all of which accounts for its nickname 'king of the fruits'.

Durian trees are typical of flowering plants that evolved alongside herbivorous dinosaurs such as *Diplodocus*, making the most of their insatiable appetite. The black seeds are surrounded by a rich, custard-like sauce that makes durians the most sought-after treat for forest animals including orang-utans, mouse deer, sun bears and even tigers. As the animals feast off the fruit, they swallow some of its seeds, which pass through their guts to be deposited far away in a pile of nitrogen-rich dung.

The survival of the durian long after their original dinosaur seed transporters died off 65.5 million years ago can be put down to the huge investment they have made in producing large fruits full of flavour that advertise their presence so spectacularly through such a powerful odour.

But the impact of the durian, or at least its ancestors, may be a great deal more significant, at least according to one theory. Edred Corner (1906–96) was a British botanist who specialized in tropical plants and spent much of his life living in Singapore. In his view ancestors of today's durian trees shed light on the 'abominable mystery' articulated by Charles Darwin as to the origin of angiosperms, or flowering plants. According to Corner, red fleshy fruits similar to today's durian were descended from cycads, trees that had already mastered the trick of using insects as pollinators (see page 101). Their durian ancestors took the idea a stage further and developed a technique for spreading their seeds that involved packaging them in well-advertised (red and smelly) fruits that would prove irresistible to passing animals.

Corner's 'Durian Theory', first published in 1949, was an attempt to associate the origins of all flowering plants with the durian lineage. Plants that evolved fleshy edible fruits were therefore the pioneers of the amazingly successful rise of the angiosperms. Corner's theory has been extensively argued, debated, rejected, supported and discussed since he first put it forward. Today it is universally agreed that flowering plants originated in tropical forests, the durian's natural habitat. Whether or not durian ancestors are the evolutionary forefathers of the angiosperms is still open to debate, and other plants, such as the ancestors of today's water lilies, pepper plants and magnolias have more recently been proposed as alternative candidates.

The group of angiosperms to which today's durian belongs is certainly among the most significant of all flowering plants. The rose-like eudicots (or rosids) are a giant group of genetically coherent plants that include many of the world's most valuable flowering trees such as rubber trees, poplars, willows, wisteria, acacias, apples, plums, elms, figs, oaks, beeches, birches, hazelnuts, walnuts, myrtles, eucalyptus, frankincense, myrrh, oranges, lemons, maples and mahogany.

If the durian really was the ancestor of all that lot then it would certainly take pole position amongst nature's most influential species. And even if its evolutionary credentials are uncertain, this most smelly species still warrants a place in the table.

Bamboo

FAMILY: POACEA
SPECIES: MELOCANNA BAMBUSOIDES
RANK: 40

Ancestral grasses that developed impressive defences against the prodigious appetites of grazing dinosaurs.

STUDYING STOOLS of fossilized dinosaur dung may not sound like one of the most appetizing pursuits in modern science, but for two American palaeontologists such patient analysis has recently begun to throw new light on the origin of one of the most important and widespread group of flowering plants on Earth – the grasses.

What Dolores Piperno and Hans-Dieter Suess found as they peered through their microscopes at dung from dinosaurs was the microscopic remains of an ancient flowering plant defence system. It was pioneered by some plants to help them avoid being eaten by the herbivores that roamed the land of what is now India some sixty-five million to seventy-one million years ago. The system worked by liberally lacing their cells with fragments of silica (like sand) that interferes with the chewing and grinding process in a typical herbivorous dinosaur's mouth.

Over time, of course, herbivores evolved ways of circumventing such defences, leading to the various types of teeth, such as molars, that evolved in mammals once the era of the grazing dinosaurs came to an end 65.5 million years ago.

Silica-laced stems and leaves are characteristic of monocot grasses that, until the findings of Piperno and Suess, were thought to have evolved only *after* the demise of dinosaurs. The earliest complete fossil evidence of grasses dates back only as far as fifty-six million years. However, thanks to the discovery of those tell-tale silica fragments, called phytoliths, in dinosaur dung, it is now thought that grasses emerged considerably earlier than previously assumed, probably during the late Cretaceous Period.

Further phytolith study reveals that most primitive grasses were ancestors of today's bamboos, among the fastest-growing plants in the world. Today a well-watered bamboo can put on a good fifteen centimetres of growth a day and reach up to forty metres in height, although one living species has been known to grow by more than a metre in a twenty-four-hour period. Back in the days of the dinosaurs, some species grew even faster – perhaps as much as several metres a day, with an ultimate height of more than sixty metres. Such prodigious growth rates were attempts to survive being grazed to the ground by greedy dinosaurs. Grasses that prospered best were those that grew the fastest. Bamboos, like all grasses, also thrived thanks to the monocot design for defence in which their all-important growth buds were tucked away safely underground (see page 133). Regardless of how savagely grasses like bamboos were grazed, their newest growth was always least vulnerable to attack.

These survival characteristics – silica, rapid growth and a submerged growth bud – are the hallmarks of what have made grasses great. As the climate has dried out over the last forty million years more than 30 per cent of the world's landmass has come to be covered in different species of grasses, many of which have become essential foodstuffs for humans following the start of farming 12,000 years ago (see On Agriculture, page 191).

Before humans began to cultivate grasses to produce wealth for settled societies, bamboos were among the world's most successful and influential plant species. As many as 1,000 types of bamboo exist today, most in East Asia, Australia, India and Sub-Saharan Africa. Huge forests of fast-growing, woody bamboos are thought to have dominated many of these landmasses before the arrival of humans.

Although the evolutionary significance of today's bamboo is clear from the analysis of dinosaur dung, it is less easy to know whether other grasses (such as wild wheat, oats and barley) originally evolved from ancient bamboos or other sister groups.

However, further evidence in favour of the bamboo's ancient lineage comes from their curious flowering habits. Bamboos flower infrequently – sometimes only once every sixty to 130 years. When they do bloom *they all flower together* sometimes regardless of geographical location. No one quite knows how they achieve this – do they possess some kind of mysterious genetic programme that acts like a clock? Are they somehow capable of sending a signal to each other? The strategy probably emerged as a response to the huge grazing pressure applied by dinosaurs seventy million years ago. By producing so many flowers and seeds all at once, some species of bamboo literally overwhelmed the appetites of herbivorous animals, ensuring that at least some of their progeny had a good chance of survival into the next generation.

As a final gesture of goodwill, the parent plants of some species automatically die off after flowering so as not to compete with their fast-growing seedlings for water, light and space. Such is the power in nature of a species' effort to ensure its continued genetic inheritance regardless of any concern for the longevity of an individual's life.

Fruit-filled forests whose tall woody trees suddenly die off all at once cause profound ecological dislocation. The effect is felt acutely in places such as the Bay of Bengal, which suffers huge surges in its rodent population whenever its bamboo forests flower, fruit and then die every thirty-five years. Vastly increased rodent populations munch their way through other crops, leading to famine. Parasites that live on rats (see page 172) cause a major upturn in the spread of diseases such as plague and typhus that are often fatal to humans.

The occasional mass suicide of bamboo forests would not matter nearly so much were it not for the fact that these fast-growing grasses are of such great social and economic importance. In some Asian societies bamboos have completely transformed human ways of life. They are used for everything from cooking (bamboo shoots) and medicine (as a natural antiseptic) to construction (roofs, flooring, scaffolding, gutters and water channels). Thanks to the tensile strength of bamboo, Chinese engineers operating over 1,500 years ago were even able to make cables for lowering drill bits into rigs for mining natural gas. Then there's the plethora of other traditional uses for bamboo that includes everything from making fences, bridges, canoes, walking sticks and furniture to musical instruments, hats, weapons, fishing rods, baskets and knitting needles.

Honey Bee

FAMILY: APIDAE
SPECIES: APIS MELLIFERA
RANK: 33

Industrious transporters responsible for pollinating many of the world's flowers and for producing pots of honey.

TODAY'S WORLD would look very different were it not for the wonder of honey bees. As many as 20,000 different species have been recorded, but it is *Apis mellifera*, the European honey bee, that has had the greatest effect, not just on human history but also on many of nature's most prolific flowering plants.

Honey bees evolved more than one hundred million years ago – in the days when ten-metre-long dinosaurs dominated the landscape, pterodactyls the size of small aeroplanes swooped down from the skies and the seas were filled with exotic ammonites and terrifying ichthyosaurs.

The bees' long evolutionary history was recently revealed by the discovery of a one-hundred-million-year-old amber fossil found deep inside a mine in northern Burma. Bees are descended from wasps and, like them, developed some of nature's first social gatherings, displaying many characteristics also seen in human civilizations, which first formed a mere 6,000 years ago.

A honey bee nest contains three types of bee. First is the queen, who lays as many as 500,000 eggs in her three-year lifespan. Most of these hatch out to become female worker bees, which share out a range of tasks such as cleaning, feeding, nursing, finding food, storing food and making the wax cells that form the honeycomb. A few, called drones, are the males and have just one function – to impregnate the queen, usually in an elaborate ritual that takes place hundreds of metres up in the air and is so physically demanding that only the strongest males can successfully mate. So much effort is required in this process of insemination that drones usually die soon after sex. This is a woman's world.

The queen communicates with her workers by secreting pheromones that are passed around the nest. Individual bees communicate through the language of dance. The 'round dance' means that food is within fifty metres of the hive. The 'waggle dance', which may be vertical or horizontal, provides specific details about the distance and direction of a particular food source. The 'jerky dance' is used to discuss whether to increase or decrease the amount of food gathering the bees need to do, depending on the hive's overall requirements.

A bee keeps guard in this fifteenth-century drawing of a hive.

In fact, honey bees are so highly developed that when they need to decide where to locate a new colony they perform the earliest known example of what in the modern human world we might call a 'democracy'. Various scouts will identify potential nesting sites and then report back the location of each one to the rest of the community by means of their various dances. Other bees check them out, before returning to the nest and dancing longer and harder in the direction of the site they think is the best. After about two weeks, the site with the strongest, most vigorous dance is the winner and the colony swarms. It has been estimated that using this form of range voting, the bees will choose the best location approximately 90 per cent of the time.

Other traits familiar in human societies even stretch to ills such as alcoholism. Bees that imbibe fermented nectar are notorious drunkards whose behaviour is similar to that of an intoxicated human. Indeed, their antics mirror the effect of alcohol on the human brain so closely that scientists now study intoxicated bees in their efforts to research the effects of alcohol abuse in mankind. Guard bees stationed outside a hive prevent such drunkards from causing trouble and – like bouncers – banish them until they have sobered up. Repeat offenders are not tolerated – their legs are chewed off as a punishment.

The co-evolution of flowers and bees during the Cretaceous Period is one of nature's best examples of evolutionary teamwork (see page 129). Without the targeted pollination services of bees, many flowering plants could never have spread themselves so widely across the world. Without these flowers, bees would never have evolved away from parasitic or predatory wasps to feed on a vegetarian diet of nectar and pollen instead.

Humans have harvested honey for thousands of years, well before the earliest civilizations, especially in Africa where hunter-gatherer people followed the paths of honey-diviner birds that led to nests full of bees. Sealed honey-pots have been found in the tombs of pharaohs and the art of beekeeping was a favourite topic among Greek and Roman writers. The oldest artificial hives found by archaeologists date back to *c.* 900 BC in Israel during Biblical times – evidence of the Promised Land of 'milk and honey'.

Bees contribute billions of pounds to today's global economy. Their value is not limited to honey production (China is today's largest producer, harvesting more than 300,000 tonnes a year) because bees are also responsible for pollinating so many of the most common and popular fruits grown for human consumption. The practice of travelling with hives of bees from one US state to another and hiring out swarms for pollination is big business. Migratory beekeeping is estimated to support some fifteen billion dollars of fruit crops in the United States alone.

There is, however, a sinister sting in the tail to the story of today's European honey bees. Since 2006 colonies have been mysteriously disappearing, reducing by as much as 50 per cent in some areas of Europe and America. If such Colony Collapse Disorder (CCD) continues, human farmers will face incalculable losses in terms of the free pollination services traditionally provided by these ancient flying insects. Concerned Western governments are now pumping millions into scientific research in an effort to discover its cause. (Is it the overuse of pesticides? Increasing urbanization? A mysterious viral outbreak?) If the agricultural lives of humans are not to be made a great deal more complicated, experts need to find a remedy – and fast.

Ant

FAMILY: FORMICIDAE
SPECIES: ATTA SEXDENS
RANK: 25

Engineering geniuses that have successfully adapted to just about every habitat on Earth.

WHEN THAT METEORITE struck the Earth 65.5 million years ago, devastating the dinosaurs, pterosaurs and so many other species that dominated the land, nature's biodiversity was its best insurance against total annihilation. Certain genetic combinations, manifested in different types of organisms with varying ways of life, found that despite the devastation all around, life went on pretty much as usual. It is common knowledge that mammals thrived in the absence of large reptilious predators (see pages 159–60) but less well known that several other animal families also came into their own after the demise of the dinosaurs. Although far smaller than mammalian species, some of these creatures have ultimately been just as significant. In particular, insects called ants.

Ants dominate life in the soil on every continent, except for frozen Antarctica. Only Iceland, Greenland and a few Pacific islands have no indigenous species. In many ecosystems, especially in the tropics, up to 20 per cent of the mass of all land-living creatures can be accounted for by ants alone, putting them at the top of today's list of nature's most populous living creatures.

Fossils of ants trapped in amber date back to the mid-Cretaceous Period (*c.* 120 million years ago), but it was only after the demise of the dinosaurs that ants spread in significant numbers. About 14,000 different species have been identified to date. Having learned to live in large societies so long ago, sufficient time has elapsed for these creatures to have explored every conceivable form of successful society. Therefore, there are few better natural parallels to human social instincts than observing the various practices of ants.

At the centre of most ant societies is a queen who, soon after she reaches maturity, indulges in a once-in-a-lifetime nuptial flight in which she mates with many males, building up a store of sperm that she uses for the rest of her life (up to thirty years in some cases) to produce fertile eggs. Once the male ants have flown and mated (or attempted to) they quickly die off. The queen then burrows into the ground and begins to lay her eggs. Most of them are fertilized and produce females which will become either future queens or sterile workers. Other eggs she leaves unfertilized, and these then grow into male clones.

Some experts believe sterility in female workers evolved because by caring for their queen (and their sisters) the chances of their genes surviving into future generations are statistically greater than if they had offspring themselves. Whatever its origins, sterility seems to act as a catalyst for selfless loyalty, co-operation and dedication – key components of an effective society.

It is not so surprising, therefore, to find that the same is true in human history. Successful social and political administrations in many historic human civilizations have been credited to the use of eunuchs, for example medieval Chinese admiral Zheng He (1371–1433). Chinese and Islamic empires traditionally put eunuchs in positions of supreme political power owing to their reputation for loyalty. In Incan society, the entire civil service was filled with staff fathered by the imperial family, providing extra genetic cohesion. By all accounts this arrangement was extremely successful. The civil war that helped bring down the empire in the fifteenth century broke out only after smallpox, brought across the ocean by European explorers, caused a dynastic vacuum (see page 26). Until the sixteenth century European civilizations, which did not generally engage in either the sterilization or fathering of a genetically related administrative class, were among the most unstable, war-torn societies on Earth.

Farming crops and animals is a habit usually thought to be exclusive to people – a way of life that began about 12,000 years ago and is often believed to set mankind apart from the rest of nature. That is not the case. Ancestors of ants that belong to the tribe *Attini*, which live mostly in South America, formed nature's original and most prolific farming communities. For many years experts believed that these leafcutter ants simply foraged for plants they preferred to eat, or perhaps used the leaves to protect their nests from rain. But thanks to the observations of a contemporary of Darwin's, British geologist Thomas Belt (1832–78), it was discovered that these ants forage for leaves to use as manure for cultivating a certain type of fungus (*Rozites gongylophora*) in neatly kept gardens buried deep inside their nests. These ants not only feed on this fungus but they also regularly pluck tufts of its mycelia to plant in new areas of their gardens, where they grow rapidly, continuing to provide the nest with a rich source of food.

So vital is the practice of cultivating food for these pioneering agriculturalists that before a new queen embarks on her nuptial flight, she tucks a store of mycelia into her body for safekeeping. When she begins the process of building a new nest, her priority is to retrieve her wad of fungi, which she uses to establish her colony's first garden – even using her own excrement as a form of manure to stimulate its growth.

Ants that cultivate crops also learned how to domesticate animals, like humans have done many millions of years later. Aphids are to ants what cows are to humans. Being specialists in sucking out the sap of plants, aphids excrete a highly nutritious juice that ants 'milk' from them by stroking the aphids' back legs with their feelers. In return for providing a source of milk, ants protect the aphids from enemies (like humans protect cows by fencing them in or keeping them in barns). So close has the association between aphids and ants become that some aphid species have even grown to mimic the face of an ant, to further facilitate the feeding process. Other symbiotic associations have arisen between ants and flowering trees such as acacias (see pages 146–7).

If farming and agriculture are social practices that evolved in nature millions of years ago, then so, too, was the insect equivalent of supporting a society through nomadic expansion and war. In the thirteenth and fourteenth centuries Mongol chief Genghis Khan and his blood relations established the largest contiguous empire the world has ever known, stretching all the way from China to the Mediterranean. Their tactics of moving in hordes from place to place, subsuming other civilizations in their path, was a way of life that originated in microcosm among types of ants, such as *Eciton burchelli*, that thrive in Brazil, Peru and Mexico.

The nests of these 'army ants' are, like those of human Mongols, a marvel of self-sufficiency. Thousands of ants link up like threads to form a chamber called a 'bivouac' in which the rest of the colony lives. When it is ready to feed, the nest breaks up, almost in an instant, and recasts itself into war-like hordes that swarm over areas of the forest floor capturing, paralysing and devouring all animal life in their path. Other insects, snakes, even birds and monkeys can be trapped, killed and eaten by these huge armies of marching, munching ants.

As in any natural self-organizing system, there are no leaders. Individuals take turns at the front of the swarm, then move back, allowing others to take their place. Group co-ordination and problem-solving is achieved through the passing of pheromones – chemical smells – which operate as a signalling language, spreading instructions from one individual to another. Like a slime mould (see page 49), these individuals are so co-ordinated that they act as if constituent parts of a super-organism, all the stronger for never having to depend on centralized control, and highly efficient at solving the puzzle of where to find the most food. Feelers are so much more powerful when they act in a concert of literally millions of co-operating pairs.

Finally, there are ant societies whose livings depend entirely on conquest. The Roman Empire and its daughter project, European overseas colonization, provide reasonable human equivalents. Legionnaire ants (*Polyergus breviceps*) are slave-raiding ants incapable of growing or even feeding themselves. Like the most affluent in Roman society, they have their food served to them by slaves. They establish their colonies either by stealing larvae from rival nests or through tactics reminiscent of a classic human *coup d'état*. A queen will sneak inside another nest, kill its ruling incumbent, and take over, instructing her new subjects to bring up her own offspring instead.

The impact of ants on the evolution of other species they come into contact with has been profound – be it symbiotic, parasitic or predatory. The societies they have succeeded in building are the oldest, most robust and versatile in all nature – far out-performing in numbers and efficacy those experimental equivalents built by humans in the last few thousands of years. With supreme success at surviving mass extinctions already proven after the meteorite strike 65.5 million years ago, these are creatures whose versatile ways of life and ability to combine as super-organisms are likely to feature near the top of anyone's list of nature's best long-term bets.

On the Rise of Reason

How the mental skills of mammals gave rise to a new evolutionary force that turned monkeys into men.

IF HUMANS are simply the product of natural selection, does it mean that their behaviour is merely a reflection of those many patterns and lifestyles found in other species? Or have people developed a different capacity from other living creatures, one in which learning, knowledge and scientific progress have taken over as a new premise for success in life on Earth?

Charles Darwin never really resolved this issue. Like many people today, he could see contradictory evidence all around. History suggests that humans can sometimes be inclined towards a symbiotic relationship with their fellow men and other species (witness the lives of St Francis of Assisi or Maimonides). But there is plenty of evidence to suggest the contrary, for example, when a ruler's appetite for selfishness manifests itself in gratuitous greed, exploitation, violence and war.

The roots of this issue reach back further than just the last few thousand years of recorded human history. Indeed, it is possible that the way of life humans lead today has its true origin more than 200 million years ago, which is when the class of creatures called mammals first emerged – that's at about the same time as the first Triassic dinosaurs (see page 109).

The mammals' survival kit

Mammals are significantly different from other types of living creatures. The selective pressures of nature, under which they evolved, equipped them with a unique set of features that go a long way towards explaining why humans behave as they do today.

The biggest influence on their early evolution came from dinosaurs and pterosaurs. Surviving in a world dominated by such successful carnivorous creatures caused mammals to adapt in a number of ingenious ways. First, and most important, was the ability to hunt at night. Reptiles mainly depend on the sun for heat. At night, when it is cold and dark, they sleep. It is then that other creatures stand their best chance of making a living. It was by adapting to survival at night that mammals gained their long-term lease of life.

To survive at night meant keeping body temperatures high enough to move around, hunt, eat and look after young. No wonder, then, that mammals are descended from creatures like the sail-backed pelycosaurs (such as *Dimetrodon*, see page 116) that had begun to develop ways of heating their bodies internally when other reptiles were still charging up from the sun.

Other improvements to their design involved staying small (less body volume to keep warm, less conspicuous); developing insulation (fur); an improved circulatory system (pumping food and oxygen around the body via a four-chambered heart); more efficient breathing apparatus (larger lungs housed in a barrel-shaped rib-cage); and the development of chewing teeth and muscular jaws (molars for pre-processing food, to allow faster digestion).

A wide-eyed bushbaby, known as the 'little night monkey', seen peeping out of a tree trunk in Africa.

Life's first true mammals were small creatures, ultimately descended from pelycosaurs. They were furry, about the size of a squirrel, hid in the day and hunted at night. Their distinguishing features were sweat glands (which originally developed to prevent their bodies from overheating) that later became adapted into food fountains from which their live young could be fed. Mammary glands evolved as a security measure so that mother mammals only had to leave their nests to feed themselves when they considered it safe to do so, not whenever their hungry children happened to demand food.

It was also apparent early on in the evolution of mammals that, in a world dominated by reptiles and birds, laying eggs was a perilous way of producing offspring. While some species stuck to using this technique (such as the monotremes), most mammals succeeded by keeping their young *inside their mother's body* until they were sufficiently grown up to be able to survive. For mammals the problem of going out to hunt for food and leaving behind a nest of eggs for another creature's breakfast therefore became a thing of the past.

Mammals also developed flexible diets – usually a mixture of fruit, seeds and insects. Flowering plants, the champions of co-evolution, responded to new opportunities for spreading their offspring by turning their seeds into nuts that could be transported, consumed and sometimes re-planted by nocturnal mammals (see oak, page 143). Insects, which became ever more plentiful on the Jurassic forest floor thanks to the rise of flowering plants, were also good fodder for nocturnal mammals, which developed the skills necessary to catch them by evolving into nimble, dextrous creatures.

Excellence in muscle co-ordination was matched by superlative senses, such as sight, sound and smell. Whiskers were an early adaptation, as demonstrated by fossils of a long extinct cat-like mammal called *Thrinaxodon*. These highly sensitive outcrops of fur were perfectly suited for developing a sense of touch and spatial presence, ideal for stealthily detecting meals in the dark or for mounting a surprise attack.

Large eyes that could work effectively in dark forests at night also made a difference to mammalian survival. Bushbabies and lemurs are mammals that have survived to this day largely thanks to their excellent night-time, black and white vision and large, bulbous eyes. Some mammals even learned how to 'see' in the pitch black. **Bats** (see page 163) are flying mammals that can build up a mental image of their surroundings based on the time it takes for sounds made by their clicking tongues to bounce back to their ears from objects such as trees, cliffs or cave walls.

High-pitched sound as a way of detecting the soft, light tread of a nearby crawling insect was another crucial component of the mammals' Jurassic survival kit. Two bones, adapted from the reptilian jaw, recessed into the skull to become parts of a mammal's inner ear (stapes and incus), substantially increasing their capacity to hear high-frequency sounds.

Mammals also developed an extraordinarily effective sense of smell. Today's domesticated dogs and pigs are beneficiaries of this trait, now deployed by humans for sniffing out everything from illegal drugs to subterranean truffles. Body hair helped scents become an important medium of communication for these creatures since they substantially increase the surface area available to volatile chemicals (pheromones) that many mammals use to send signals. It explains why humans, terrestrial mammals that have lost most of their body fur (see also page 186), still maintain hairs in many of their smelliest places …

Information technology

As the designer of any advanced computer system knows, data inputs such as video feeds (eyes), audio streams (ears), smelling sensors (noses) or even pressure gauges (whiskers) are effective only if there is sufficient *centralized* processing power literally to make sense of them. The evolution of the mammalian brain, with its advanced computing power and more extensive cerebral cortex (the outer layer of the brain), was therefore an essential and direct result of finding a way of surviving a world dominated by dinosaurs.

Brains are what have given the **elephant** its extraordinarily long memory (see page 168) and the **whale** its ability to communicate via low-frequency sound-waves over thousands of miles (see page 165). Brains have given dogs the capacity to learn new tricks and monkeys their mischief. Mammals can be distinguished from other types of living creatures because, quite simply, they have bigger-than-average brains.

Before brain power became an evolutionary response to enhance a species' chances of survival, animals simply relied on *instinct* to dictate their behaviour. Such traits first developed in Eukaryotic multicellular creatures to avoid being eaten or going hungry in the seas (see pages 71 and 74). These instincts are the outward expressions of inner genetic dynamics that drive the actions of everything from fish, insects and amphibians to reptiles, birds and dinosaurs. Nature's myriad patterns of life, from the symbiotic relationship between ants and acacias (see page 146) to the parasitic wasp larvae and their hapless arachnid hosts (see pages 128–9), rely on instinct not deliberation and thought. Even social insects, creatures that build the most sophisticated societies on Earth, are controlled by the genetically driven instinctive behaviour of a natural self-organizing system, not by the centralized planning of a ruler's rational mind (see ant, pages 154–6).

So it was that with the evolution of mammals came the beginnings of an alternative operating system for the working and organizing of life on Earth. The most extreme expression of this rise of reason, of problem-solving, deliberative thought and self-awareness, is still very much alive today – it belongs to us: modern man (see also On Rivalry, page 367).

Breaking free

While living with dinosaurs provoked evolutionary innovation in the form of mammal survival skills (warmbloodedness, senses, dexterity, giving birth to live young, and bigger brains), it was their mass extinction 65.5 million years ago that caused these lifestyle innovations to spread rapidly around the globe.

Within ten million years of the meteorite strike that wiped out most large dominant reptiles (except crocodiles, which somehow made it through), mammals had radiated on to every continent, even Antarctica, like rats freed from a trap. What were once small, furry, squirrel-like creatures that skulked fearfully in the trees or burrowed in the ground, now adapted to a new environment where it was safe to roam outdoors, even during the day.

Smallness for mammals had no intrinsic survival value in this bright new, dinosaur-free world. In fact, bigness was often best. Hoofed, herbivorous ancestors of today's rhinos, hippos, horses, camels (see page 248) and elephants (see page 168) evolved separately as the continents slowly drifted apart. Some groups became extremely numerous, for example Artiodactyla (even-toed ungulates) such as deer, antelope, buffalo, yak and bison, from which today's domesticated cattle, goats, pigs and sheep are ultimately derived.

Aepycamelus – a tall giraffe-like camel that lived in North America about twenty million years ago.

These were creatures that developed specialized teeth and innovative digestive systems that host symbiotic micro-organisms in the gut that break down hard-to-digest cellulose. Food regurgitated into the mouth as cud and chewed again before delivery to the animal's main digestive chamber allowed herbivorous mammals to grow large and sustain themselves as warmblooded creatures by extracting sufficient energy from difficult-to-digest grasses and plants (see bamboo, page 150).

Herbivorous mammal populations were generally kept in check by less numerous but stronger, more violent carnivores (see pyramid of life, page 103). Ancestors of today's lions, wolves, tigers and boars evolved alongside now extinct creatures such as the huge four-metre-long *Andrewsarchus*, a kind of hoofed tiger with an enormous head and long crocodile-like jaws. Ancestors of these beasts also gave rise to an order of mammals called Cetacea that returned to the seas and evolved into the dolphins, porpoises and whales (see page 165), some of which became the largest living creatures of all time – bigger even than the dinosaurs.

While some mammals grew extremely large, others stayed small, adapting to a climate that cooled as the continent of Antarctica headed as far south as it is possible to go, taking its current place at the bottom of the world. From about forty million years ago, this once tropical and lush land of vegetation was smothered by a bleak blanket of ice more than one and a half kilometres thick.

Climate changes driven by continental drift meant that from about thirty-six million years ago the Earth's climate grew steadily colder. Water gradually became locked up as ice at the Earth's poles, rainfall levels reduced and grasslands came to replace many of the world's forests.

Rodents are small creatures that thrive in grasses, eating mainly seeds and nuts. Today they are represented by creatures such as mice, **rats** (see page 172), squirrels, guinea-pigs and porcupines, which between them account for more than 25 per cent of all mammal species. Rodents have not always been small, however. Fossils have been found of an extinct South American species, the giant capybara, that once grew as large as a donkey.

Compared with insects, birds or flowering plants, mammal species are not particularly numerous – there are an estimated 4,300 species alive today. But mammals can boast a spectacular diversity of lifestyles, from the pygmy jerboa, the size of a small human finger, to the blue whale, as big as a nuclear submarine, and from moles that bury themselves underground and live in complete darkness to bats that have learned to fly blind at night.

Climate change, geographical isolation and a lack of other predators (such as dinosaurs) are common explanations as to why mammals became so successful and exhibit such diverse lifestyles.

Recently, however, another intriguing theory has begun to emerge that also gives life's most ancient genetic forms – viruses – a potentially pivotal role in the mammals' winning evolutionary formula.

Infection connection

Of the eight groups of mammals known to have existed over the last 220 million years (of which only three groups remain today), by far the most successful have been the placentals. This group of creatures delay giving birth to their live young far longer than other types, such as marsupials (kangaroos and wallabies, for example). They are able to do this because they have a valve-like organ called a 'placenta' that grows out of the embryo into the wall of its mother's womb to regulate the exchange of oxygen, food and nutrients between mother and child. The strategy has proved highly successful because it allows mammal embryos plenty of time for their large, complex brains to develop in the highly protected environment of their mother's womb.

But delaying birth in this way represents extra problems for the immune systems of placental mammal mothers which, like those of all vertebrate creatures, have been programmed over millions of years to reject and kill off organisms that invade their bodies, a central component in the fight against infection and disease. Hosting a genetically distinct embryo (see meiosis, pages 47–8) is no different for an adaptive immune system from tackling an invasive parasite, be it a virus, bacteria, worm or protozoa.

It is thought by some experts that a solution emerged from a series of ancient viral infections. Recent analysis of the genetic makeup of placental mammals shows an unmistakable tendency for their various species to become colonized by retroviruses, genetic replicators that copy themselves into their host's genes (see page 14). Each placental species seems to have its own unique suite of such viruses, preserved and inherited in their genes as so-called 'junk DNA'. Although such genes appear to have little function in everyday life, they have been found to be responsible for helping to establish the placenta during the first few days of pregnancy by producing proteins (common to invasive viruses) that suppress the mother's immune system.

Standing on two feet

The emergence of hominids between five and seven million years ago presents perhaps the biggest evolutionary conundrum of all. Here are species that, in a mere blink of evolutionary time, developed such a powerful new way of life that the whole world has literally been re-shaped by their presence.

Since the widespread acceptance of Darwin's big idea that humans evolved from apes that lived in Africa, dozens of theories have been proposed as to what forces of nature were at play in humanity's rise to predominance.

Fossils of what are regarded as the earliest human ancestors were first discovered by South African palaeontologist Raymond Dart in 1924. It wasn't until 1974, when further fossil evidence was found, that these creatures, called **Australopithecus** (see page 176), were widely accepted as belonging to the world's first two-legged apes. These direct descendants of gorillas learned to live in open grassland where, for reasons still hotly debated, they swivelled up on to two feet, and began to walk upright, freeing their front limbs as arms and hands.

Was it so they could carry food to put into storage more effectively? Was getting flatter feet just a natural adaptation to squatting for food on the forest floor? Was walking on two feet a necessary skill for wading through rivers? It has even been proposed that the desire to stand up on two feet came from a kind of elaborate sexual display on behalf of males wishing to show off their manhood, forcing females to stand up to protect their modesty ...

Whatever the cause, walking upright on two feet led to a spiral of evolutionary changes key to the process of turning primate monkeys into people. Chimpanzees and bonobo apes, humanity's closest genetic relatives, continued to live their lives up trees while the *Australopithecus* ancestors of humans learned to walk upright along the ground. With their hands now free they could craft tools and missiles that helped them become top predators and the only creatures able to kill effectively at a distance. Precise hand-to-eye co-ordination and fully mobile ball-and-socket joints inherited from ancestors who spent their lives swinging in the trees gave this lineage the edge it needed to fashion a completely different way of life.

This puffing chimpanzee was used to illustrate Charles Darwin's book *The Expression of the Emotions in Man and Animals* (1872).

Becoming big-headed

Within the space of just a million or more years, the brains of these ape descendants swelled into species that experts regard as recognizably human. Fossils of *Homo habilis*, dating back 2.1 million years, show an unmistakable increase in brain size from 450 millilitres to 700 millilitres. A later species, **Homo erectus** (see page 179), shows a further significant increase in brain size to 1,100 millilitres. The even more recent Neanderthals (*Homo neanderthalenis*, see page 400), a species that only died out about 25,000 years ago, had heads up to 1,400 millilitres in volume, at least as big as our own species, **Homo sapiens** (see page 182).

What was the evolutionary trigger that caused such a dramatic increase in brain power? What's the benefit of a modern human brain four times the size of nature's nearest genetic rival? Was this large cerebral cortex really necessary for the survival of such bipedal species? What were the consequences of creatures with such big brains for the rest of life on Earth?

Reason versus instinct

In the dynamic global environment of the last sixty million years the mammal brain has evolved a unique capacity for deliberative thought, problem-solving and self awareness. Strategies for securing the inheritance of ever larger brains, for example by giving birth to live young as late as possible, caused placental mammals to become the most rationally minded group, an evolutionary pathway that eventually led all the way from mice to men.

As a result rational brain power began to compete head-on with the world of genetic instinct. With the rise of *Homo sapiens* the local, self-organizational world of genes was confronted by the centralized planning and processing power of the modern human mind. In the last 100,000 years rational behaviour, originally honed by man's mammalian ancestors to hunt in packs, began to be focused on directly manipulating the process of genetic evolution, a line that can be traced from ancient hunter-gathering and Neolithic farming to today's modern genetic science. Now the effects are all around, from the creation of laboratory mice that can sprout human ears out of their backs to the deliberate transplanting of drought- and disease-resistant genes from one species into another (see apple, page 344).

Life's age-old evolutionary instinct for the immortality of a genetic line, regardless of the survival of any one species or way of life, now sits restlessly alongside the human brain's rational desire to live its individual life as long and as comfortably as possible (see On Rivalry, page 367). Space shuttles and supertankers are the products of collaborative rational planning, not self-organizational evolutionary design. Today's contest for supremacy between the competing evolutionary systems of natural selection versus artificial selection – one instinctive, one deliberate – originated with the rise of reason and the domination of mammals. It has reached its zenith in man.

Bat

ORDER: CHIROPTERA
SPECIES: DESMODUS ROTUNDUS
RANK: 75

Flying musicians that have successfully adapted to most available habitats.

AUSTIN, the state capital of Texas, has a number of claims to fame. Not only is it one of America's great university cities, but for anyone who loves jazz, Saturday night on Sixth Street is an unforgettable experience. The place is filled with music, spilling out from bars the length of the town.

About ten blocks to the south another spectacle lies in wait. Although music lovers won't find quite the same thrill here, naturalists will, and there is something of a link with music, too. Underneath Congress Avenue Bridge lives a colony of about *1.5 million* bats – the largest in the United States. They stay there during the summer and migrate to Mexico for the winter. More than 100,000 tourists come to see the sight each year.

Bats are the only flying mammals. There are gliders, like flying squirrels and colugos, but sustained flight seems to have evolved just once in the mammal world, representing evolution's fourth and most recent sortie making use of the combined forces of wings, muscles and air resistance to achieve aerial mobility (see also dragonfly, page 103; *Quetzalcoatlus*, page 120; and *Archaeopteryx*, page 122).

The earliest bat fossils date back nearly sixty million years, with a major population increase occurring about fifty million years ago during a period of sudden and dramatic climate change when global temperatures soared by as much as six degrees in just 20,000 years (see *Azolla*, page 93).

As the world got warmer, insect populations soared, ideal for creatures like bats that feast on them. Fossils show clearly that by this time, bats had already learned how to fly. With such an abundance of insect meals to devour and their consummate skill in 'seeing' in the dark using high-frequency sonar, bats developed into some of the most successful creatures on Earth.

Bats inhabit every part of the world except the polar regions. More than 1,000 types are known to science today, representing nearly 20 per cent of all mammal species, the second most diverse group after the rodents (see page 172). Some are as small as a bumble bee, others have wingspans that reach out for more than 1.5 metres. The most ancient lineage, microbats, mainly eat insects, while the more recently evolved megabats mostly feast on fruits such as figs and play a vital part in many tropical ecosystems by pollinating plants and spreading seeds.

By far the most powerful ecological impact of bats is as a king predator of insects, keeping invertebrate populations in check throughout many parts of the world. The secret of their success is sound, which takes us back to the origins of music and all that jazz.

Birds make sounds as a way of attracting a mate from the right species. Some herbivorous dinosaur families (such as the hadrosaurs) are thought to have made

noises as a means of warning others in their herd about predators. But sound became more like music in mammals, with their increasingly complex brains (see also the giggling rat, pages 174–5; singing whale, page 166; and trampling elephant, page 169). Bats used extra cerebral processing power to become some of the most expert mammal musicians of all.

Humans don't tend to think of bats as musical creatures because most of the sounds they make lie far beyond the scope of human hearing (between 14,000 and 100,000 hertz – people can hear only up to 20,000 hertz). Some bats make continuous tones of various notes. In others the tone modulates and can be emitted as a harmonic chord.

Bats learned to make music because it helped them hunt in the dark when they were safest from attack by predators. Differently pitched sounds help bats judge the whereabouts and proximity of solid objects, using time delay and pitch change as cues to build up a mental image of the outside world without the need for light or sight. This technique has a name – echolocation.

A bat's song comes from its larynx. After its ears pick up the echo of its song, its brain calculates the time difference between each ear to work out where an object is located. Differences in pitch between the sound when emitted and the echo also tell the bat if the object is moving or not – very useful when trying to find flying insects in poor light. So skilful are these creatures at 'seeing' by sound that some species are able to distinguish between two separate targets less than half a millimetre apart – all in the pitch dark.

Bat species are endangered today as a result of habitat loss, owing to human urbanization. The use of pesticides on fields also poisons insects that are then eaten by predators such as bats (see page 99). Conservation efforts have been under way since the 1960s, and even include attempts to convert British World War II pill boxes, originally designed to defend against German invasion, into bat shelters.

A more recent hiccough in the human–bat relationship is the growing realization that some bats are, like rats and mosquitoes, vectors of disease. A few bat species have diversified their tastes and become bloodsucking vampire bats (*Desmodus rotundus*, for example). They bite other mammals (such as cows) with a slash from the upper teeth or by nipping off a small piece of skin. The bat then feeds for up to half an hour from the same wound while anti-coagulants in its saliva prevent the wound from clotting.

Rabies is a virus spread by vampire bats. It passes from bat saliva into the blood system of a mammal host (such as a dog or human), causing it to salivate and burst into fits of biting, tactics designed to help spread the virus via saliva. Bats are also thought to be vectors of other diseases such as ebola, a deadly haemorrhaging virus that affects humans and is currently annihilating populations of African gorillas. SARS, the often fatal respiratory, influenza-like disease that nearly caused a human pandemic in Asia in 2003, is also thought to be carried by bats, which act as a persistent reservoir species for the virus.

As the only mammals that can fly and with more species than almost any other group of mammals, the extraordinary lifestyles of bats, honed by their musical pitch in the blackness of the night, make these creatures amongst the world's most remarkable aviators.

Sperm Whale

ORDER: CETACEA
SPECIES: PHYSETER MACROCEPHALUS
RANK: 59

A deep-sea swimmer that has learned to survive the onslaughts of man.

'CALL ME ISHMAEL' is one of the most famous first lines in American literature. Herman Melville's nineteenth-century, pre-Civil-War classic tells the story of a sea captain, Ahab, and his vendetta against an aggressive male sperm whale called Moby Dick that had bitten off his leg. Ishmael, the narrator, is a crew member of the whaling ship *Pequod*, which sets off in pursuit of the whale but is eventually sunk by the creature after an epic three-day mid-ocean duel. Man's attempt to control nature lies at the heart of Melville's tale and in this story mankind is the loser.

The story behind Moby Dick was based on fact, not fiction. In 1820 an American whaling ship, the *Essex*, was rammed and sunk by a large sperm whale 2,000 miles off the west coast of South America. Eight crew members survived. One of them wrote about the episode and his account gave Melville the context for his classic tale.

Scientific researchers have recently been trawling through the records of other acts of aggression between sperm whales and humans and have found that the *Essex*'s fate was not unique. A similar incident occurred in 1851 when another wooden whaling ship called the *Ann Alexander* was rammed and sunk. After the introduction of steel-hulled ships and explosive harpoons in the second half of the nineteenth century, the scales in the contest between man and whale became tipped in mankind's favour. Since the mid eighteenth century whales, among them blue whales, the largest creatures that ever lived, have been subject to severe human predation.

There are two types of whale – those with teeth (the dolphin and sperm whale families, for example) and those without (filter-feeders, such as the baleen and blue whale). Both are derived from a land-based common ancestor that also led to the Artiodactyla, an order of hoofed mammals that includes today's pigs, camels, deer, giraffes, sheep, goats and cattle. Fifty-three-million-year-old fossils of a creature called *Pakicetus,* found in Pakistan, are thought to represent the whale lineage shortly before it returned to the water. Genetic analysis suggests that the hippopotamus is the closest living land relative to the dolphin and whale.

Evolution changes direction without any sense of historical progression. Bats are mammals that evolved the skill of flight even though their ancestors had previously adapted to a life on the ground. Whales and dolphins are descended from hoofed wolf-like animals that later 're-adapted' to live as swimmers in the sea. For these creatures at least, returning to the seas fifty million years ago must have seemed a safer bet than staying on the land, which was by then becoming overrun with mammals of ever-increasing size and numbers. Large birds of prey came to dominate the skies after the demise of the pterosaurs, perhaps providing another evolutionary pressure on these creatures to try survival by submergence.

Over millions of years the nostrils of these creatures migrated to the top of their heads to become blow-holes. Bones that were once their legs, hooves and pelvis became redundant and sank deep into their bodies, helping to streamline them for a life spent swimming. Some whales also evolved a ball of oily fat in their heads to help focus the production and detection of sounds underwater, which in many species served the equivalent function of echolocation in bats (see page 164).

Relatively large brains, with the capacity for using rational thought as a tool for survival, explain why whales (like elephants, see page 168) are among the most successful creatures on Earth. Most whale species are highly social. They communicate either by using a series of clicking sounds or by singing long, repetitive twenty- to thirty-minute songs, mostly voiced by males in the winter months.

Some whales use 'culture' and showmanship to win the favour of females during the mating season. These are the less aggressive, 'toothless' baleen whales, such as the humpbacks that sing complex, haunting songs. Although no one is 100 per cent sure of the purpose of whale song, most experts believe it forms part of the various rituals used by males to try to win the approval of potential female mates.

The sperm whale sports an enormous head, filled with precious spermaceti.

Charles Darwin dedicated a section in his *On the Origin of Species* to what he called 'sexual selection' (see *Archaeopteryx*, page 122). Antics in toothless whales range from singing and spy hopping, to lob-tailing, tail-slapping, flipper-slapping, charging and parrying. They are classic examples of how in nature males show off to impress prospective mates. In these species, females choose their mates on the basis of which is the best-performing male. This genetic instinct derives from a female's aspiration to have male offspring that inherit good performance skills, therefore increasing their chances of attracting females in future generations and so passing their genes on (see also *Homo erectus*, page 179).

But in some whale species brute force is what counts most when it comes to making the best impression. The biggest, strongest, most successful males take part in head-to-head combat from which only those that are successful go on to mate with females. Such behaviour explains why, over the course of many generations, male offspring have genes selected for size and strength leading to sexual dimorphism – a condition in which one sex is typically much larger than the other. Male sperm whales are often up to *twice as large* as females.

Sperm whales are historically the most successful whales probably because their highly aggressive mating rituals have led to prodigious male strength. The sperm whale's ability to dive to depths of up to three kilometres and withstand enormous water pressure is down to this robust design. Diving this deep means the food on which they live (mostly octopus and squid) is normally out of reach of human fishermen. Therefore, the sperm whale's food supply is largely secure despite the increasing impoverishment of marine stocks by modern man (see cod, page 212).

Another adaptation resulting from a male sperm whale's violent mating ritual is its characteristically swollen head, which is filled with a thick, white oil. This feature is now thought to have evolved at least in part as a battering ram to deliver blows and absorb the shock of head-on attack during duels for damsels (focusing sounds for echolocation and providing buoyancy are other uses).

It was an unfortunate by-product of the evolutionary process that high quantities of this oil, known as spermaceti, proved so attractive to human hunters for use in everything from making candles to lighting lamps. A large sperm whale could yield up to *three tonnes* of spermaceti, once its head had been cut up on a whaling ship. Human hunters would scoop out the oil into barrels before taking it to market onshore. At its height, in the late eighteenth century, the American whaling fleet numbered about 360 ships and employed as many as 5,000 crew who produced 45,000 barrels of spermaceti annually (compared with 8,500 barrels of conventional whale oil). Spermaceti was also an important lubricant for machines used in cotton mills during the early Industrial Revolution and it was the main fuel for public lighting in the United States until the 1860s when it was replaced by lard oil and later petroleum.

Sometimes sperm whales fought back – as the crew of the *Essex* found out to their cost. With the deployment of explosive harpoons, however, the contest became one-sided. By 1880 nearly a quarter of the estimated 1.1 million sperm whales worldwide had been killed. Stocks bounced back, however, as petroleum and other products replaced the more expensive and hard-to-acquire spermaceti. But following World War II whaling became popular again and sperm whale numbers plummeted by as much as 33 per cent. It is estimated that at least 700,000 sperm whales were slaughtered for their precious oily heads between 1946 and 1980. Human sentiment began to change only when scientists found that whales were cultured, intelligent creatures that could sing, click and, in their own way, talk, leading to the advent of the worldwide whaling ban that began to take effect in the late 1960s.

Unlike many other whales (such as the blue whale), sperm whales are not on the list of most endangered species. There are many hundreds of thousands living in all the world's oceans. Although still categorized as 'vulnerable', these are resilient beasts. Their strong, deep-diving survival skills and their inherited instinct to take on new predators – even mankind in the days when a fight could be fought fairly – are impressive. *Moby Dick* is not just a great piece of American literature, it pays homage to a species that, at least for now, has won a remission from man and whose extreme deep-ocean feeding habitat puts it at lower risk of extinction than most.

Elephant

FAMILY: ELEPHANTIDAE
SPECIES: LOXODONTA SP.
RANK: 44

Lumbering giants with big brains, emotions and an impressive ability to learn and never to forget ...

GEORGE ORWELL (1903–50) was a journalist and author who is most famous for writing about his fearful vision of the future in a book called *1984*. Less well known is his life-changing encounter with an elephant while serving as a British army officer in Burma in 1936.

Locals were being terrorized by the animal, which was out of control. The elephant had recently come into 'musth' – a condition of sexual arousal that can, for short periods of time, make such creatures uncharacteristically violent. This particular beast had broken its shackles and run amok in the local bazaar. Worst of all, it had trampled a native 'coolie' (labourer) to death.

Orwell, a liberal-minded pacifist and animal-lover, was the only officer available to bring the situation under control. For him, the consequences of restoring calm to the enraged local population were appalling: *'As soon as I saw the elephant I knew with perfect certainty that I ought not to shoot him ...'* But, being a white man with a gun and the only officer 'in charge', doing what the crowd willed and expected of him denied him any realistic choice: *'White man mustn't be frightened in front of "natives" ... there was only one alternative.'*

An Indian elephant – the same species as the creature encountered by George Orwell in 1936.

By the time Orwell took aim, the elephant had recovered its poise and was calmly grazing in a nearby field. A mysterious, terrible change came over the creature's face after the first bullet struck home: *'He neither stirred nor fell, but every line of his body had altered. He looked suddenly shrunken, stricken, immensely old ...* ' It took three shots in all to fell the beast. *'At last he sagged flabbily to his knees. His mouth slobbered ... He trumpeted for the first and only time. And then he came down, his belly towards me with a crash which seemed to shake the ground even where I lay.'*

It then took more than an hour for the creature to die.

Elephants are among the strongest, most emotional and highly intelligent creatures on Earth. Their brains are larger than that of any other living land mammal (including humans) with a mass of just over five kilograms. Correspondingly, their capacity for memory, learning, emotions, play, problem-solving and socializing is thought, by many, to be the most advanced of any non-human species.

Two types of elephant exist today – African and Indian – although many others have lived but become extinct over the last few thousand years. Woolly mammoths, a related species specially adapted to intense cold, died out soon after the last ice age melt as a result, it is thought, of a combination of climate change and over-zealous, spear-throwing *Homo sapiens* (see page 182).

Like humans, elephants have a prodigious capacity to learn as they grow up. Most mammals are born with brains that are 90 per cent of their final size. Elephants are born with just 35 per cent of their final capacity, the other 65 per cent developing during the first ten years of life (it is slightly more for the human brain, which puts on 72 per cent). Thus the elephant's ability to modify instinctive behaviour by learning through experience has made it (along with dogs, whales, chimpanzees and, of course, humans) special in nature, giving it the advantage of a more adaptive survival strategy. Rational thought offers the ability to 'learn' during a single generation in order to develop new patterns of non-instinctive behaviour. That's why from an evolutionary point of view, in a rapidly changing world, a big brain can make good survival sense.

Large brains with a capacity for developing off-the-cuff survival skills may also exhibit a capacity for foresight. An elephant can use its trunk as an arm for making wooden tools. Some have been observed digging out water-holes. They then plug these holes with a ball of chewed-up bark and cover them with sand to prevent the water evaporating and so preserve them for future use. The same elephants have later been observed returning to their 'artificial' water-holes at a later date, taking out the plug and helping themselves to a refreshing drink. Captive elephants have even been known to remove their shackles, and then to make their escape only once they have looked around to ensure no one is watching. Meanwhile, wild elephants communicate over vast distances using sensors in their trunks and feet that pick up signals in the ground sent as vibrations by other more distant elephant groups. These low-frequency infrasounds travel more effectively through the ground than through the air.

Creatures with large brains usually take years to learn from experience. Like humans, therefore, elephants rely on strong social support systems to care for the young. A mother elephant has to eat huge quantities of vegetation each day to keep her supply of milk flowing, so she needs the support of other adults to help in the tuition and care of her offspring. Strong emotional bonds, indicated by an enlarged region of the elephant brain called the hippocampus, are instrumental in maintaining the integrity of such groups. Devotional behaviour, such as burying the bones of their dead in branches, leaves and sand, has been observed in elephants, as has grief for the loss of loved ones, expressed by standing mournfully over graves and returning to visit ancient burial sites. Humans are the only other species in which such rituals are known, beginning with the Neanderthals (*Homo neanderthalensis*, see page 400).

The group of placental mammals to which elephants belong, Afrotheria, emerged in Africa about 105 million years ago. Along with their extinct ancestors (such as stegodons, mastodons and mammoths), these creatures have had a dramatic impact on the environment of the continents where they have traditionally dwelt – Asia, Africa, Europe (Crete) and the Middle East. As the strongest animals on land, these creatures can, literally, rip out trees with their trunks. Deforestation by elephantine herbivores creates space in the tree canopy through which new shafts of light can penetrate, providing the all-important conditions in which seedlings can thrive. When modern humans deforest an area, they build on it or farm it, thereby preventing trees from regrowing. Yet, when nature's 'lumberjacks' – elephants – unearth tracts of forest they give it ample time to recover before returning many years later.

Just as brutish elephant strength supports a healthy forest ecosystem, their copious quantities of faeces support other lifestyles on an altogether smaller scale. Large dollops of elephant dung provide the ideal foundations for the nests of termites, insects descended from cockroaches (see page 400). These nests, which can contain upwards of *two million individuals*, strike an extraordinary sight, dotted in towers sometimes more than two metres high across the African plains, each one seeded in the wake of a passing elephant's waste.

Elephants have had no less an impact on human history. Ancient Indian, Persian and Phoenician armies were usually fortified by many thousands of elephant cavalry, most popularly by Indian chief Chandragupta Maurya. His 9,000 elephants were, according to Greek historian Plutarch, instrumental in deterring invasion by Greek adventurer Alexander the Great in the fourth century BC. Two hundred years later, Hannibal, the Carthaginian general, deployed a now extinct species of smaller African elephant (*Loxodonta africana pharaohensis*) to cross the Italian Alps in his bid to shatter the Roman Empire. So enduring has the military contribution of elephants been that they were still being used by the British army in World War I to haul artillery across munitions yards.

Elephants have traditionally been hunted by humans, not so much for their meat, but for their precious ivory tusks. Elephants use these extended, modified teeth to dig for water, uproot trees or, occasionally, as weapons. For thousands of years ivory has been one of the world's most popular craft materials used to make everything from decorative ornaments and religious artefacts to pipes for smoking opium, piano keys and billiard balls. This easy-to-carve material has been a highly prized commodity for Asian merchants in particular, much of it traded along the ancient silk route that connected markets from Africa and Europe to the Far East.

Objects previously made from ivory are now more often constructed from plastic. Intensive lobbying by animal conservationists led to a worldwide ban on the sale of ivory in 1989, following fears that both African and Indian elephants would soon be hunted to extinction unless action was taken. African elephant populations have since begun to recover. As many as 500,000 elephants are thought to be alive in the wild today, their populations growing at an annual rate of about 4.5 per cent. Human efforts to co-ordinate the worldwide ban on hunting elephants (and whales, see page 167) demonstrate the capacity for creatures with large brains to act 'altruistically'.

Elephants exhibit this trait, too. Joyce Poole, an American elephant conservationist, reports how a ranch herder's leg was broken when a female elephant knocked him over with her trunk. On seeing that he was injured, the elephant:

> *' ... gently moved him several metres and popped him under the shade of a tree. There she stood guard over him through the afternoon, through the night and into the next day. Her family left her behind, but she stayed on, occasionally touching him with her trunk.'*

Charles Darwin, in his *On the Origin of Species*, stated unequivocally that it was contrary to the laws of natural selection for any creature to behave altruistically, that is *'for the exclusive good ... of another'*. Many biologists have assumed that selfless acts of charity on behalf of other living things cannot therefore be natural behaviours. Yet the power of rational thought – which operates not over generations but within a single lifetime through that individual's ability to learn – does not necessarily obey this rule. Mammals with larger-than-average brains (such as elephants, whales, dogs, monkeys and humans) do seem to have within them the capacity to override genetic instincts and behave altruistically (see the story of Gelert, page 323), be it an elephant protecting a vulnerable human or *Homo sapiens* lobbying for an international ban on the hunting of elephants for ivory.

The Burmese elephant shot by that British officer in 1936 couldn't control its instincts, nor could its executioner, George Orwell, use his powers of reasoning to stand up to the pressure of a crowd of locals baying for blood. The contest between behaviour by experience versus instinct – or head over heart – is something elephants and humans have in common. Perhaps that's why Orwell felt compelled to write about the incident as movingly as he did and why, at least for him, this extraordinary creature made such a profound impact.

Hannibal and his war elephants cross the Alps.

Rat

ORDER: RODENTIA
SPECIES: RATTUS NORVEGICUS
RANK: 32

Roguish rodents whose intelligence and adaptability have been mostly misunderstood.

Rats! They fought the dogs and killed the cats,
And bit the babies in the cradles,
And ate the cheeses out of the vats,
And licked the soup from the cooks' own ladles,
Split open the kegs of salted sprats,
Made nests inside men's Sunday hats,
And even spoiled the women's chats,
By drowning their speaking
With shrieking and squeaking
In fifty different sharps and flats.

From a version of *The Pied Piper*, by Robert Browning, 1842

SEWERS, FILTH, VERMIN and disease are just a few of the many negatives that are conjured up in the minds of most people in the Western world when they think of rats. The historic association goes back to the medieval bubonic plague (see pages 138–9). Rats never featured in the original story of the Pied Piper of Hamelin, who apparently stole 130 German children in 1284. By the sixteenth century, however, their place in the tale was firmly established. By this time these vermin infested every overcrowded street in Europe's burgeoning cities.

Rats belong to the order Rodentia, which comprises more than 40 per cent of all living mammal species and includes mice, squirrels, chipmunks, porcupines, beavers, hamsters, gerbils, guinea-pigs, chinchillas and groundhogs. They are distinguished by their two front incisor teeth, which grow continuously throughout their lives. Therefore survival for these creatures means constantly gnawing through anything they come across – from fully grown trees to plastic-coated electric wires – just to keep their teeth in check.

Rat populations increased with the spread of the world's great grasslands, which in turn arose as a response to the world's ever-cooling climate over the last thirty million years. These are creatures that can eat insects or fruit, but mostly thrive on seeds. So when humans indulged in their first experiments with farming crops about 12,000 years ago, and began to store their agricultural wealth in granaries and barns, it was as if rats had won life's lottery. Ever since, the success of rats has mirrored the rise of human civilizations. In Britain there are an estimated eighty-three million rats, twenty-three million more than humans. No one knows the global figure, although some estimates suggest there may be as many as a hundred million in the sewers beneath New York city alone.

Success for rats wasn't just down to hanging off the coat-tails of humanity. Rats are among nature's most prodigious reproducers. Gestation takes just twenty-one days from conception to the birth of a litter that may number as many as ten young. Even more extraordinary is that females are ready to conceive again within just twenty-four hours of giving birth. With a potential output from a single female of more than one hundred offspring per year, numbers can rise dramatically.

There are more than 300 rat species alive today, although two predominate. The black rat (*Rattus rattus*) and the brown rat (*Rattus norvegicus*) both originated in Asia and have spread around the world, often independently of man. Black rats reached Australia thousands of years ago by rafting or swimming, rapidly becoming the predominant placental mammal. Being small, shy creatures that are not fussy about what they eat, rats have frequently been the uninvited guests of human travellers who, during the last few thousands years, have transported them all over the world along with their precious supplies of food. Black rats reached Europe by AD 600 accompanied by traders who were following the silk route. Brown rats arrived later, perhaps by about AD 1600, but have since become the most successful, dominating most habitats except those in tropical regions.

Sometimes sailors have owed their lives to their uninvited guests. A survivor from the original crew that set off on Ferdinand Magellan's epic voyage to circumnavigate the world in the early sixteenth century (Magellan himself never survived the journey) described living on powdered, wormy biscuits stinking of rats' urine. Nevertheless, Magellan still complained that he and his companions could not get enough rats to eat ...

Human–rat relationships vary a great deal from culture to culture. Generally, rats are prized in the East and persecuted in the West. Today many poor people in Asia, and especially in Cambodia and Vietnam, depend on rats as a rich source of protein in their diet. In Ghana rats are both farmed and hunted for their meat. Traditional Indian tribes such as the Irula, of Tamil Nadu in southern India, who work in rice fields, smoke rats out of their burrows not only to protect the rice crop but also so they can cook and eat the animals. Lord Ganesha, a popular Indian deity, is often depicted with a rat at his side. The Karna Mati Temple in northern India, which is dedicated to him, is jam-packed with rats which, it is believed,

The black rat, which has been more recently eclipsed by its brown cousin, *Rattus norvegicus*.

are destined for re-incarnation in the next life as holy men. Pilgrims are therefore encouraged to eat and drink grain and milk that has been touched by the rats as a blessing.

As rats from the East spread westwards, attitudes towards them hardened thanks to the rise of Judaism and its successful offshoot Christianity, which regarded them as unclean. Touching – let alone eating – rats was strictly taboo in the Judaic culture of the Middle East: *'Any creatures that teem along the ground are vermin: they shall not be eaten'* (Leviticus, chapter 11, verse 42).

By the sixteenth century rats had became associated with diseases such as plague, which is roughly when they appeared as embellishments to the original Pied Piper of Hamelin story. The charismatic musician's primary purpose in the modified tale was not to steal children but to catch rats. In Victorian England rat-catching became big business. The creatures were caught to be sold to London gamblers, who ran as many as seventy rat pits. Punters would take bets on how many rats a bull-terrier could kill in one minute. The world record was held by a dog called Jacko who, on 29 August 1862, polished off sixty rats in two minutes forty-two seconds – an average kill time of one rat every 2.7 seconds.

Today, the laboratory rat (*Rattus norvegicus*) is prized by scientists because it is so easy to study, drug and genetically modify for the benefit of research into disease. Such experiments generally cause less public concern than those that involve larger mammals, such as monkeys. Few people in the West care much about what happens to rats, with the exception of those who keep domesticated varieties, also derived from scientific research, as fancy pets.

One surprising benefit of observing and breeding laboratory rats is an increasing realization by experts that there are similarities in behaviour between all mammals, be they rats or humans. Conscious thought, social communication, childish playfulness, laughter and joy are attributes usually associated with people. It's one thing to assume that our closest primate relatives, such as chimpanzees, also share such characteristics; the idea that vermin could have similar traits has seldom entered the human mind (again, except maybe for those who keep rats as pets).

Indian god Ganesha with his two wives, two attendants, lion and rat.

However, recent research indicates that rats do laugh. It is symptomatic of tactile creatures that tickle each other, and roll around in playful jest – just like human children. Such behaviour is now thought to be common in young mammals as a way to help youngsters weave their way into the social fabric of their kind. Tests have revealed a distinctive series of vocal chirps at 50 hertz (above the threshold of human hearing) during rough-and-tumble play between juvenile rats which, researchers claim, indicates the *'early origins of laughter and joyful social processes that are carried out by the mammalian brain'*.

Other research suggests that rats are able to think for themselves. Trying to determine the extent of self-awareness among non-human species is made difficult by the simple lack of any common system of communication between humans and other animals. However, an intriguing set of tests has been conducted in which rats were presented with sounds of varying durations. They were 'asked' if an individual sound was long or short by being given a large reward when they got the answer right and no reward for getting the answer wrong. They were, however, given a small reward for not answering at all, which sometimes happened when the sounds were of medium length and the rats found it hard to be sure of the correct answer. *'Our research showed that the rats know when they don't know the answer to a question'*, says Jonathan Crystal, associate professor of psychology at the University of Georgia. Such results suggest levels of consciousness in non-primate mammals that were previously thought to be restricted to creatures much more closely related to humans.

The common dismissive perception of rats as unclean, detestable vermin underplays their enormous success and significance. Some biologists have even dared to predict a future path of evolution. They have placed the long-term prospects for rats far above those of humanity. If some pandemic or other disaster were to wipe out humans, rats are well placed to adapt. With their varied diets, fast reproduction rates, strong social communities and apparent sense of self-awareness, it has even been proposed that new rodent species could some day evolve into a replacement race of highly intelligent beings (see also crow, page 378).

Pilgrims at Karna Mati Temple praying amid colonies of rats.

Australopithecus

FAMILY: HOMINIDAE
SPECIES: AUSTRALOPITHECUS AFARENSIS
RANK: 23

A fortuitous swivel on to two feet that eventually led to extraordinary consequences for the planet, life and people.

CHOOSE A SINGLE moment or event that is the most significant to the entire history of humanity – what would it be? The discovery of electricity? The rise of religions? How about the invention of writing? Or, going even further back, the birth of spoken language itself? There is, in fact, a development more significant that any of these. It probably happened sometime between four and five million years ago and occurred within the primate family called the Great Apes (Hominidae). This transformative evolutionary adaptation originated in a species, ancestral to modern humans, which did simply this: it learned to walk on its own two feet.

The two biggest physiological differences between man and his closest genetic cousins, chimpanzees and bonobos, are that humans have bigger brains (roughly four times larger) and that humans walk on two feet, whereas chimps and bonobos have stayed on all fours.

The decision to swivel on to two feet is one of the most significant evolutionary milestones in what it means to be human. For a long time its impact was masked by a notorious scientific hoax. Piltdown Man was a collection of human skull fragments found near a village in Sussex, England, between 1908 and 1915. These bones, it was claimed, dated back at least twenty million years. For decades they were seen by some as proof that man was not after all descended from apes but belonged to its own separate evolutionary line. More important to others was the idea that fossils found in Edwardian England showed clearly that imperial Britannia led the way in human progress.

A reconstruction of the head of an Australopithecus.

After forty years of acceptance by most of the scientific community, Piltdown Man was finally revealed as a forgery in 1953 thanks to the research of Oxford anatomist Joseph Wiener. He showed conclusively that these bones were nothing more than the merged remains of a modern human and ape skull. The Piltdown perpetrators have never been revealed.

But the debacle, even after its exposure, left the lasting impression that in human evolution large brains evolved before the ability to walk upright. As a result, it was thought that humans began to walk upright owing to their large capacity for mental thought. They did it because, well, they deduced that it was a good idea ... But no. Recent fossil discoveries of the bones of bipedal apes have now provided incontrovertible proof that walking on two feet evolved long before big brains.

Australopithecus, an apeish creature that pioneered bipedalism, had a brain about the same size as that of a modern chimpanzee (about 400 millilitres). Therefore, it is far more likely that the innovation of learning to walk upright itself contributed towards the development of larger brains rather than the other way around.

Australopithecus lived in Africa between about 4.5 and 2.5 million years ago and, since its bones were first discovered by Raymond Dart in 1924, a wide range of skulls, skeletons – even footprints – have been unearthed by palaeontologists. The most famous is a 40 per cent complete skeleton, nicknamed Lucy, painstakingly uncovered in Ethiopia in 1974 by a team of palaeontologists led by American anthropologist Donald Johanson. The shape of the feet, pelvis and spine of this 3.2-million-year-old *Australopithecus* shows beyond all reasonable doubt that her kind walked upright. Volcanic footprints found in Kenya at Laetoli were most likely made by the same, or similar, species about 500,000 years before.

Why swivel on to two feet? Unfortunately, there is still no scientific agreement on that critical point. There are a number of competing theories. Clearly, there are certain advantages to a life on two feet, rather than four. Stretching up for food that would otherwise be out of reach is one. Looking out for danger in grassland areas where there are no trees is another. Wading through rivers is a third, since it is easier to ford deeper water by walking upright than on all fours.

Others think that the gorilla-like ancestors of *Australopithecus* rose on to two feet, Tarzan-like, to ward off predators. It has even been suggested that their two-footed antics could have been a sexual display by males who wished to impress females with the size of their manhood (apehood?), while females rose up to protect their reproductive regions from rearward attack. Some think walking on two feet began as a fashion – something 'cool' designed to impress females (or perhaps pioneered by females to impress males) that then caught on. Once some of the advantages of living on two legs rather than four kicked in, natural selection in favour of those with the best attributes for walking upright took over.

Perhaps the most significant advantage of walking on two feet is the one highlighted by Charles Darwin in his book on human origins, *The Descent of Man*, published in 1871: *'Man could not have attained his present dominant position in the world without the use of his hands, which are so admirably adapted to act in obedience to his will.'*

By freeing his arms and hands from the task of walking they could then be used for other purposes such as fetching and carrying food for storage to help him survive severe weather conditions. When wading across streams his hands could grip rocks and trees for safety. Hands not designed for walking could become more dextrous and be deployed for making tools such as clubs and spears to kill fish and other animals for food.

Walking upright is therefore an evolutionary magic bullet – an adaptation that has had a most profound impact on the planet, life and people. Its pioneering species was *Australopithecus*. Until recently it was thought that hand tools originated with the descendants of *Australopithecus*, a line called *Homo habilis.* Traditionally these species are classified as the first true hominids because their brains, at around 650 millilitres, were significantly larger than those of either modern chimpanzees or ancient *Australopithecus*. But in 1996 a research team led by Ethiopian palaeontologist Berhane Asfaw and American Tim White discovered primitive stone tools dating back 2.6 million years scattered next to the remains of a previously unknown *Australopithecus* species. They named it *Australopithecus garhi*, meaning 'surprise' in the local Ethiopian language.

This find suggests that *Australopithecus* gained major survival advantages as a result of walking upright and freeing its hands to make tools. What extraordinary evolutionary consequences lay in wait from this single adaptation! Like two-legged dinosaurs hundreds of millions of years before (see page 126), these bipedal mammals were soon well placed to dominate life on land. But, unlike the dinosaurs, they did it not by brute strength, but by using their hands and with what came next, much larger brains.

By about 2.5 million years ago various species of *Australopithecus* – some slender omnivores, others more solidly built vegetarians – gave way to descendants, *Homo habilis*, whose increasing skill at crafting tools made a vital contribution towards the evolution of more advanced hand–eye co-ordination, better problem-solving abilities and, ultimately, larger brains. Throwing weapons they had fashioned allowed them to hunt at a distance without endangering their own lives, ensuring that the genetic impetus of their two-legged lineage, pioneered by Lucy and her ilk, powered ever on.

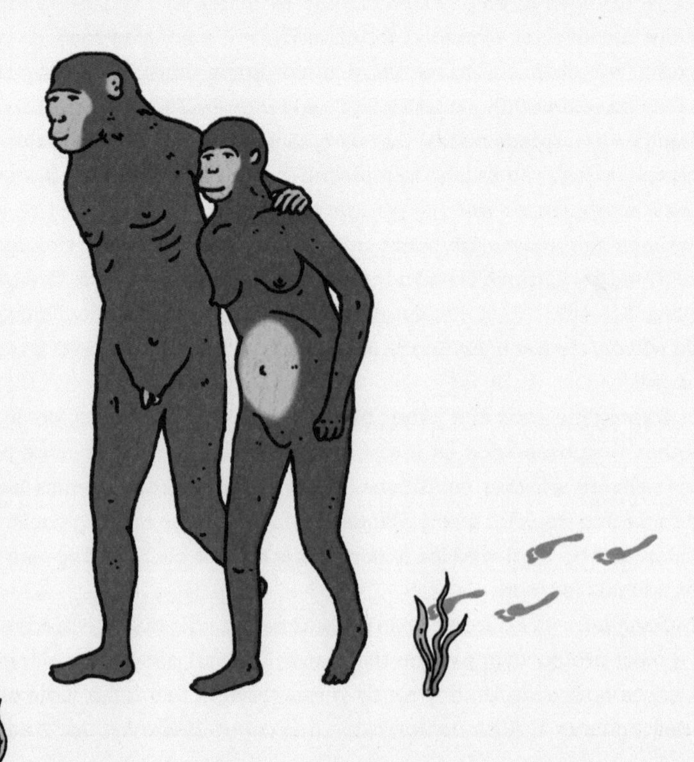

Homo Erectus

FAMILY: HOMINIDAE
SPECIES: HOMO ERECTUS
RANK: 13

The longest-surviving human species with an unprecedented growth in brain power.

CHARLES DARWIN dedicated an entire chapter in his *On the Origin of Species* to the topic of instinct. The behaviour of all creatures, he said, operates according to the same laws of natural selection that have governed the evolution of the shape and form of physical characteristics. From the collaborative bee hive to the parasitic cuckoo, instincts drive the habits of creatures in their struggle for survival. The most successful pass their patterns of behaviour on to future generations, enhancing the chances of survival for their species.

Which is all very well until conditions rapidly change. Instincts, however well honed, cannot prepare any species for the sudden arrival of a new, unknown predator or for a dramatic shift in living conditions. Many generations of natural selection are required for new instincts to become dominant in any species. For small, fast-breeding creatures (such as insects) with short generation cycles, that may not be such a problem. But for larger animals that live longer it could take centuries for new instincts to evolve, by which time they could easily get wiped out.

Brain power is the mammals' evolutionary answer to the problem. Its greatest attribute is that a creature can adapt its behaviour *within the space of an individual lifetime* without having to wait for more appropriate instincts to evolve over many generations. Mammals such as elephants, whales (even rats) can *learn* from their parents, peers and their surroundings, equipping them with the ability to override, modify or supplement instinctive behaviour, if circumstances are so dynamic as to require such changes in order to survive.

It is the capacity for rational behaviour in many mammals (and some birds, too – see page 378) that has allowed modern humans to train creatures to do non-instinctive things, as any trip to a traditional circus will show. A creature that stands out as the most prolific in advancing its potential for learning and reasoning was the direct ancestor of modern humans, and the descendant of *Australopithecus* (see page 176).

Homo erectus was a species of human being that could boast a string of superlatives and has by far the best track record of any human species to date in terms of survival. He originated in Africa about 2.6 million years ago, yet died out only around 70,000 years ago. In contrast, modern humans (*Homo sapiens*, see page 182) have lasted for just 160,000 years so far – less than 7 per cent as long. *Homo erectus* was also the original hominid tourist, the first to explore the world outside its continent of origin. About 1.8 million years ago *Homo erectus* migrated out of Africa to populate parts of Europe and Asia, even reaching as far as Indonesia and China, where numerous remains of its skeletons, skulls and bones have been found (Java Man and Peking Man, for example).

Homo erectus was also the first human species to enhance its environment through the controlled use of fire; evidence comes from the discovery in Africa of charcoal remains dating back 1.5 million years. The remnants of fires have also been unearthed in northern Israel dating back to around 800,000 years ago, and more recently in France and Spain, confirming that this species included the world's first barbecue chefs. Cooked meat releases energy more efficiently and more quickly than raw meat, providing better nutrition to satisfy the greater energy requirements of these people's increasingly big brains. Fire also protected *Homo erectus* against wild beasts. This made them king of the predators, eclipsing all other species with their superlative survival skills. Rival humans such as *Homo habilis* who did not use fire are thought to have died out about 500,000 years after *Homo erectus* first appeared.

Without doubt the most notable evolutionary adaptation forged by these people was the dramatic increase in the size of their brains. Early *Homo erectus* skulls show a cranial capacity of about 800 millilitres, roughly twice the size of that of *Australopithecus* (see page 176). Later specimens show how this had increased to as much as 1,100 millilitres (approaching the average 1,400-millitre capacity of modern humans).

One possible cause for both the increase in brain size of this species and its migration out of Africa was climate change. Over the last three million years global temperatures have gyrated wildly between long fits of severe cold and shorter periods of interglacial mildness. Only recently has it been discovered that conditions can sometimes flip from hot to cold and vice versa literally within the space of a human lifetime. Feedback loops in the Earth's temperature regulation systems (such as ice caps, carbon dioxide and methane levels) mean grasslands and forests can be reduced to scrubland sometimes in a matter of decades, not millennia, as was once thought.

The Younger Dryas period of sudden cooling, which followed the last glacial period *c.* 11,400 years ago, is one well-studied example. Some experts now believe similar sudden climate changes provoked *Homo erectus* to migrate out of arid Africa in search of wetter, more fertile hunting grounds. Essential survival strategies such as crafting better tools for hunting, igniting and controlling fires, eating cooked meat and even working out how to build rafts for coping with rising water levels might never have arisen through generations of slow adaptation driven by the selection of genetic instinct alone. Instead, they all required the use of substantial, impressive, fast-acting rational brain-power.

Problem-solving skills, communication between individuals in hunter-gatherer bands and the construction of weaponry were all vital for the survival of this species in the dynamic, rapidly changing environmental conditions of an ice age. The potential for intellectual reason, which had been gradually evolving in mammals since the demise of the dinosaurs, came into its full bloom with the rise of two-legged *Homo erectus*. Working with his hands led to the further enhancement of his cognitive skills. Over time, *Homo erectus* individuals with the best mental skills for working out how to survive in changing environments passed on these attributes to their future generations. And so, over not so much time, average cranial capacity grew.

An additional explanation for the dramatic increase in human brain size was proposed by Charles Darwin in his second most famous work *The Descent of Man*, published in 1871.

Sexual selection is the second evolutionary force (after natural selection) that Darwin believed may have helped humans enhance their reasoning powers. With freely available hands, males could prove to females their worthiness as mates by 'showing off' (see also *Archaeopteryx*, page 122; and *Homo sapiens*, page 184). Females may have been most impressed by men that could excel in design and technology skills, such as toolmaking or boat building, and survival techniques, such as lighting fires. These attributes were then more likely to be passed on to the offspring of such well-skilled mates.

Likewise, males may have been impressed by females with desirable domestic skills such as being great cooks or entertainers who danced and sang well. Any daughters they bore with these qualities would be more likely to attract future males. Good toolmakers that bred with excellent cooks, and fancy dancers that mated with brilliant boat-builders would proliferate. Poorly performing individuals would be generally regarded as unattractive to either sex – who wants their children? According to Darwin, this probably led to a pattern of breeding that favoured enhanced mental reasoning, and performance and handicraft skills. Of course, large males that excelled in hunting (or warding off rival suitors) were highly desired by females too, and their genes were also likely to be passed down the line – which accounts for why *Homo erectus* males were typically 30 per cent larger than females (see also sperm whale, page 166).

An artist's impression of *Homo erectus* based on 800,000-year-old-fossils found in Java, Indonesia.

No one knows to what extent *Homo erectus* were really cultural beings and if they could communicate or talk in the sense that modern humans do. However, the extraordinary increase in brain size during their two-million-year existence on Earth was certainly caused by a powerful interplay of selection forces, be they the brilliance of brain power needed to survive rapid climate change or the courtship contests of sexual selection.

Thanks to the success of *Homo erectus*, humanity came to enjoy a much broader global distribution. It also created a lineage that ended up with a brain-to-body-weight ratio nearly four times larger than any other mammal. Other animal predators were no longer significant threats to humans such as these, with their advanced hunting techniques, sophisticated tools and command of fire. From the time of *Homo erectus,* humans became the dominant land-living species and as a consequence the rest of life on Earth, and even the evolutionary process itself, was increasingly subject to the command and control systems of early man.

Homo Sapiens

FAMILY: HOMINIDAE
SPECIES: HOMO SAPIENS
RANK: 6

The beginning of a new evolutionary era in which living things thrived or failed according to their usefulness to a single species: man.

MY APOLOGIES. A thoroughly human author writing about his own species for an exclusively human readership does not bode well for objectivity. But, owing to the lack of any common language with other species and the exclusive ability of *Homo sapiens* to read and write, it is an unavoidable inconvenience that any account must bear.

Modern humans first emerged in Africa, about 160,000 years ago. They evolved from the stock of *Homo erectus*, the longest-surviving species of humans, whose brains had grown out of all natural proportion (see page 179). By about 25,000 years ago *Homo sapiens* was the only surviving species of human on Earth.

According to recent genetic analysis, a near-catastrophic event, about mid-way through its history, reduced the population of *Homo sapiens* to no more than a few thousand individuals. Speculations as to the cause of this near-extinction range from viral infection to the giant volcanic eruption of Mount Toba 70,000 years ago that poisoned Africa's habitats. Large brains with a cranial capacity of *c.* 1,400 millilitres may have proved vital to its survival. During catastrophic events advanced mental reasoning powers truly come into their own. Quick thinking, manual dexterity and increasingly complex toolmaking skills are likely to have made the critical difference between life and death.

At about the same time the behaviour of *Homo sapiens* started to change significantly. The species began to develop a unique capacity for complex spoken language, learning, education and other forms of what is now called intelligence. No one knows for sure when human language first appeared because, unlike writing, speech leaves no physical record. However, highly sophisticated toolmaking techniques date back around 70,000 years. Spoken language communication would probably have arisen by then since it is difficult to see how such people could have passed on their craft-making skills from one generation to the next without some kind of verbal communication. Extreme survival conditions may also have put greater emphasis on speech as an aid to hunting as climates changed rapidly during the Earth's oscillating bouts of glaciation.

Vitruvian Man – Leonardo da Vinci's famous fifteenth-century drawing shows how human limbs move in perfect circular symmetry (see also page 339).

One possible explanation for the rise of language comes from the same sexual selection pressures that are thought to have increased the rational thinking powers and brain size of *Homo erectus*. With populations of *Homo sapiens* falling to just a few thousand, competition between men to mate with the few remaining women and produce offspring is likely to have become extremely intense. It would have been

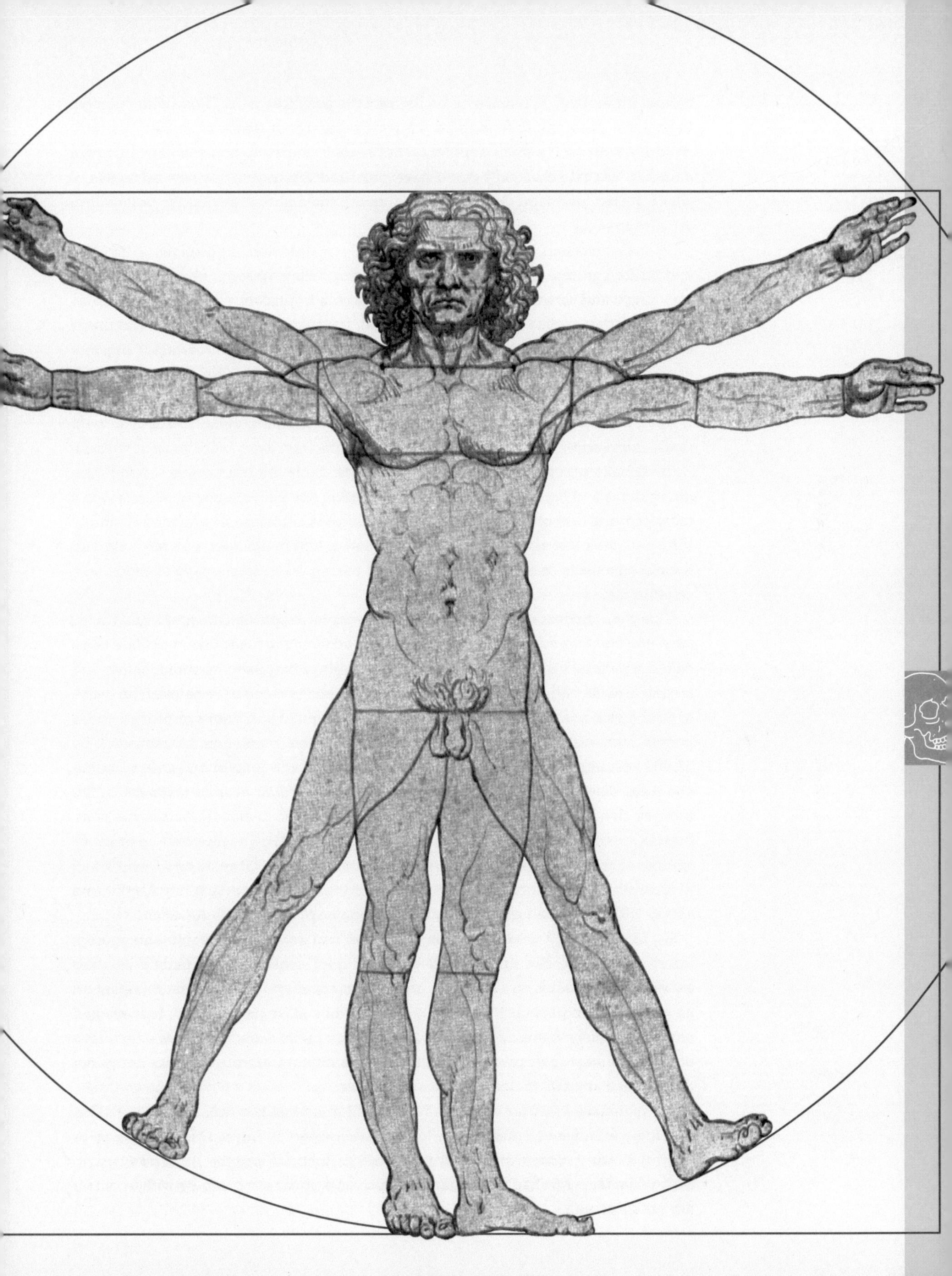

natural for females to choose carefully from the available men. Those with the most impressive skills would tend to be the most attractive or desirable fathers for their children. With such a small population bottleneck, courtship contests (or, as Darwin called it, 'sexual selection') could have provided the trigger for new advances in which verbal language skills became a male's most potent device for persuading females to mate.

Verbal courtship displays have frequently manifested themselves in spoken and written culture throughout recorded history. Plays, poems, songs and stories about love and romance appear to have supreme importance in human culture, as popular today as they were in the medieval/troubadour age of chivalry hundreds of years ago. Mostly, they are written and performed by men in a bid to impress women, who are frequently influenced by such displays as they determine their partnership choice.

With such small human populations 70,000 years ago, those males that were really impressive linguistically, the most charming, perhaps, were most successful with verbally receptive women. Their linguistic skills therefore passed into future generations and the human capacity for language has become mammalian reason's most impressive evolutionary product yet. Complex language, as opposed to simple birdsong, may therefore be best explained as a highly efficient way for male and female humans to find out about each other during the critical phase of choosing a suitable mate (see also pages 339–40).

The same process of sexual selection that may have spawned bigger brains in his ancestor, *Homo erectus*, could also have helped *Homo sapiens* take what has been called a cultural 'great leap forward'. Sophisticated linguistic, communicative and technical skills helped populations of *Homo sapiens* to recover – whatever the cause of their sudden fall in numbers – and, as environmental conditions improved, these people soon began to spread out of Africa, like their *Homo erectus* ancestors. By 50,000 years ago they had reached the Middle East and fanned out across Europe and Asia, displacing other human species, such as *Homo neanderthalensis*, in the process. Some built rafts and canoes, which they used to paddle across the seas, populating continents such as Australia and the islands of Polynesia where no previous species of humans had ever lived before. Others, about 13,000 years ago, found a way to cross from the eastern edge of Asia to the Americas, via modern-day Alaska, and within 1,000 years had roamed down to the southern tip of South America.

By about 12,000 years ago *Homo sapiens* had reached all continents except frozen Antarctica. The arrival of these intelligent, well-armed bipeds with their exceptional capacity for reasoning and communication had a dramatic impact on many other living species, especially those that lived in places where humans had never previously ventured. First in Australia, then in the Americas, mass extinctions of large animals took place soon after human hunters arrived. Instinct could not equip these animals to deal with the sudden arrival of such a new, unknown threat. Large mammals in Africa and Asia had lived for at least two million years in close proximity to humans, enough time to have developed an instinctive fear, hiding or fleeing to avoid close contact. But mammals in Australia and the Americas had no such advantage. Nor had they sufficient reasoning powers to compete with or outwit the brains of man.

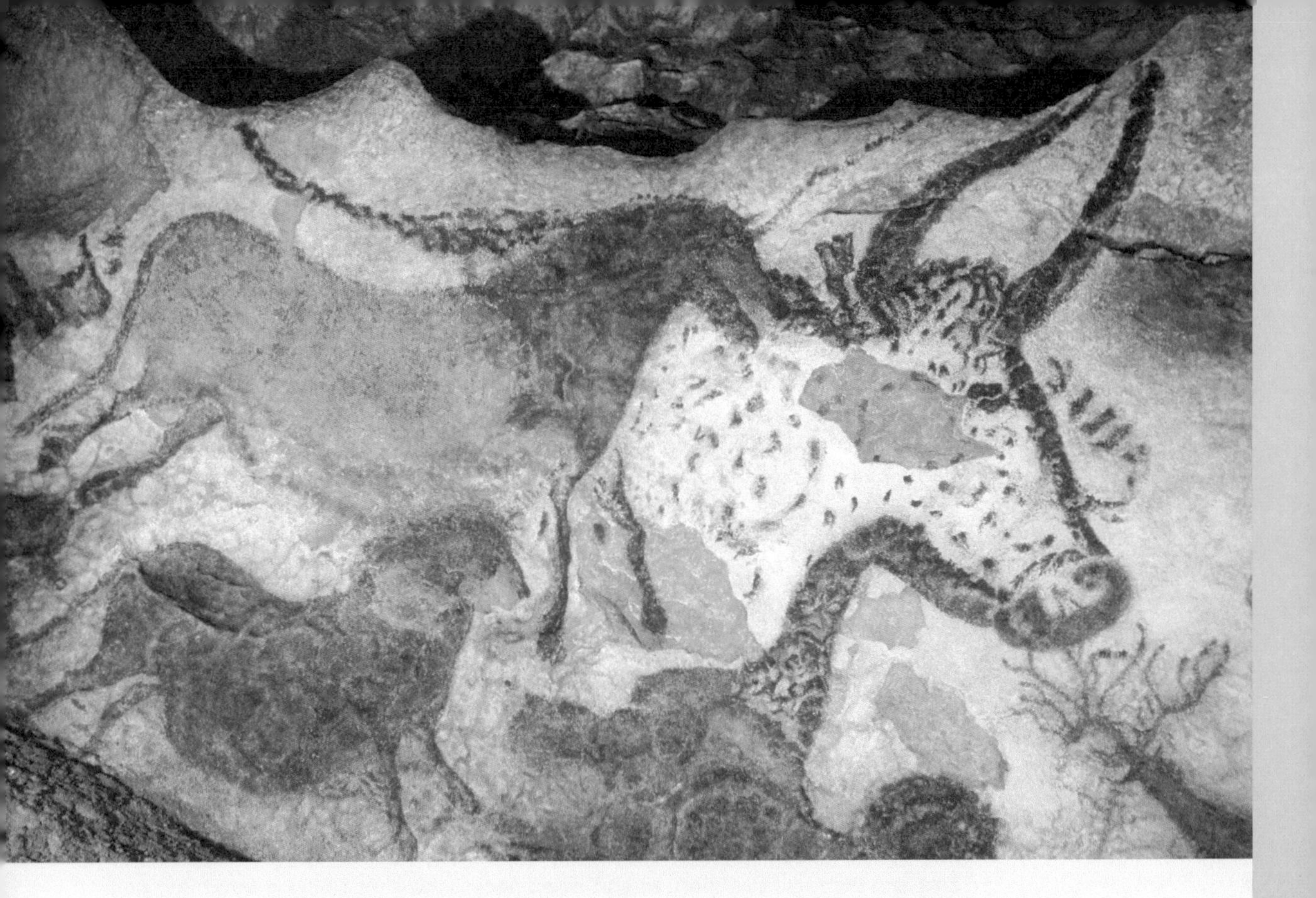

Cave paintings of animals at Lascaux, France, created by prehistoric people to contact the spirit world.

Soon after the arrival of *Homo sapiens* in Australia, only one out of sixteen groups of large mammal survived. In North America, thirty-three out of forty-five groups became extinct. In South America, forty-six out of fifty-eight groups perished. The decimation of many of the top American predator species (such as the American lion, cheetah, sabre-toothed cat and cave bear) led to huge increases in the populations of herbivorous mammals. Starvation followed as food ran out and, together with drought, this led to the extinction of many creatures, such as mammoths, mastodons, glyptodonts, North American llamas and all five species of native American horses.

As human populations migrated across the globe, evidence of their advanced culture emerged in the form of cave paintings that venerated animals, carefully crafted figurines dedicated to a fertility goddess and ceremonial burials that indicated that this was an increasingly 'self-aware' species. The reasoning powers of *Homo sapiens* were now sufficiently robust to begin to seek answers to questions about the origins and purpose of life on Earth (such as Australian aboriginal Dreamtime stories), life after death (burial ceremonies) and even morals for distinguishing between right and wrong behaviour (such as totems and taboos).

Other physiological traits further distanced this species of human being from his immediate ancestors. So large were the brains of infantile *Homo sapiens* that a child had to be born much earlier than the offspring of other mammals, so that its head could squeeze through the mother's birth canal. These vulnerable human offspring required the constant attention, protection and care of adults. Social groups also proved effective for the transferral of cultural values (especially language and crafts) and other essential life skills, which were taught by older generations to the young. Such birth dynamics may also explain why there is less sexual dimorphism in *Homo sapiens* than among its ancestors *Australopithecus* (see page 176) and *Homo erectus* (see page 179). Males are just 15 per cent larger than females, on average, indicating a greater inclination towards monogamous relationships and nuclear families in which both parents invest time in raising their offspring.

Male and female *Homo sapiens* evolved into the only species of primate not covered in thick body hair. It was an adaptation triggered, some experts believe, by the female's preference for males with exposed skin. Clean, unblemished skin was indicative of a man in good genetic shape, free of fleas and lice. Sexual selection meant such traits were passed down to future generations. Some physiological differences between the sexes remained. For example, facial hair in males (beards) may have developed as a visual cue to ward off other males during courtship contests.

But by far the biggest impact of these mostly hairless, two-legged hunters with language, art and a curiosity about their place in the world began to be felt about 10,000 years ago. As a result of dramatic swings in global temperatures, traditional hunting grounds began to disappear under rising seas and wild game became scarce. As a consequence humans had to use their brain power to find new ways to feed themselves.

Natufian people, who lived in the area of the Middle East now called Lebanon, retreated into the hills and began sowing the largest seeds of wheat they could find in well-prepared soil. This was the first known attempt by humans to grow crops for harvesting months later. They discovered that grain could be stored in makeshift huts and then used as food, as and when necessary, after being ground into flour and baked into bread.

Humans in several parts of the Middle East, North Africa and Asia also figured out, through trial and error, that some wild animals could be tamed and bred into forms that were more suited to the needs of human beings. After being bred in captivity over several generations, such creatures became compliant, their genetic instincts of fear and aggression – vital for survival in the wild – became suppressed, turning them into the domesticated breeds of goats, sheep, cows and pigs that we recognize today. Such animals could be kept on farms – or in herds to support pastoralists – providing regular supplies of milk, cheese, hides, furs and meat. Now *Homo sapiens* had its own 'artificial' solution to hunger management that was more resistant to the effects of climate change and the scarcity of hunting grounds. Such methods were painstakingly pioneered by such 'Neolithic' (new stone age) people, using calculated rational techniques learned through experimentation that were passed on and perfected via their verbal culture from one generation to the next.

The rise of farming had a huge impact on the planet, life and people. So much food could now be kept in store that people could settle in permanent dwellings and many of them were able to explore ways of life beyond the tasks of food provision. The rise of artisans, priests, rulers and soldiers led to the development of large human settlements, well-guarded cities, empires and complex civilizations. Others stuck to the more traditional, nomadic lifestyle, taking their flocks and herds with them for food, trading between settlements – or carrying out raids, when the opportunity arose. They began to used domesticated animals, such as horses and camels, as personal transport systems, or to carry goods or wage war (see pages 242 and 248).

Humans learned how to convert their spoken language into a permanent written form using marks and scratches. Such codification further helped pass on knowledge, such as the best time to sow crops as well as keep accurate accounts of trade and exchange. Thanks to cultivated food, animal husbandry and the application of rational thinking, human populations grew from approximately *five million* individuals *c.* 12,000 years ago to approaching *seven billion* by AD 2009.

Human societies differ in one important respect from their counterparts in the insect world (see pages 152–6): they are products of central planning and authority, the work of thinking minds, not of instinctive self-organization that arises through genetic inheritance. Such a difference introduced a radical new selection pressure into the process of evolutionary change.

In this world of 'artificial selection' species thrive or die not as a result of their fitness for any given ecological niche but rather because of their usefulness to man. Over the last 12,000 years, global glory has followed those species that have successfully adapted themselves to this new evolutionary paradigm. Isolation has come to species that have failed to prove their worth, many of which have been driven to extinction. A few groups have soldiered on regardless, remaining representatives of those species strong and fit enough to resist the clamour of humanity's rational demands (see On Rivalry, page 367). These, for now, continue to evolve according to the laws of nature – not of man.

After Man

From 12,000 years ago to the present day ...

On the impact of species that thrived in the presence of modern mankind

See what a lot of nice new friends I've made already

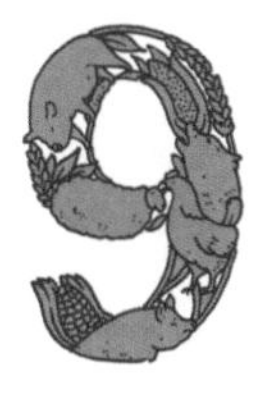

On Agriculture

How the birth of farming helped create a new top tier of world-conquering species, thanks to the co-evolution of humans and a few select animals and plants.

TRADITIONAL HISTORY seldom considers the impact of a range of living species that have, in their own way, had a far greater impact on the planet, life and people than human contributions such as politics, wars and inventions. In fact, it is thanks only to a few select non-human life-forms that modern human civilization exists at all.

Top of the list of such 'super-species' are those animals and plants that have adapted to provide food and nourishment to suit human appetites. Such species fascinated Charles Darwin so much that he focused on them at the beginning of his *On the Origin of Species* in a chapter entitled 'Variation Under Domestication'. Darwin's interest was stirred by common farm and household animals such as cows, horses, pigeons and dogs, and everyday kitchen garden produce such as potatoes, cabbages, strawberries and gooseberries. These were living species that had, to Darwin's mind, undergone change as a result of selection by humans: '*Our domestic races show adaptation in their structure or in their habits to man's wants or fancies.*'

If the power of selective breeding was the cause of variation in domestic species, then perhaps other, more natural selective forces were at work long before mankind emerged? It was from the study of domesticated varieties that Darwin's more general evolutionary theory about natural selection emerged.

Over the last 12,000 years, certain plants and animals have demonstrated a stunning ability to modify their ways to please human beings. In the process they have become supreme at enlisting humans as farmers, transporters, householders, gardeners and chief defenders against attack. This symbiotic co-evolution of humans and a select few domesticated species can be divided into four distinct phases – farming, globalization, disease and addiction.

Spuds up: how the British Ministry of Food put potato propaganda on the menu during World War II.

The farming revolution

No evidence of farming exists before the glaciers retreated. Before then, humans – including *Homo sapiens* – lived as hunter-gatherers, feeding only when they needed to by killing game or collecting wild fruits, berries and seeds. They lived mostly a nomadic lifestyle, wandering in bands of between ten and twenty individuals and sharing all they needed, carrying as little as possible to lighten the load.

However, during the last 12,000 years humans have almost completely abandoned their traditional ways of life. Now most live in vast cities and settled communities surrounded by individual material possessions. Populations have increased from roughly five million before the birth of farming to about seven billion today – a 1,400-fold increase. In many places layers of asphalt and concrete have smothered and sterilized the earth to support huge conurbations, and rivers have been dammed to protect cities from flooding and to provide hydroelectric power. Meanwhile, lakes and marshes have been drained and trees hacked down to create open fields on which to grow food or graze animals.

This extraordinary transformation has come about thanks to surprisingly few varieties of crops and animals. These were species that responded vigorously to the changing lifestyle habits of humans who, with their freely available hands and large, inquisitive brains, began to experiment with selecting, planting and cultivating crops as well as enclosing, herding and breeding animals.

The first human experiments with artificial selection – the deliberate cultivation of plants and breeding of animals with particular traits that could further human wellbeing – took place between 12,000 and 3,000 years ago. They occurred independently in a number of fertile river valleys, including those of the Tigris and Euphrates rivers in the Middle East; next to the river Indus (modern-day Pakistan); alongside the Yellow and Yangtze rivers in China; along the upper reaches of the Nile in Egypt; beside the shores of lakes in Central America (modern-day Mexico); and on hillside terraces in north-west South America (modern-day Peru).

In these places the deliberate cultivation of just a few plant species, such as rice, wheat, maize, millet, lentils, barley and peas, made a big difference to the way people lived. Each helped plot a distinctive path for ancient human history in Europe, North Africa and the Middle East (for example **wheat**, see page 199), China and India (for example **rice**, see page 230) and the pre-Columbian Americas (for example **maize**, see page 234).

These crops demonstrated an extraordinary alchemy. Over generations of cultivation they changed dramatically from their natural, wild state into forms that better suited human needs. Early Neolithic farmers selected individual plants with the biggest, most nutritious seeds that were easy to harvest because they stuck firmly to the stalk. Such criteria were quite different from what natural selection had previously demanded over millions of years. Plants that produced small seeds, easily blown by the wind, were those that thrived in the pre-agricultural past.

A similar transformation occurred in the breeding of animal species. Natufian people in the Lebanon began the first experiments with domesticating dogs, to help them hunt and guard against wild beasts. The hundreds of breeds of dogs we have today are all descended from an original wild species of grey wolf variously bred by humans over the last 12,000 years into different varieties suited to helping people hunt game, herd sheep, guard homes and livestock or just to enjoy for the companionship they provide (see page 323).

When humans began cultivating plants and breeding animals those species with the broadest genetic variety thrived. For example, grasses such as wheat had some varieties with large seeds that could be selected for high levels of protein and carbohydrate and turned into bread or puddings to sustain hungry humans. Foodstuffs such as **olives** (see page 209) that were easy to grow, rich in oil and didn't quickly degrade became part of the staple diets of other human communities.

Genetically pliable animals also benefited from human selection. Aurochs – large mammalian herbivores that evolved in India about two million years ago – had sufficiently diverse genes to allow human farmers to breed varieties that were able to live in captivity as well as produce gallons of milk and copious quantities of meat. Modern domesticated **cows** (see page 222), of which there are an estimated 1.3 billion living today, are their descendants. They are now the most prolific large animal species on Earth.

Other domesticated species include **sheep** (see page 218), goats (see page 401), **pigs** (see page 215) and horses (see page 242). Although not always farmed for food and drink, each of them has played a vital part in the establishment of modern human diets in the last 12,000 years. In return, they are looked after by humans, protected from predators, inoculated against disease, fed and watered.

All change

Thanks to agriculture, many people gave up the traditional hunter-gatherer way of life, choosing instead to live in settlements (towns and cities). They no longer regarded land as common hunting ground and they built shelters in which to store their grain, and barns in which to keep their animals.

The world's first walled cities, Jericho and Çatal Hüyük, emerged in the Middle East about 8,000 years ago. They had populations in the thousands – each family living with their own individual house and personal store for agricultural produce. Here lie the origins of the concept of private property, a right most people take for granted today. It is a bizarre thought that the idea of individuals owning anything goes back only 12,000 years at most and happened entirely thanks to the genetic versatility of certain species of animals and plants.

These domesticated species produced so much food that people living in settlements could become full-time artisans, soldiers, rulers, administrators, priests, merchants and slaves. All these specialized social roles were made possible because of the adaptation of a number of essential agricultural species that provided human societies with surplus supplies of stored food, all year round.

The use of dogs to herd animals helped some farming cultures stick to their ancient nomadic ways, but instead of hunting they took domesticated livestock with them. Grazing animals that convert plant material into milk and meat provided an ideal solution to the problem of feeding human populations in areas with poor-quality soils unsuited to growing crops. Their lifestyle, known as 'nomadic pastoralism', dramatically altered the course of human and environmental history through the mass migrations of people such as the Bantu in Africa (see page 142) or the Mongols in Asia (see page 245). Sometimes nomads abandoned their herds along the way to become settlers (such as the Israelites who settled in their 'Promised Land' of Canaan). Today, only a few societies cling on to their nomadic herding way of life, since it takes ten times more land to support pastoral people than it does settlers dependent on high-yielding crops.

Both agricultural and pastoral modes of existence have had major environmental impacts. Large-scale deforestation is needed to produce fields for planting crops and to provide open pasture for grazing animals. Britain was mostly forest before the rise of farming and pastoralism. But, by the time of William the Conqueror's Domesday survey in 1086, 85 per cent of the land was bereft of trees. Exmoor, Dartmoor and the North Yorkshire Moors were all ancient woodlands felled by Neolithic people in the last 5,000 years. Such was the impact of a few genetically malleable plants and animals that could be customised to suit the changing appetites of modern man. Today 99.9 per cent of all people have adapted to living in larger, mostly settled societies – all of them dependent on farmed food as fuel for a mostly urban way of life.

The globalization of food

The second great achievement of easy-to-domesticate plant and animal species was globalization. Two major waves of human migration provided the

mechanics of transportation and transplantation for farmed crops and livestock – both took place over the last 2,000 years with dramatic effects on all history.

The first came with the rise of Islam from the seventh and eighth centuries AD onwards. Muslim traders connected the rich agricultural produce of south and east Asia to Europe and around the Mediterranean. North Africa and Spain were transformed by crops from the East that were transported and successfully cultivated using irrigation systems, wheeled ploughs and slave labour. **Sugarcane** (see page 202), rice (see page 230), lemons, apricots, cotton (see page 252), artichokes, almonds, figs, bananas (see page 364) and aubergines were all imported by Islamic merchants for their nutritional value or utility. These species were successfully established in Europe as new staple crops thanks to their ability to adapt to a range of climates and habitats.

By 1481 the agricultural riches of Spain, well watered by Islamic systems of irrigation, had fallen completely into Christian hands. Within fifty years, European merchants had found their own way to circumvent Islamic traders by mastering the Atlantic trade winds. Portuguese explorers pioneered voyages by sea around the Cape of Good Hope to the spice-rich markets of the East, while Spanish and Italian opportunists headed west in search of an alternative, quicker route.

Each time they took with them life-support kits primarily in the form of livestock such as pigs, sheep, goats, horses, chickens and cows. When the Spanish arrived, quite unexpectedly, in the unknown 'New World' of the Americas, their successful conquests restocked the American continents with large animals, replacing with domesticated breeds those that had gone extinct shortly after the arrival of *Homo sapiens* 12,500 years before (see page 185).

European settlers in the Americas also brought with them crops such as sugarcane, coffee, bananas, cotton, oranges, lemons, oats, wheat, olives, apples, rice, rye and turnips. In this way crops from Asia ended up in the Americas via a two-stage process. Muslim traders brought them to Al Andalus (Islamic Spain) following which Christian conquistadors took them across the Atlantic.

Oceanic trade routes also brought a rich source of exotic crops from America back to Europe. **Potatoes** (see page 206), tomatoes, sunflowers, tobacco (see page 289), strawberries, vanilla (see page 346), rubber (see page 256), pumpkins, avocados, beans, nuts, cacao (see page 274), maize (see page 234), and pineapple were species painstakingly domesticated over thousands of years by Native Americans. From the seventeenth century, however, they were being transplanted wherever appropriate climates could be found in the cultures of Europe, the Middle East and Asia. The global success of such species benefited from human communities that were becoming increasingly interconnected by population growth, transport, conquest and trade.

That's why today oranges are cultivated in Florida, potatoes in Ireland and Iowa, tomatoes in Italy, rubber in Africa, coffee in Brazil; and why cattle are reared in Texas, and sheep – hundreds of millions of them – in Australia and New Zealand. These are all species now cultivated in mass quantities on the opposite side of the world from their original habitats, largely thanks to their climatic adaptability and their genetic ability to morph into varieties that appeal to human appetites.

The domestication of man

The gradual co-evolution of two unrelated groups – such as orchids and moths (see page 131) – usually involves some kind of genetic modification that benefits them both. Even though humans have been farming for only a relatively short period (12,000 years is a mere snapshot in humanity's 2.5-million-year history) genetic changes to humans provoked by the species they have come to depend upon (and which now depend on humans) have already affected the course of recent history.

The most obvious evolutionary changes have occurred with the spread of pandemic diseases. Measles and smallpox (see page 24) originated as persistent infections in cows but jumped the species barrier around 10,000 years ago to become deadly infections in humans, subsequently killing hundreds of millions of people. Influenza (see page 16) jumped from domesticated chickens and pigs. Other diseases such as malaria and sleeping sickness are spread by bloodsucking insects. They feed on herds of cattle, which act as host reservoirs for these diseases that then infect humans (see pages 134 and 140).

Humans who have lived with farm animals over thousands of years have gradually gained immunity from such diseases. These traits have been passed to subsequent generations via genetic inheritance.

For human populations that had no experience of living in close proximity to farm animals, the consequences of pandemic outbreaks were much greater. Smallpox is estimated to have killed as many as *twenty-five million* Native Americans – 90 per cent of the population – following the arrival of Europeans and their animals in the sixteenth century.

Invading Europeans suffered less since they were farther along the path towards genetic immunity – at least until the discovery of a smallpox vaccine in the twentieth century, which removed the threat altogether (see page 27). In a very real sense, therefore, European humans were in the process of being 'domesticated' by certain farm-animals and their parasites into a breed that was immune to this particular infection. Indeed, immunity from smallpox was 'selecting' humans in favour of those that tended to animals such as cows over those who, because of their hunter-gatherer lifestyles, did not.

Gulping down copious quantities of milk sounds like an unlikely second example of how direct genetic change has taken place in humans that live in close proximity to farm animals. But before the domestication of goats, cows and sheep, humans were able to digest milk only in infancy. Lactase, the enzyme that digests lactose, a sugar found in milk, naturally disappears from the human gut soon after infants are weaned off their mother's breast.

However, milking a cow vastly increases the amount of energy that the animal can deliver to humans over its lifetime – as compared with killing it for the one-off supply of meat. Finding a way of tolerating milk is therefore a significant survival advantage to humans. Civilizations such as the Romans devised a method for removing indigestible lactose by fermenting milk into cheese (see *Lactobacillus*, page 386). But this was less practical for many nomadic people who needed to drink from their draught animals while on the move.

These people, therefore, drank unprocessed milk, regardless of the digestive consequences. Over generations their offspring have developed the ability to continue to digest lactose into adulthood. Populations in Britain, Germany, Switzerland and Scandinavia are highly tolerant of drinking unprocessed milk since they are descended from Neolithic pastoralists that roamed the land from eastern Europe and the Middle East several thousand years ago. Yet among Native Americans 95 per cent of the population are intolerant of milk since their continent was cleared of nearly all large mammals by a combination of human hunting and climate change more than 12,000 years ago (see pages 184–5).

Once again, species of farm animals have, in effect, 'domesticated' humans, all part of the process of co-evolution that so often characterizes

Cheese-making in Renaissance Italy – the traditional way to avoid the problem of lactose intolerance.

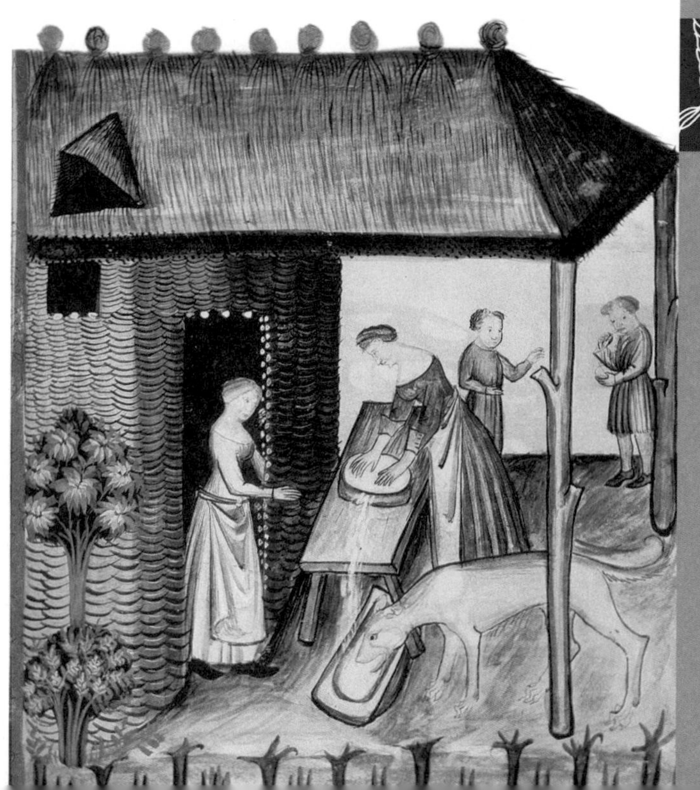

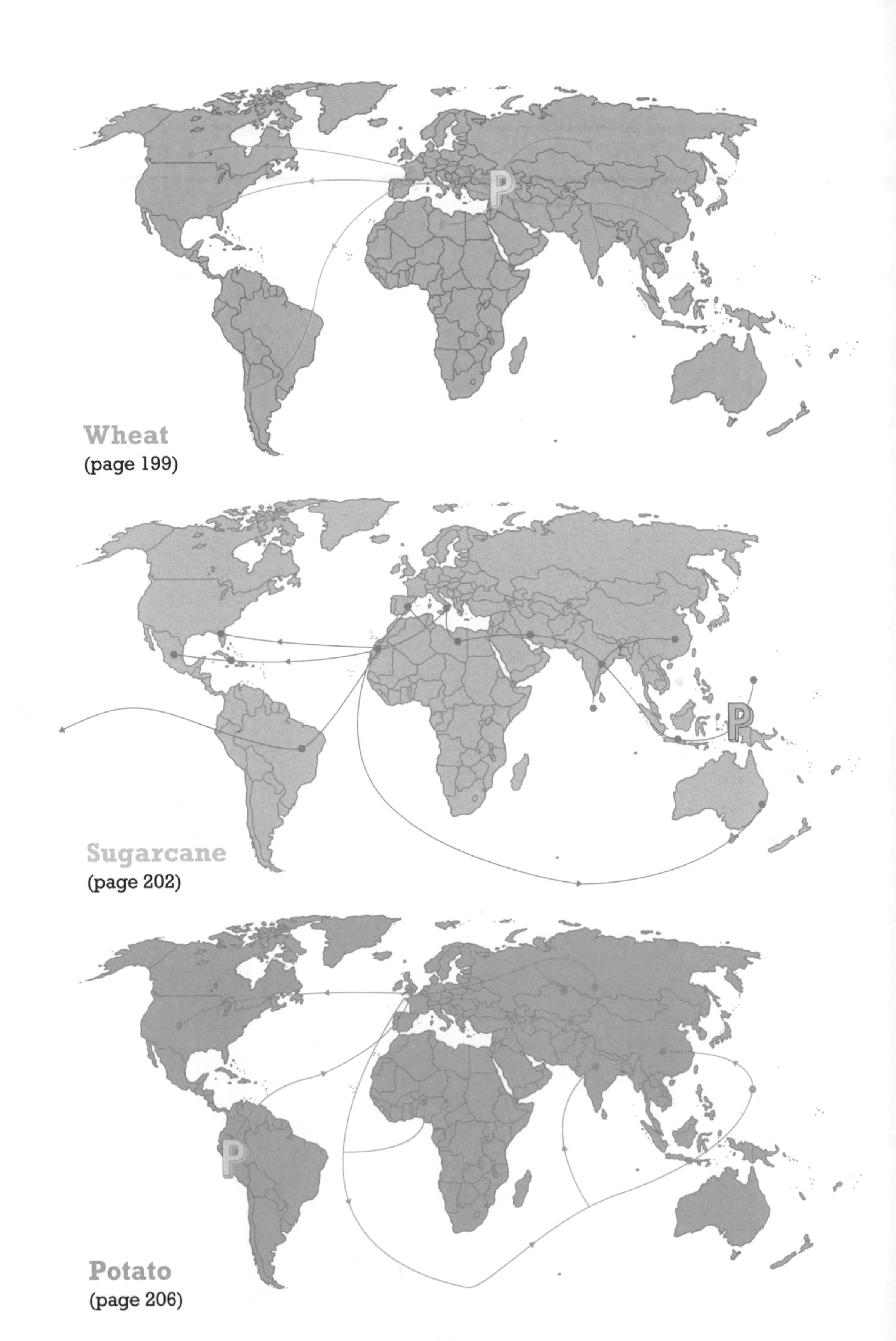
Wheat
(page 199)
Sugarcane
(page 202)
Potato
(page 206)

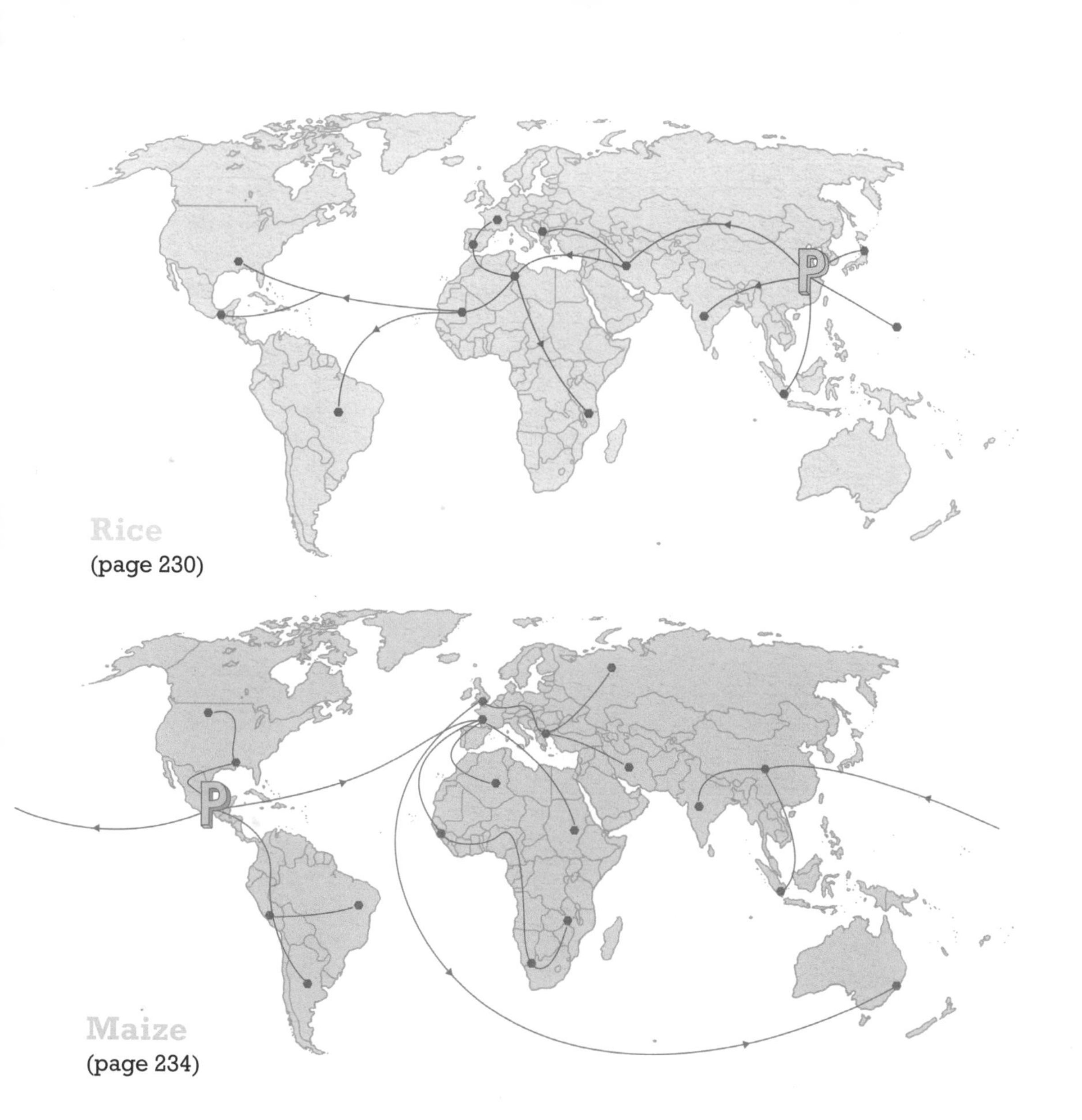

The spread of global agriculture

How certain crops have evolved with human civilizations, boosting population levels and causing the worldwide cultivation of a few super-species.

P Point of origin

● Major cultivation centre

mutually beneficial relationships between species (see symbiosis, pages 44–5).

Addictive behaviour

It is well known that some plants are able to influence human behaviour through their ability to modify the appetites of human consumers. Cacao (see page 274), tobacco (see page 289) and poppies (see page 308) are famous examples.

However, our basic agricultural diet has its own in-built system of addictions, too. That ancient Neolithic civilizations were besotted with bread is clear from their worship of a fertility figure – a Mother Goddess. Thousands of figurines and images representing her have been found next to bread ovens by archaeologists working around the Mediterranean coast (see pages 199–200).

Human tastes grew more exotic thousands of years later in wealthy societies such as the Roman Empire whose ruling class was addicted to grapes (see page 296), for fermenting into fine wines, and olives (see page 209), perfect for trading, cooking and eating. When early-modern European societies first tasted chocolate, sugar, coffee and tea – products that came from colonial expansion eastwards and westwards from the sixteenth century onwards – their palates quickly adapted, stimulating increased demand.

Throughout the nineteenth century the European appetite for beef from Argentina and lamb from the Antipodes created a worldwide addiction to eating red meat. European tastes even for American crops such as maize and potatoes, originally shunned in European courts as food fit only for peasants, have been transformed over the last 200 years. The humble spud is now the most popular vegetable in the Western world (see page 208).

Addiction to farmed foods from across the world has created such consumer demand that, from the beginning of the twentieth century, teams of Western scientists and business communities have collaborated globally to make food provision their top priority. While attempting to prevent hunger in parts of the world threatened by starvation (such as Mexico in the 1940s and India in the 1960s), rich nations became besotted with eating ever larger quantities of fruit, vegetables and meat, which have created global markets and hence big opportunities for large corporations to make succulent profits (see also bananas, page 364).

The biggest break came with the invention of artificial fertilizers, the brainchild of Germans Fritz Haber and Carl Bosch in the early twentieth century (see page 41). Nitrogen-based compounds, made by converting natural gas and air into ammonia, are now used on crops that feed more than 40 per cent of the world's population – that's about 2.5 billion people.

Genetically modified plant species have now been developed, such as dwarf wheat and rice, that can grow fat on artificial fertilizers without collapsing under their own weight, feeding more people and preventing famines from Central America to India in what has been dubbed a 'Green Revolution'.

An unforeseen consequence of these artificial interventions is that world populations have increased by five billion in the last eighty years alone (see page 371). Such a massive increase is matched by a similarly exponential rise in the populations of, and land devoted to, a select band of well-protected, fertilized, nurtured and watered species such as cows, pigs and sheep that are required to feed the enormous human race. In such a way is the relationship between humans and their domestic animals as symbiotic as the farming of fungi and aphids in an ants' nest (see page 155).

Artificial pesticides keep insects, fungi and viruses at bay; genetically modified strains help crops grow where nature on its own could never have situated them, and antibiotics protect animals from infection, even when their cramped living conditions would normally incubate population-controlling disease. The combination of humanity's big brains, burly hands and insatiable appetite with the genetic versatility of a few highly adaptive crops and animals accounts for a co-evolutionary *tour de force* that has truly transformed the landscape and ecology of the modern world.

Wheat

FAMILY: POACEAE
CULTIVAR: NORIN 10
RANK: 19

A highly nutritious genetically modified grass that keeps millions of people from starvation but which could not survive in the wild.

A UNITED STATES law, passed on 14 December 2006, declares unequivocally that Dr Norman Borlaug (born 1914), an American agronomist, has 'saved more lives than any other person who has ever lived'. More than one billion people alive today owe their existence to this man, it is claimed, yet only a tiny fraction will ever have heard of him. Recognition of his extraordinary contribution to the modern world is loud and clear: the Nobel Prize for Peace (1970), the Presidential Medal of Freedom (1977), and America's highest civilian honour, the Congressional Gold Medal, awarded in 2006. What miracles, you may rightly wonder, has this man performed to deserve such unprecedented praise?

In a word, it was wheat. Not just any old variety, but the pioneering of a new hybrid called Norin 10 which, since the early 1940s, has increased the yield of crops grown in places such as Mexico and India by more than ten times their historic average. Worldwide fears of mass starvation caused by the relentless rise of human populations have been staved off by the cultivation of this carefully crossed grass species that grows on short stalks but has very large seeds. The dwarf-like characteristics of Norin 10 mean it does not topple over despite the weight of its bloated grains. Fed with plentiful supplies of water, artificial fertilizer and chemical pesticides, Borlaug's ingenious wheat pulled back many regions of the world from the brink of disaster in the mid twentieth century. Countries such as India and China are now able to feed themselves thanks to his wheat, despite their massively increasing populations.

Dr Norman Borlaug pictured in a field of his own specially cultivated wheat in Mexico, 1970.

Wheat is a species of grass that has truly conquered the world. Its rise to predominance began about 12,000 years ago when humans in the eastern part of modern-day Turkey began to cultivate varieties of grasses, tending to select those with the largest, most nutritious seeds from their stores. The discovery of how to grind grains of wheat into flour and then bake bread is one of history's most important but least recorded stories (see yeast, page 293). Over the last few thousands years, bread has sustained and inspired humans all over the globe. Just think of the many different types that characterize various human cultures: baguettes, bagels, pitta bread, naan bread, croissants, focaccia, sandwiches, chapati and dumplings ... to name just a few.

For early farmers the transformation of wheat flour into a sumptuous leavened loaf was a miracle that lay at the heart of their religious worship. Excavations at Achilleion, a Neolithic settlement near Farsala in eastern Greece, have revealed several ancient temples (*c.* 6000 BC) dedicated to a pregnant goddess. Hundreds of figurines have been found next to bread ovens that were equipped with platforms on which these statues stood, turning them into shrines and altars. This was the deity that represented Mother Earth with all her glorious bounty, worshipped wherever grain was ground or bread prepared.

Thousands of years later, the association of bread with divinity was not lost on the scribes of the Jewish Torah (Christian Old Testament). They revealed how God sent manna from heaven, a type of bread that was ground up and baked into cakes, to sustain the Israelites on their epic journey out of Egypt to the Promised Land. Two thousand years ago a Jewish carpenter's son called Jesus became, in the eyes of many, the personification of what it took to sustain humanity. He was, he said, the bread of life. To this day millions of Christians celebrate the life of Jesus by eating bread that symbolizes his holy body.

Bread-making has not only permeated religious cultures. Its impact was just as significant in the world of science and engineering. Cogs, sails, mills and wheels were devised by humans from ancient Greece (33 BC) to China (*c.* AD 200), originally for the purpose of pounding and grinding grain into flour. It was only later that the concept of harnessing wind and water power was applied to sawing wood, pumping water, making paper or driving looms. Today's ubiquitous renewable source of electricity, the wind turbine, is therefore adapted from a technology that has its origins in the processing of domesticated wheat.

British history received a huge jolt thanks to the highly combustible property of powdered flour. The premises of one Thomas Farriner, a baker in Pudding Lane, London, exploded on the night of Sunday 2 September 1666, leading to the most devastating fire in the city's history, destroying the homes of at least 80 per cent of its 80,000 inhabitants. Many of the houses that were decimated in that fire were roofed with thatched straw, a similarly combustible material made from wheat stalks.

The pivotal role played by wheat and bread in the history of human civilizations lives on in the modern English language with phrases like 'breadwinner', 'bread basket' and the use of the word 'dough' as slang for riches. Scientists keep trying to come up with discoveries that they hope will become 'the greatest thing since sliced bread ...'

The amount of land now allocated to wheat cultivation is a staggering 600 million acres – terrain that was once mostly forest or savannah. According to at least one eminent historian, wheat has *'diversified more dramatically, invaded more new habitats, multiplied faster and evolved more rapidly without extinction than any other known organism ...'*

Two major factors help explain the enormous impact of wheat. Gluten is a protein that occurs in far greater quantities in wheat than in any other grain or cereal crop. This is what makes it possible to knead ground wheat seeds (flour) with water into an elastic, sticky ball of dough. With the addition of a little yeast (see page 293), the dough ferments and the carbon dioxide gas that is given off gets trapped by the elastic gluten, expanding the mixture into a capacious, neatly formed, aromatic loaf.

Wheat's second remarkable feature is the ease with which it naturally combines with other grasses to make new species. Modern wheat is actually a combination of at least three different grass species whose genes have quite naturally spliced themselves together to make wheat as we know it today. These are einkorn, emmer and *Aegilops squarrosa*. Such genetic agility helped establish mutant varieties with large, firm seeds that were subsequently selected by the first human farmers as promising sources of food.

Wheat is a perfect example of a new species that has evolved through *acquiring* the genomes of others (for other examples, see page 64). All living things have in their genes evidence of the combination of different genetic forms, be it through infection, mutation or symbiotic association. But plants seem to have a greater propensity for such antics in nature than other life-forms. Modern wheat, with its six chromosomes from lineages of three separate species, has no single common ancestor, much to the chagrin of some of Charles Darwin's modern scientific disciples.

If variety was the key to wheat's rise to global predominance then the modern uniformity of wheat today makes its heyday look increasingly vulnerable. Unlike ancestral varieties, today's world-conquering crops are entirely dependent on humans for their propagation and survival. The seeds of Norin 10 are too large and too firmly attached to the stalk for wind power to act as a dispersal mechanism. What's more, wild wheat seeds gradually sow themselves mechanically by using their 'awns' (bristles) like oars that become erect and then draw together, drilling the grain into the earth. Modern varieties rely completely on humans, usually using seed drills, to manage the process for them. Therefore, ingenious evolutionary techniques for burying fallen seeds two centimetres into the soil have now been lost.

Cloned wheat cultivars, such as Norin 10, which are typically grown as a single variety, are highly susceptible to infection by parasites, fungi or viruses since they lack nature's most effective defence against disease – biodiversity. As a result, humans have to spray wheat crops with chemical pesticides to combat the threat of harvest failure, mostly regardless of its effects on other parts of the terrestrial ecosystem.

Theoretically a virus or parasite could evolve a way of infecting wheat in a manner that humans could not control (see *The Death of Grass*, page 358). Its reward would be a truly global feast. Without wheat billions of humans would inevitably starve. If catastrophe were to strike the human race, then the survival of Norin 10, now one of the most populous types of plant on the planet, would likewise wane through its own inability to reproduce.

Modern man's dilemma is how to carry on feeding his ever-growing population without fundamentally endangering the balance of the world's ecosystems through either chemical poisons, genetic wizardry or shrinking biodiversity. No one is as wise to this conundrum as Norman Borlaug who, in an echo of Thomas Malthus just over 200 years ago, has predicted that the world's current food supply must double by 2050 to avoid human catastrophe:

> *'Future food production will have to come from higher yields. And although I have no doubt yields will keep going up, whether they can keep on going up to feed the population monster is another matter. Unless progress with agricultural yields remains very strong, the next century will experience sheer human misery that, on a numerical scale, will exceed the worst of everything that has come before.'*

Sugarcane

FAMILY: POACEAE
SPECIES: SACCHARUM OFFICINARUM
RANK: 29

An addictive sweetener that has risen to become the most widely cultivated plant in the world with many modern consequences.

PLANET EARTH is overrun by a species of vengeful plants that go to war with the human race. That's basically the plot of one of the twentieth century's most celebrated science fiction dramas, the apocalyptic *The Day of the Triffids.* Yet John Wyndham's concept is not so very far from the truth. Sugarcane is a type of grass that has, over the last few thousand years, been instrumental in enslaving millions of humans and intoxicating billions more with its sweet, irresistible flavour.

The idea that a plant could have a rational intention or will to inflict damage on another species, such as man, belongs exclusively to science fiction. Rather, it is the increasingly interwoven fate of modern humans with these spectacular blades of grass that has had such a dramatic impact on the planet, life and people, especially over the last 500 years.

Sugarcane, like wheat, has a complex natural history. It is thought to have been the product of the genetic fusion of several species of wild grasses, making its descent especially hard to pin down. The plant grows quickly, up to six metres tall, and in sections, rather like bamboo (see page 150). First contact was made by people living on the island of New Guinea, in Polynesia, about 10,000 years ago. They discovered that they could cut down this plant's stems and, by chewing its ends, enjoy a fix of sugary, syrupy juice – the plant's natural store of energy intended for use as power for its next year of growth.

By selecting those plants with the sweetest stems, these people cultivated canes for thousands of years until eventually a fifth of a sugarcane's stem contained pure sucrose. After grinding out the juice, and concentrating it until it became supersaturated, a substance at least as magical as bread crystallized out of the thick syrup. We call it sugar.

So important was the growing of sugarcane to Polynesian people that they believed it was responsible for the very birth of humanity. A mysterious, beautiful young woman who stepped out from behind a dense cluster of sugarcane leaves coupled with a fisherman called To-Kabwana. Their children became the founding members of the human race ...

Polynesians, with their outrigger canoes, were seafarers. They took their sacred sugary treat with them to Indonesia and the Philippines before reaching the northern tip of India by 8000 BC, where sugar eventually became a fundamental part of the religious history of southern Asia. One day, shortly after his enlightenment *c.* 2,500 years ago, the Buddha was resting under the shade of a tree by the roadside. Two passing merchants saw that he was a holy man and gave him a piece of peeled sugar to eat. This was the first food Buddha had received for seven weeks. So grateful was the Buddha that he anointed these men, Trapusa and Bhallika, and they became his first disciples.

Sugar spread with Buddhism as a holy food down through India to Sri Lanka. Westerners got their first taste in *c.* 325 BC when an officer in the army of Alexander the Great described eating a milky rice pudding sweetened with sugar. By AD 50 sugar had reached China, which was the first civilization to experiment with turning sugar into 'rock candy', before being taken overseas by Buddhist missionaries to imperial Japan.

Sugar's biggest break came in the eighth century, however, with the rise of Islam. Of all the many products that swirled from East to West with the whirl of Mohammed's people, sugar and gunpowder were surely the most significant. Islamic sugar mills, irrigation systems and expertise in its cultivation ensured that excellent conditions for growing and processing sugar were pioneered in areas around the Mediterranean such as Egypt, Sicily and the Near East. These people learned how to cultivate the crop by cutting back the canes to a stump, then letting them re-grow. This way a single plant could be harvested up to ten times before its productivity waned. Mills originally designed for grinding grain were adapted to extract the juice from sugarcane.

The crop's global conquest then progressed via Christian Europe during the twelfth century. Cyprus, Crete, Italy and Spain were all pressed into sugar production by Crusading states eager to avoid trading with their infidel Muslim rivals. Christian cultivation reached its climax on the island of Madeira thanks to

Back-breaking: a nineteenth-century drawing shows black workers harvesting sugar in the Antilles in the Caribbean.

Henry the Navigator, a Portuguese crown prince, sugar's most notorious patron. He was the first to combine the mass production of sugar with the enslavement of black Africans, dispatched to toil on plantations owned by European whites. Initially his plan was to use Muslims, captured from Africa, to work in Madeira's sugar fields. However, they persuaded him to grant them freedom in return for sourcing native Africans, who were, they said, the children of Noah's son Ham, whom the Bible had condemned to be 'servants of servants'.

When Henry agreed terms in 1444, it was the start of a most inauspicious pact. Mercenary Muslims and commercial Christians worked together to bind Sub-Saharan Africans into slavery to grow sugar for export. Later that year the first 235 'Negroes' were shipped out of Lagos to work on the cultivation of sugar in Madeira. Over the next 400 years more than twenty-five million Africans were transported westwards in slave ships to work in European colonies of the Caribbean, Brazil and North America, mostly to feed Europe's growing addiction to sugar.

Sugar is so ingrained in the modern human diet that to suggest it is unnatural for people to eat it sounds bizarre. But sugar as extracted from sugarcane in its refined form of *sucrose* was not a significant part of the traditional human diet. Thanks also to the discovery of sugar beet, another plant that can yield sugar in less sunny climates (see page 401), sugar intake in countries such as England has risen from an estimated two kilograms per person per year in 1700 to more than thirty-five kilograms today – a seventeen-fold increase. The same picture is mirrored throughout developed nations.

This recent human addiction to sugar is a modern menace that is having an impact not so very different from the triffids of Wyndham's tale. His killer plants caused people to go blind. But sugarcane and sugar beet have a more insidious method. They work in cahoots with addictive substances in drinks such as tea (see page 282) or snacks like chocolate (see page 274). The combined effects of sweetness and the kick of caffeine can easily become too addictive for even the most disciplined human appetites to resist (see On Drugs, page 268). Sugar initially reached wealthy European people from the seventeenth century onwards through their fancy for sweetened cakes, puddings and pastries. Meanwhile, it was the habit of adding sugar to tea that ensnared the working classes during the nineteenth century, as it still does to the present day.

Why has sugar been so successful? Part of the reason lies in its extraordinarily powerful system of photosynthesis. Unlike most plants, sugarcane uses a more modern method for converting sunlight into energy, called C_4 photosynthesis. This technique is thought to have evolved independently in several plant lineages about thirty million years ago, but became increasingly common as climates altered six million years ago. In hot tropical conditions, sugarcane is able to convert as much as 8 per cent of the sunlight it receives into sugars. By comparison, some plants achieve as little as 0.1 per cent. Without this level of efficiency, the first Polynesian farmers could never have cultivated such rewardingly concentrated strains.

Sugar, in its most common form *sucrose*, also has a profound effect on the human body. The full consequences of the high doses now imbibed by humans are still the subject of much research, dispute and debate. A simple but significant impact has been the birth of tooth erosion. Before humans began eating refined sugar there is little, if any, record of dental decay as we know it today. Yet after sugar became widely available in the nineteenth century, poor-quality teeth became a huge problem, to the extent that some men were refused entry to the allied armies in World War I – they didn't have enough teeth even to chew their rations. Sugar is so easily digested that bacteria in the mouth (*Streptococcus mutans*) can metabolize it in seconds, creating acid as a waste product that quickly rots teeth. Sugars found in vegetables and fruits are not so easily digested, making the mouth a less favourable environment for such bacteria to flourish. Unlike modern humans, wild animals suffer little tooth decay.

The sudden introduction of high quantities of sugar into modern diets has had other, more pernicious, effects on the human body. Recent research indicates that high levels of sugar intake in children may be related to a decreased sensitivity to dopamine, a drug released by the brain that is connected to an animal's ability to learn. Sugar is also believed to be addictive; stop eating it and withdrawal symptoms soon set in, leading experts to conclude that an inability to control excessive sugar intake is *'in the realm of an addictive disorder, rather than a failure of willpower'*. Dr Candace Pert, a professor at the Department of Psychology at Georgetown University, Washington, is in no doubt about the highly addictive qualities of sugar. *'Relying on sugar to give us a quick pick-me-up is analogous to, if not as dangerous as, shooting heroin,'* she says (see poppy, page 308). Of course, when sugar is fermented by yeast into alcohol its addictive qualities are far better understood (see grape, page 296).

The power of sugar is not limited to its effect on the human brain. Modern epidemics ranging from diabetes to obesity are on the list of ailments increasingly being connected to our modern fixation with high doses of sugar. So addicted has the world become to this ultimate species of success that sugarcane is now the most widely grown crop in the world – even more so than wheat – with worldwide production estimated in 2007 at *1.5 billion tonnes*.

Sugar certainly has parallels with the triffids in Wyndham's tale: its sweet success over the last few hundred years has also resulted in the subjugation of many billions of people, be it through slavery, addiction or disease.

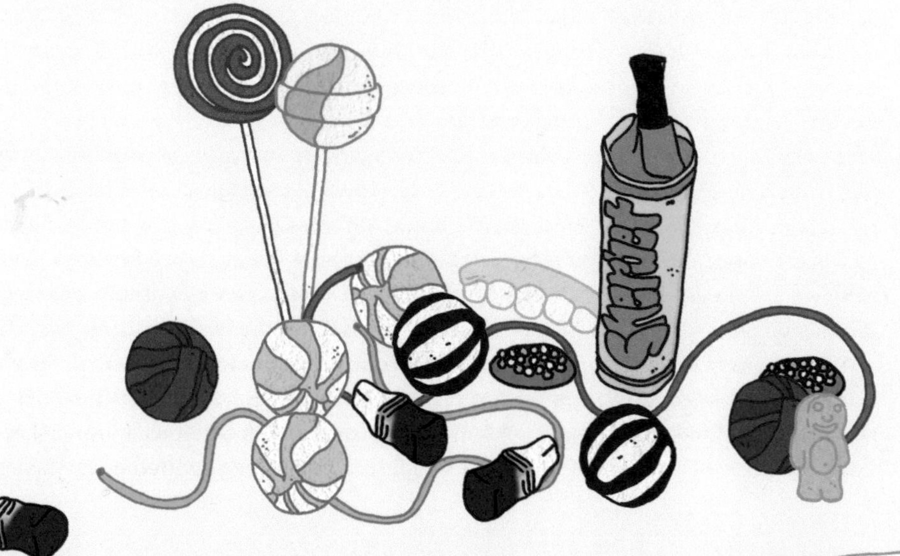

Potato

FAMILY: SOLANACEAE
SPECIES: SOLANUM TUBEROSUM
RANK: 43

A naturally toxic tuber that has successfully adapted into one of mankind's most nutritious and popular modern foods.

VINCENT VAN GOGH painted a scene of peasants eating potatoes by the light of an oil lamp because he wanted to depict the life of ordinary, poor people who had 'earned their food by honest means'. His picture, painted in 1885, reflects a social stigma attached to potatoes that delayed their rise to global dominance for more than 200 years.

Few people today think of potatoes as being fit only for animals or the very poor, since they are now one of the world's most popular vegetables. From fish and chips and bangers and mash to *pommes de terres à la dauphinoise*, the versatility of these tubers is unique. Even India and China, traditionally rice-growing nations, have finally embraced the potato, the vegetable of choice in the West, because of its high nutritional value and its capacity to grow just about anywhere. In 2007, for the first time ever, China became the world's largest potato producer.

Potatoes are the fifth most widely grown food crop today, after sugar, wheat, rice and maize. Yet not so long ago they were objects of derision. Their cultural legacy is buried in our language. Terms of abuse still used in several European languages include: 'potato head' for stupidity, 'couch potato' for laziness, and 'sack of potatoes' for clumsiness. So the potato's rise to modern glory is all the more remarkable in the history of human/vegetable relationships, given the colossal cultural barriers put in its way (see also cinchona, page 316).

Being naturally poisonous was not a great start. Quite why or how South American natives began to selectively cultivate these members of the deadly nightshade family over 5,000 years ago is a mystery of agriculture that will probably remain forever hidden. These are plants whose leaf, stem and nutritious reproductive food store (in the form of underground tubers) were designed by natural selection to be highly resistant to even the most persistent of herbivorous attacks.

Potato-eating peasants, illustrated by Vincent van Gogh (1853–90).

But people living high up in the Andes in what is now Peru desperately needed a crop that could sustain them and would grow at altitudes of up to 4,000 metres. After years of labour (probably a female endeavour) they somehow focused the potato gene pool into growing large quantities of underground tubers that were not only safe to eat but so nutritionally rich that, when eaten in sufficient quantities, they could provide all the nutrients needed for human wellbeing and survival. The potato is still the only vegetable in the world that furnishes, in one staple, all of life's essentials. It's a list that includes carbohydrate, fat, protein, vitamins (B and C) and minerals such as calcium, iron, magnesium, potassium, sodium and zinc.

The sheer range and number of potato varieties found in South America is itself evidence of the gargantuan effort made by these early Andean farmers to find a food that would work as a high-altitude, one-stop shop for survival. Thousands of wild potato varieties persist in the mountains of Peru today, perhaps the relics of ancient, long-lost selective experiments conducted by humans in their efforts to transform a well-armed plant laced with poison into a compliant vegetable that tastes great whether boiled, mashed, roasted or fried.

By AD 1500 when Spanish conquistadors arrived, they found that at least half the natives ate nothing but potatoes. But just when transportation in the form of European adventurers offered this vegetable its biggest chance for global spread, a surprising hurdle, which had nothing to do with taste, digestive intolerance or nutritional value, was placed in its path. European invaders weren't interested in these nutritious, easy-to-grow tubers because (a) they were not solid and shiny like gold and silver; (b) food eaten by native savages could never be, in their eyes, fit for civilized European consumption; and (c) potatoes suffered from a crisis of bad branding: they grew underground, were predominantly eaten by animals and were suspected of having something to do with the occult since they were so closely related to *belladonna*, the deadly nightshade plant, notorious as an ingredient used in a medieval witch's brew.

The potato plant, made palatable only after painstaking artificial selection.

It is the first of several ironies associated with the history of the potato that Spanish conquerors in the 1530s had absolutely no idea that the real source of treasure they sought lay not with precious metals and gems but with *chicha* – the curious potato vegetable diet of the enslaved, uncouth, savage natives.

More than 200 years elapsed before the potato became widely accepted in any major European country. Until then it was grown mostly as a botanical curiosity. Various rumours abounded about the potato's curse – everything from it being a powerful aphrodisiac (therefore experimented with by Henry VIII of England, desperate for an heir) to a common source of leprosy. Most people agreed that root vegetables corrupted the blood, spread disease and were generally unfit for human consumption. Anyone who ate potatoes, therefore, was making a direct class statement about themselves.

Despite such prejudice, this vegetable struck back. Ireland was the country that became its adopted European home. The Catholic peasantry, ostracised by British Protestant society, found in the potato a source of food that was relatively easy to grow, tolerant of the Irish weather and which provided plentiful fodder for humans and animals alike. In the two centuries leading up to 1840, the Irish population soared from one million to eight million, a rise that reflects the successful establishment of the South American tuber as the country's new staple crop.

But, starting in 1847, a devastating famine and blight destroyed Ireland's potato crop, killing more than one million people through starvation, and triggering the mass emigration of millions more to North America and Australia (see also pages 51–2). How big is the dose of irony that the global transplantation of one of nature's most genetically diverse plants was undertaken with so few varieties that a single protoctist pest *Phytophthora infestans* could, within a couple of years, wipe out an entire human potato-growing generation's crop.

The Irish diaspora took the potato-growing and -eating habits of the Irish to the US, which has since become the West's biggest potato-consuming culture. The rise of this fast-food nation in the twentieth century, with its hunger for burgers and fries, is mainly explained by the nutritional value packed into potatoes that can grow just about anywhere from the plains of Idaho to the mountains of Peru.

Even in class-conscious Britain, the potato finally broke through the stigma associated with root vegetables. Would the rise of towns such as Manchester and Birmingham, which provoked such political upheavals in the nineteenth and twentieth centuries, have been possible were it not for the establishment of the potato as a foodstuff for feeding the stock of urban workers? Would the industrial revolution in Britain ever have succeeded without the contributions of Irish navvies, a labour force whose availability had arisen entirely thanks to the destruction of their homeland's vital potato crop? Could Europe's national armies that fought from Napoleonic times through to World War II possibly have been sustained without the contribution of this quick-growing crop from South America? Seen in this context, political milestones such as the rise of popular sovereignty, representative parliamentary democracy – even universal suffrage in the early twentieth century – can also be attributed to the increasing popularity of the potato from the 1850s onwards.

By the twentieth century the potato had smashed its way through the last vestiges of European prejudice to take its place at the world's top table. It was a symbiotic relationship – the potato depended on humans for global cultivation and people relied on potatoes to nourish their ever-increasing numbers.

After the United Nations declared 2008 as 'the year of the potato', the worldwide conquest of this versatile vegetable was officially complete. This global organisation now trumpets the potato as a cheap, plentiful and highly nutritious foodstuff that is essential for sustainance, as human populations climb ever upwards.

The potato's future success is predicated on it being so green. Not green in the sense that greenish potatoes can be a sign of unwelcome toxicity, but in the fact that potatoes are ideal for economies geared towards sustainability. When food prices rocketed in 2008 as global demand outstripped available supply and speculators bought up tradable commodities in a rush to secure their positions, the price of humble potatoes stayed stable. The reason is that, once harvested, they do not travel (and therefore trade) particularly well because they quickly perish. Therefore these are vegetables well-suited to a new sustainable form of agriculture in which production and consumption are reunited in an effort to reduce the effects of food miles and crop transportation on increasing global carbon emissions.

Olive

FAMILY: OLEACEAE
SPECIES: OLEA EUROPAEA
RANK: 80

A multi-purpose, self-preserving fruit that launched the ancient Mediterranean world on a path towards modernity.

THERE IS a saying, attributed to the most holy book of Jewish religious law, the Talmud, that it is easier to raise a legion of olive trees in Galilee than to bring up one child in the land of Israel. What this proverb reveals about olives is extremely pertinent to the story of this Mediterranean fruit that helped shape the modern Western world.

Olives grow naturally in Asia Minor. They were among the first fruits to be cultivated by Neolithic people following the last ice age because they tolerate craggy soils and don't require manual cultivation. Other crops, such as barley, were also an important source of food for those living on the north-eastern shores of the Mediterranean. But by *c.* 2000 BC many areas were already showing signs of soil degradation as a result of intense irrigation and farming. It was the souring of the earth that led to the fall of the Sumerian civilization in the Middle East.

Olives, as one of the only easy-to-grow sources of nutrition, therefore began to gain the status of a divine gift. From Minoan Crete to Mycenaean Greece and Israelite Canaan to their Philistine and Phoenician neighbours, ancient cultures around the Mediterranean hallowed these fruits as a sacred link between the human and spiritual worlds. Saul, King of Israel, was anointed by God's prophet Samuel with a horn of olive oil. The Garden of Gethsemane, where Jesus took refuge to pray

Jesus and the twelve apostles in the Garden of Olives at Gethsemane, depicted in a sixth-century mosaic.

on the night of his betrayal, was an olive grove. Even the name Christ means, literally, 'anointed' one, and the Mount of Olives, in Israel, is still the world's holiest burial site for Jewish people today.

The association of olives with divinity reached its zenith in ancient Greece from about 700 BC. A dark age of wars lasting nearly 800 years had previously preoccupied this part of the world as horse-wielding pastoralists battled with the settled civilizations that had arisen around the river valleys of the Nile, Jordan, Tigris and Euphrates. Then, from *c.* 700, a more peaceful period of ancient history began. Its success had a great deal to do with olives, since growing them was as much an incentive for peace as owning horses was a temptation to wage war (see page 242).

A newly planted olive grove takes from eight to twenty years to mature and produce fruit. Once established, however, these trees produce crops year after year, sometimes lasting for more than a millennium. Olive groves need little labour other than cutting back in the winter and vigorous shaking when the fruit is ripe, to gather the olives into nets for processing. Craggy, acidic soil does not bother these trees, which was ideal for people wanting to settle away from fertile, alluvial river valleys.

Olive oil was a rich source of Mediterranean wealth, and growing groves sometimes provided an incentive for making peace.

Such a lifestyle, however, chiefly depended on living in a climate predominated by peace. Hostile attacks that destroyed mature groves were devastating to a way of life founded on trading olives because it took at least a generation to re-establish the crop. Preservation of olive groves therefore became a big incentive for resolving disputes via discussion in the ancient world. Trade, compromise and negotiation were the hallmarks of a new style of human society stimulated by olive-growing and still symbolized today by the olive branch as a sign of peace.

Olives weren't simply nutritious to eat. Their oil, extracted in giant stone presses, was as valuable to the ancient world's economy as crude oil is to the Middle East today. Not only could olive oil be used to light lamps and as fuel for heating, but it was also ideal for cooking, making soaps and cleansing the body.

The olive's other natural bounty comes from a symbiosis with a strain of bacterium called *Lactobacillus plantarum* (see page 387). As harvested olives are left to ripen in the sun, this bacterium's fermentation process produces lactic acid, a natural preservative, which allows the fruit to be stored for up to a year without going off. Nothing could be better as tradable produce for people who relied on importing other products (chiefly wheat from Egypt) for their welfare. Civilizations such as the Minoans, Phoenicians and later on the Greeks all grew wealthy on the trade of self-preserving olives.

Not having to work in the fields all year round gave people in olive-growing cultures the time and incentive to build fleets of ships to further their bartering of olives for grain. The seafaring Phoenicians spread olive-growing all around the Mediterranean, to North Africa, France and Spain. Meanwhile, by *c.* AD 650, the Lydians, from Asia Minor, pioneered a non-violent mechanism for commerce using minted silver coins. This system of exchange, based on tokens, was more easily established by people whose worlds had a surplus tradable product. Such commodities provided an additional form of economic security.

Trade in olives also led to the need for skilled craftspeople such as potters. Amphorae were vessels fired to store and transport olive oil. Artisan, merchant, trader, banker and shipbuilder were therefore all professions that arose in Greece to support the olive trade. The contrast with China, where rice-growing shaped society into a more rigid tripartite structure – farmers, soldiers and bureaucrats – could hardly have been more striking (see page 230).

The other precious by-product of an economy based on growing and trading olives was time. Since people were no longer shackled to farming all year round, they could engage in politics (democracy, even) or pontificate about the laws of nature and the mechanics of the universe. Thales, Aristarchus, Pythagoras, Plato and Aristotle were just a few of the brilliant thinkers who were fortunate enough to live in a society where food production could be managed by the few on behalf of the many. An olive-based economy therefore underpinned the freedom to investigate and articulate new natural, scientific and rational perspectives on man's relationship with nature. Charles Darwin's father, Robert, made an equivalent contribution by providing his son with sufficient private income to liberate his mind from the drudgery of earning a wage. Such investments paid off well.

With the rise of the Roman Empire, olives were no longer quite as dominant as they once had been between the time of Minoan Crete (*c.* 2,200–1600 BC) and the demise of Greek independence (*c.* 160 BC). By 30 BC the bread basket of Egypt was firmly in Roman hands and Europe's dependency on crops grown on one side of the Mediterranean Sea to be exchanged with grain grown on the other was replaced by an enforced one-way flow of treasure in the form of taxes and slaves to Italy.

But the olive's religious and cultural associations lived on – particularly as a sign of peace and divinity. As European maritime explorers spread across the world from the sixteenth century onwards, they took their biological and cultural survival kits with them. The Jesuits brought olive seedlings to California, ensuring they were equipped with the necessary ingredients for holy sacraments such as baptism, confirmation, ordination and marriage (see also grapes, page 298). From there olives spread to South America. Other European expeditions took olives to South Africa, Australia and New Zealand.

Olives are still one of the most widely grown crops today – Italy is the world's largest producer, weighing in with an average annual harvest of about four million tonnes (out of an estimated worldwide harvest of seventeen million tonnes). The olive's easy-to-cultivate, unfussy habit and natural resistance to disease makes it one of the least ecologically damaging crops of all, which is why, perhaps, olive oil is valued as one of the natural world's most healthy, nutritious products. Such is the hallowed status of the fruit today, cultivated since ancient times, that without it the philosophy, culture and lifestyle of the modern Western world might never have taken seed.

Cod

FAMILY: GADIDAE
SPECIES: GADUS MORHUA
RANK: 57

A robust saltwater fish that fuelled European expansion but fell foul of over-hunting and the rise of modern technology.

'Unless the order of nature is overthrown, for centuries to come our fisheries will continue to be fertile.' Canadian Ministry of Agriculture, 1885

'Today it is estimated that offshore cod stocks are at 1 per cent of what they were in 1977.' Canadian Broadcasting News, 2 July 2007

IT IS EXTRAORDINARY to think that a creature as common as cod once was is now threatened with extinction. For more than 1,000 years, humans living in the northern hemisphere have relied on this species of highly desirable fish to sustain their economic and nutritional welfare. Yet today stocks are so low that the Atlantic cod has joined the inglorious ranks of the world's most vulnerable species.

Viking fishermen found copious stocks of the cod species *Gadus morhua* when they started to explore the Atlantic seas from the ninth century onwards. Between AD 985 and 1011, five Viking expeditions discovered Iceland, Greenland and the east coast of America (Newfoundland). By catching and drying these large, omnivorous, fleshy fish, the intrepid Norse explorers had a nutritious chewy meal that they could depend on, which didn't easily go off.

Later, Spanish fishermen from the Basque region discovered that by bathing cod in salt they could increase its shelf life still further. Unlike most other fish, cod is unusually low in fat (typically 0.3 per cent), making it easy to salt. Its high protein content (80 per cent when dried) meant that, even in the pre-refrigeration age, this food could be successfully transported thousands of miles back to Iberia and sold on to markets in the Mediterranean for a good price. What's more, after being reconstituted in water, this dried, salted, well-preserved fish tasted just as good, if not better, than when freshly caught.

Fishermen in Europe had a ready-made market for their easy-to-trade cod largely thanks to the medieval Catholic Church. Only 'cold foods' such as fish were permitted to be eaten on traditional days of fasting – every Friday (to commemorate Christ's sacrifice on the cross) and throughout the forty days of Lent.

Cod congregate in areas where warm and cold waters meet – attracted by plentiful food supplies stirred up by ocean currents. From *c.* 1100 to 1500, German tradespeople from the Hanseatic League controlled a cartel to monopolize supplies harvested from the North Atlantic seas off the coasts of Norway and Russia. Set against them were the Basque fishermen who made a beeline for stocks off the coast of Newfoundland where the Gulf Stream meets the Labrador Current. There, on a series of shallow banks stretching down the eastern seaboard, supplies of cod were so abundant that by the late fifteenth century it was said they could be scooped up in buckets dropped over the sides of boats.

By the sixteenth century, British sailors headquartered in Bristol had discovered the location of the Basques' treasure trove. French fishermen soon joined them, and the ports of St Malo and La Rochelle sprang to life on the back of the West Atlantic cod fishing trade.

When in the spring of 1621 a band of early-seventeenth-century Puritans from Britain began to colonize the north-east coast of America, they landed at a place they came to call Plymouth, to remind them of home. According to a chronicle written by their leader William Bradford (1590–1657), one reason they chose that place to settle was *'chiefly for the hope of present profit to be made by the fish that was found in that country'*. Despite initial difficulties at settling in, they were not left disappointed. Within twenty years, the Massachusetts Bay Colony was bringing some 300,000 cod a year to the world markets.

It was thought until relatively recently that nothing other that a complete collapse in the 'natural order' could possibly dent supplies (see quotation opposite). Off-cuts could be sold to labourers, vitamin-rich liver oil to health-conscious middle-class European families, and premium fleshy white fish to markets throughout America, Europe and the Mediterranean. As a result, the American city of Boston grew rich off the profits of the cod trade, symbolizing its wealth with a gilded cod that hung down from the ceiling of the city's prestigious town hall.

Only the most exceptional wild species have been able to thrive despite the combined onslaught of increasing human populations and industrialized food processing that emerged during the nineteenth century. Cod stocks fared well enough until the arrival of steam fishing ships at the beginning of the twentieth century, despite relentless harvesting by man. When these vessels were later upgraded to more powerful diesel trawlers, from the 1930s, it became possible to drag fishing nets along the bottom of the ocean floor, scooping up all forms of sea life. Valuable cod were harvested in their millions, while the marine ecosystem was devastated, most often beyond repair. Meanwhile, freeze-drying, a technique for processing frozen food pioneered by American inventor Clarence Birdseye

A fine figure of a fish, drawn for an encyclopaedia in 1859 when the seas were still abundant in cod.

Aye! Aye! How Captain Birdseye sold frozen fish-fingers made of cod to consumers in the West.

(1886–1956), further expanded the markets for cod in the form of fish-fingers, now made available to inland areas far away from the coast.

Warning signs that cod stocks were on a slippery path to decline came from a series of three so-called 'cod wars' between Iceland and the UK from 1958 to 1976. Desperate to preserve the wealth of its only viable material export – fish – Iceland extended the six-kilometre fishing zone around its coastline to a 322-kilometre zone in 1976. At the same time, enormous bottom-trawling factory ships equipped with onboard sonar systems for detecting shoals of fish with pin-point precision were added to fishing fleets all over the world. Modern factory ships, capable of staying at sea for months at a time, sucked up and processed so much cod (and all manner of other, less economically valuable fish) that by the 1990s global fish stocks faced complete collapse.

In July 1992 the Canadian government declared a total fishing moratorium on the Grand Banks off Newfoundland owing to the decimation of Atlantic cod, stocks having dwindled to less than 5 per cent of their historic norm. European and Russian fleets have since focused their attention on taking a negotiated quota of remaining stocks in the Barents Sea. Confirmation of the global plight of Atlantic cod came in the year 2000 when *Gadus morhua* was officially placed on the list of endangered species by the World Wildlife Fund (WWF).

Farming cod in dedicated lakes is one possible substitute for harvesting them from the wild. However, viruses are notoriously efficient at infecting fish in farms owing to a lack of genetic diversity. Meanwhile, releasing farmed stocks back into the wild to boost native populations risks further devastation. Fish bred in an unnatural habitat typically lack the survival instincts necessary to prosper in the wild. If these were to mate with wild species, what remains of the cod gene pool could become even more vulnerable to extinction by predators.

The Atlantic cod is a fish that successfully sustained humans from the northern hemisphere for more than 1,000 years, provoking economic development, the exploration of the Atlantic seas and even the European settlement of North America. Today, however, it is on the verge of complete collapse and there is no guarantee that, despite recent fishing restrictions, it will ever re-establish its ecological niche.

Pig

FAMILY: SUIDAE
SPECIES: SUS SCROFA
RANK: 48

An unfussy mammal most usually kept in unnatural conditions but which has proved more than capable of re-adapting to life in the wild.

WHAT PLEASES PEOPLE about a pig? They are not bred for their colour, friendliness or as a source of labour, transport or milk. Nor is there any selective advantage for a pig with brains – even though these are among the most intelligent mammal species and quick to learn. Pigs are popular with most people for one reason only – they taste so good. The Romans summed it up in a saying: *'The race of pigs is expressly given by nature to set forth a banquet ...'*

Today, pork is the most widely eaten meat in the world. More than a hundred million tonnes of pig meat is farmed, mostly in intensive warehouses, despite taboos in at least two major religions, Judaism and Islam, against its consumption. A wide variety of traditional dishes are associated with eating pork – from bangers, burgers and bacon to ham, gammon and pork pies. But by far the most prolific consumers are in Asia. More than half of all pork is reared and consumed in China.

Compared with our treatment of other domesticated species, the relationship between people and pigs is a lot more sour than sweet. Pigs were among the first

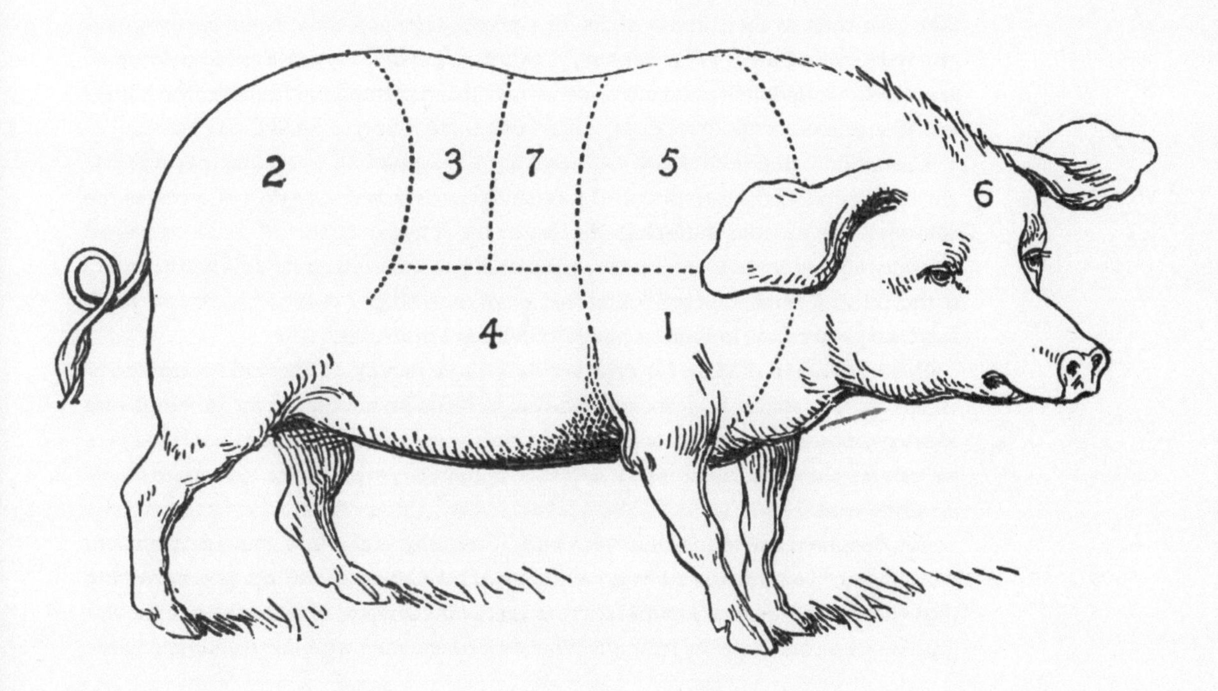

An early-twentieth-century diagram showing how to cut up a pig without wasting a sausage.

creatures that *Homo sapiens* discovered could be kept in captivity and bred for human consumption. Genetic studies reveal two main areas of initial domestication – in Asia and the Near East – beginning about 9,000 yeas ago. Swineherds seem to have roamed across Europe and taken the idea westwards. A third area of domestication in western Europe soon followed.

• Traditionally the life of a pig wasn't so bad. These are natural foraging creatures with an undiscriminating diet. Like humans they are omnivores, and will eat almost anything from dead insects and worms to tree bark, roots, fruits and flowers. Their most powerful tool is their snout, which not only houses a superlative sense of smell but is used to turn over the ground. Pigs are often employed as assistants in a treasure hunt for truffles, underground fungi that are considered a delicacy by humans in various parts of the world.

Humans that tended to pigs (swineherds) traditionally allowed their herds to forage in their natural woodland habitat, enabling these social animals to look after their young and move about in family groups. But rearing pigs this way requires a great deal of space because of the potential environmental devastation resulting from their omnivorous diet. Pigs also lack sweat glands so in hot climates they wallow in mud as a way of keeping cool. This only adds to the damage they cause if kept too closely confined, which is why a typical pig sty looks so bare and dirty. With widespread deforestation since Roman times, as a result of growing civilizations, demand for pork combined with a lack of space has led to a radical change of approach in the way pigs are kept.

Today pigs are mostly reared indoors on intensive farms. Hundreds of animals are kept in individual stalls where they can be fed on a carefully controlled diet of grain laced with antibiotics, vitamins, hormones and dietary supplements to ensure they mature as fast as possible and fatten up sufficiently to turn a good profit. Females are typically kept in crates, too small for them even to turn around, where they give birth to their litters of five or six sucklings at a time. Their offspring are removed immediately after weaning. Their living space is kept deliberately small as a conflict-avoidance measure since in this most unnatural environment, hostilities between animals often break out, ranging from tail biting to genital mutilation.

Castration, tail docking and ear notching are human interventions designed to alleviate some of the symptoms of intensive farming, but they do not address the causes of the obvious suffering of pigs, or the 'misbehaviour' of these creatures. Despite regulations to try to improve pig welfare, there is currently little enforcement of the rules. A recent survey found that more than 80 per cent of pig farms across five European countries were engaging in illegal practices.

Historic human disdain for pigs is echoed in a legacy of linguistic adaptations. Words such as swine, hog, sucker and runt indicate how little is ever said in favour of the pig. Some cultures (possibly originating in ancient Egypt) have declared pigs so 'impure' (Leviticus, chapter 11, verses 7–8) that they are unfit for humans to touch, let alone to eat.

Yet, despite these taboos and such verbal and physical abuse, humans have over thousands of years invested huge amounts of time, money and effort into rearing pigs in ever-increasing numbers. From the point of view of the pig's genes, this represents a considerable triumph. What's more, humans were directly responsible

Little room for manoeuvre: hogs huddled on an American pig farm.

for substantially increasing the range of these creatures far beyond their natural habitats across North Africa, Europe and Asia. Spanish conquistador Hernando de Soto (1497–1542) regarded these animals as being so important for survival that when he ventured to Florida in 1539 with 620 men and 220 horses, he also took with him 200 pigs. During his three-year progress through present-day Georgia, South Carolina, Tennessee, Alabama, Mississippi, Louisiana and Texas many of the pigs escaped, seeding populations of wild boar, now called Razorbacks. These creatures still roam the southern United States, where they are hunted by humans for sport.

In a similar vein, Yorkshireman Captain James Cook took domesticated pigs with him when he arrived in New Zealand in 1773 on his ship *Resolution*. These breeds also escaped and became wild, or feral, pigs. Meanwhile, another variety, the Kune kune, originally descended from Chinese pigs, reached New Zealand with the islands' first human settlers, the Polynesian Maoris. The impact of these wild pigs on that island's habitat has been profound, damaging crops and pastures, and killing lambs. Human hunting is now deemed necessary to keep populations in check.

Today, the supremacy of farmed domesticated pigs as a source of human nutrition is unsurpassed. In China consumption of pork increased by 25 per cent between 2002 and 2007 alone and shows no sign of slowing. As 211,000 humans are, on average, added to the world's population every day, the success of pigs, like that of most domesticated species, is inextricably tied to the fate of humans. Yet, unlike many other domesticated creatures, pigs have proved themselves sufficiently adaptable, intelligent and unfussy about where they live and what they eat to thrive successfully in the wild (see also camels, page 248). Pigs are even placed in the top one hundred most invasive species, according to the *Global Invasive Species Database* run by the University of Auckland, New Zealand. It's an impressive accolade, indicative of past, present and potential success.

Sheep

FAMILY: BOVIDAE
SPECIES: OVIS ARIES
RANK: 26

One of the world's most populous domestic species that has adapted to provide human civilizations with essential sources of fibre and food.

GABRIEL OAK, hero of one of Thomas Hardy's most famous novels, *Far from the Madding Crowd,* was a most unfortunate shepherd. When his sheepdog over-zealously herded a flock of about 200 animals through a weak fence and over a cliff, where their bodies piled up, lifeless, hundreds of feet below: *'All the savings of a frugal life had been dispersed at a blow; his hopes of being an independent farmer were laid low – possibly forever ... '*

From that moment on, Gabriel's destiny relied on making a good marital match, setting the scene for Hardy's heart-wrenching Valentine's Day tragedy.

The wellbeing of country people traditionally revolved entirely around the lives of their animals. They were completely dependent on their livestock for milk, wool, skins and meat. Hardy's novel, written less than 150 years ago, depicts a pastoral Britain still shaped by the symbiotic relationship of humans with just a few species of domesticated animals. Gabriel's sheep represent the earliest example of the co-evolutionary dynamic between humans and farm animals that first emerged about 12,000 years ago. As it developed, humanity's relationship with sheep became one of the most significant economic partnerships in the history of civilizations.

Sheep were the first creatures selectively bred to help improve the human way of life for the simple reason that they proved to be nature's most easy-to-manage animals. Their pliability in the face of unorthodox human expectations explains how a thinly dispersed wild animal that probably originated in Asia has now become a ubiquitous creature that numbers more than *one billion* individuals today. Sheep now thrive in vast numbers on every inhabited continent of the world.

The sheep's success lies in its gene pool. Over thousands of years, this creature's genes have been focused by pastoralists to produce a suite of products that have changed world history. Wild sheep do not produce much wool. Instead, their bodies are covered with hairs. The fleece of pre-domesticated sheep provided just a thin undercoat of thermal insulation.

In a world full of humans, the most easily controlled animals are those that naturally live in communities with social structures where the majority submits to the authority of a few dominant individuals. If the leadership role is taken over by humans, the species can easily be domesticated. Other characteristics for successful domestication include being non-territorial. The natural compliance of sheep has served them very well indeed.

Farmers from the southern Himalayas were among the first to capitalize on these attributes. After many generations of living in close proximity, sheep that prospered in the company of humans were those with the thickest fleeces. Humans preferred these animals because they could pull out clumps of wool with their bare hands and compress the fibres to make felt. They later learned how to shear off the fleeces and spin the fibres into yarn which, using ancient basket- and fence-making techniques, they wove into cloth. Wool rapidly became one of humanity's most widely traded and valuable commodities.

What oil is to the smooth functioning of the modern global economy, wool was to the ancient world. The Sumerians, widely regarded as the first people to build substantial urban settlements, controlled massive flocks of sheep, which provided the wool to clothe most of their people. Wearing warm garments made it possible from about 3000 BC for the Sumerians to live in higher, more mountainous areas, taking their flocks with them, their way of life fanning out across the Mediterranean and into Europe.

Sheep are uniquely easy and cheap animals to farm. They do not require feeding in winter (unlike chickens or cows), they can graze in pastures all year round without needing shelter and they are able to survive well in cold mountainous areas and in rough, uneven terrain. In return they provide an enormous range of useful commodities: ewe's milk can be drunk or made into cheese. Sheep meat is a major source of nutrition in many parts of the world in the form of either lamb or mutton. Lanolin is a naturally occurring grease extracted from the sheep's fleece that is used in everything from cosmetics (such as Oil of Olay) to industrial lubricants. Finally, wool is a material that has clothed humans from China to the Americas, at least until the rise of artificial fabrics after World War II.

The winged ram Chrysomallos soars through the air. Later its golden fleece was strung up in a tree and guarded by a dragon until Jason and his thieving Argonauts stole it.

Sheep have played a central part in human folklore, too. One of the most famous surviving Greek myths tells the story of a man called Jason and his team of sailors, called the Argonauts, who were sent by the gods on an epic adventure to retrieve a golden fleece from a winged ram. Archaeologists have found evidence that woollen fleeces were indeed used for panning for particles of gold in rivers, perhaps the source of the famous story. Alternatively, it may have been the trading of woollen fleeces in exchange for gold that inspired the legend, since wool was the perfect commodity for barter, exchange and trade. Wool is light to carry, slow to depreciate and has endless, practical value for clothing.

Sheep were central to the economy of medieval Europe. Monastic orders that originated in France and Italy spread to England following the Norman Conquest in AD 1066. Forests were felled by Cistercian monks to provide pasture for their grazing sheep. Thanks to a papal dispensation that freed these monks from paying taxes, a profitable arrangement was established in which wool grown in the UK was traded in Continental markets to be spun and woven in Flanders and northern Italy. From the eleventh century onwards wool became the most significant international currency for trade with Islam and, in an echo of Jason's adventure, it was swapped for gold from the south Sahara Desert in the markets of Champagne, now in northern France.

Families such as the Medici in Florence became fabulously rich, patronizing the arts and triggering what is known to history as the Italian Renaissance. Meanwhile gold from the profit of the wool trade filled the coffers of Cistercian monks who became so immensely rich that, by the sixteenth century, the English monarch Henry VIII used his marital dispute with the Pope as an excuse to avail himself of their fabulous establishments and wealth. The money he gained helped pay for England's first powerful naval fleet – the initial step towards a mighty British Empire which, at its height, encompassed 35 per cent of the world's landmass.

When Spanish monarchs Ferdinand and Isabella took southern Spain from the Arabic Moors of North Africa in 1492 they inherited highly prized flocks of wool-bearing sheep called Merinos, a breed introduced from the Near East by Arabic tribes in the twelfth century. So fine is the wool from Merino sheep that it has become the most successful breed in the world, dominating flocks in Australia and New Zealand.

Its sumptuous wool secured enormous wealth for Spanish monarchs who jealously guarded their flocks to prevent them being smuggled to rival nations – a crime punishable by death. Wealth earned from trading fine wool paid for overseas expeditions in search of the spice-rich markets of the East, but which resulted in the discovery of the New World. Spanish settlers took their sheep with them to the Americas. Consequently, Argentina has become the most populous sheep-grazing nation in the Americas with sheep populations that peaked in the mid twentieth century at around sixty million.

Meanwhile the global domination of domestic sheep, led by Merinos, continued to flourish. German principalities eventually got their hands on the precious breed after Ferdinand VI of Spain sent a flock to his cousin Prince Xavier, the elector of Saxony, in 1765. These were crossed with Saxon varieties to produce even finer fleeces. As a result, German wool became the next most sought-after variety in the world.

So powerful was the allure of Saxon sheep that one woman, Eliza Forlong (1784–1859), spent much of her life trekking on foot all the way from Scotland and through Germany selecting and buying the finest sheep she could find for export to the new British colonies of Australia and New Zealand. Thanks in part to her efforts, Antipodean sheep herds came to have the finest fleeces in the world, even surpassing German breeds. Within fifty years of the arrival of the first convict settlers from Britain, domesticated sheep had been established as the engine of the Australian and New Zealand economies. By 1830 there were over two million sheep in Australia. Today the continent hosts more than a hundred million, second only to China's 160 million. New Zealand has a more modest forty million but, given its small human population, that represents a staggering ten sheep per person. In the UK the ratio currently stands at about two people per sheep.

Sheep made their biggest modern economic impact on the British-led industrial world. Incentives to find ways of speeding up and automating the manufacture of woollens lay behind some of the industrial revolution's most powerful inventions. John Kay's technique for weaving cloth, the flying shuttle (pioneered in 1733)

inspired other inventors such as James Hargreaves to find a way of supplying looms with more thread. His spinning jenny (invented in 1764) could spin yarn ten times faster than traditional medieval-style spinning wheels. Later that century ingenious minds adapted water power – and later steam power – to automate the weaving of cloths. Mills were opened wherever natural conditions best suited their location – usually next to fast-flowing rivers – turning the English Pennines into a worldwide centre for cloth manufacture. With demand to feed mills with raw materials outstripping supply, the Highlands of Scotland were cleared of their traditional lairds and replaced by a mostly absent English aristocracy, which dedicated their empty land to housing permanent flocks of millions of sheep. The Highland Clearances of the late eighteenth century paved the way for mass immigration of Scots to the Americas, Australia and New Zealand, a fact not forgotten by those still clamouring for Scottish political independence today.

By the end of the nineteenth century Thomas Hardy's England was undergoing an irreversible transformation from a mostly pastoral society based on rearing sheep to a series of linked urban metropolises. Wool from India, Australia and New Zealand supplemented the never-ceasing appetite of Britain's steam-driven 'satanic' mills.

Long-lasting, waterproof, insulating woollen fibres are less important today than they once were thanks to the growing ascendancy of artificial fibres in the last sixty years. Nylon, acrylic and polyester have come to replace wool, cotton and linen in many parts of the modern material world. But sheep have versatile genes. Today they are reared more for their meat than for their wool, and eating lamb is on the rise. Sheep are the only domesticated species that has not succumbed to the economies of scale of intensive farming. These creatures are so simple to farm that families all over the world are still able to scrape a living through the sale of wool and meat, providing their flocks have access to sufficient supplies of rough and ready pasture.

Sheep have been fundamental to the rise and sustenance of human civilizations, past, present and future. That's why the dog that chased Gabriel Oak's sheep over the cliff edge was summarily executed the following day, at twelve noon, for his heinous over-exuberance. It was, said Hardy:

> *' ... another instance of the untoward fate which so often attends dogs and other philosophers who follow out a train of reasoning to its logical conclusion, and attempt perfectly consistent conduct in a world made up so largely of compromise.'*

Cow

FAMILY: BOVIDAE
SPECIES: BOS PRIMIGENIUS
RANK: 17

Versatile animals essential to the development of many civilizations but which now pose a serious threat to the global environment.

JACK'S MOTHER was beside herself with fury when her young son announced his latest and greatest deal. On the way to market the boy had been persuaded by a traveller to sell the family's precious dairy cow in exchange for a handful of beans. His mum was right to be cross. The cow was their main source of income and should have fetched a high price. How could she have known that the beans would later grow into a giant beanstalk that would help fearless Jack retrieve the treasure of a miserable giant that lived high up above the clouds?

This nineteenth-century English fairy-story reveals the intimate relationship that has evolved between humans and cows. It's an association that stretches back at least 8,000 years when wild aurochs, which originated in India about two million years ago, began to be domesticated by humans in the Near East. Domesticating a beast like the aurochs was no mean achievement. In the wild these animals sought safety from attack from wolves by gathering into vast herds, fleeing at the first sign of an approaching predator. Contrast this with the scene of a pastoral milkmaid, approaching confidently with her three-legged stool. She pats the cow on the back, places her stool on the ground beneath its giant udders and proceeds to squeeze them rhythmically until the compliant cow lets down her milk and the maid's pail is filled.

No one really knows how such an extraordinary transformation in animal behaviour was actually achieved. Domestic dairy cows are the same species as wild aurochs, but over thousands of generations of living with humans these creatures proved pliable enough to provide man with a wider range of goods and services than any other living creature. The success of cows is reflected in their present populations. Estimated at 1.3 billion individuals, these are the most well kept, highly bred, populous domestic farm species of all time.

'Cattle' is a word derived from the Latin *caput*, meaning head, which originally referred to moveable property. The modern word 'capital' and the medieval word 'chattel' – meaning possessions – are closely related. The association of cattle with personal wealth therefore goes to the heart of why these animals were so very important to the development of many early human civilizations, especially those in Europe, Asia and Africa.

It is thought that the first domestic cattle emerged in Mesopotamia. They were farmed by the same people who pioneered the husbandry of sheep for their wool and milk (see page 219). Perhaps they persuaded the wild aurochs to approach their human settlements by leaving supplies of salt and water nearby as bait. As the animals ate the salt, these Neolithic humans may have approached and stroked them, teaching them to have nothing to fear from the presence of humankind.

Charles Darwin revealed that instinct is an inherited trait, and once fear was bred out of their system, compliance in cows took its place.

Domestication resulted in smaller breeds because they were easier to house when conditions got too cold or too dangerous for farmers to leave their animals outdoors. However, cows kept inside at night required food to be brought to them. For these reasons, as any modern farmer will attest, keeping cows is altogether more hassle than farming sheep.

Cows became so important to the wellbeing of some societies, for example in India, that they are regarded as sacred animals – provided by the gods as a source of milk and labour – and far too precious to be slaughtered for meat. In Asia, castrated male humped cattle, commonly known as oxen, have been the primary source of automotive power for pulling carts and ploughs for thousands of years. *'Protection of the cow means protection of the whole dumb creation of God'* declared India's early-twentieth-century independence leader Mahatma Gandhi. *Ahimsa*, the Hindu, Jain and Buddhist doctrine of non-violence to living things, was largely derived from, and most powerfully symbolized by, the sacred status of the domesticated cow.

Different breeds of cow came to have huge significance in other civilizations, too. China's ability to feed its massive rice-dependent population owes as much historically to the muscle power of its herds of patient, mild-tempered water buffalo

The sacred cow, depicted here as the mother of the world, her body studded with deities.

as to its pioneering development of the wheeled metal plough. There are more than 145 million buffalo still collaborating with humans in China today. Tibetan civilization is largely dependent on a single breed of shaggy, thick-coated domesticated cow known as the yak. There are an estimated five million of these pack animals that provide milk, meat, hides, hair, underwool, dung (for fuel) and blood, which is drunk as a source of protein. Butter from yak milk is even used as a source of fuel for Buddhist lamps. Tibetans will never deliberately kill a yak, although they are experts at ensuring nothing from the carcass of one that has died from old age ever goes to waste. Hair and hide are turned into ropes, saddlebags, blankets, bags, tents, footwear, bottles and boats.

In the Middle East, Europe and Africa, the idea of eating beef was never taboo. Cows were the perfect multi-purpose domesticated animal that could be self-transported wherever there was dry land. In Britain, wild aurochs were mostly replaced by domesticated breeds before the Roman era. The last European specimen died out in Poland in 1627.

A seemingly unstoppable radiation of cow herders that began in the Near East in about 5000 BC and had spread throughout Asia and Europe came to an abrupt halt in Central Africa. A microscopic parasite called *Trypanosoma brucei,* which travels in the saliva of a bloodsucking fly called the tsetse (see page 140), causes the fatal cattle disease nagana. The tsetse-fly zone, which straddles Central Africa like a belt, prevented the farther southwards spread of cattle-herding for as long as 5,000 years until a breed called N'dama finally developed immunity. Eventually by about 400 BC, herders progressed, overrunning most of the remaining hunter-gatherers, eventually establishing themselves into kingdoms such as at Great Zimbabwe along the banks of the Zambezi river.

Bantu herds-people mastered the art of milking using their own special cow language, which consisted of a series of screams, shouts and loud whistles mixed with tender words of admiration. Special breeds with specific skins, coats and markings were selected and bred by African tribal leaders for thousands of years, as if they were heraldic emblems from a proud family of European knights. Dingan (1795–1840) was a Zulu chief who made a big statement with his precious herd of 2,424 whitebacks.

Cattle were fully exploited as a source of wealth in other parts of the world during the European colonization of the Americas and Australia. Columbus brought cows with him from the Canary Islands on his second transatlantic voyage in 1493. South American herds were seeded by the arrival of one bull and five cows with Portuguese settlers who landed in Uruguay in 1555. By the eighteenth century there were so many herds roaming through Argentina that the government declared any individual could help themselves to up to 12,000 cattle each.

By the eighteenth century, specialist breeding techniques, pioneered by Englishman Robert Bakewell (1725–95), had begun to revolutionize cattle herds. Rather than developing breeds suited to specific locations or lifestyles, Bakewell deliberately inbred animals to increase their yields for producing beef and tallow (for making candles). He selected shorter-legged individuals that matured early and that put on plenty of fat, stating that his aim was to produce a breed with *'the greatest value in the smallest compass'*. In so doing, Bakewell began a trend of breeding animals to maximize commercial profit. Success was swift. As a result of his techniques, the average weight of a castrated British bull destined for slaughter more than doubled between 1700 and 1786, from 170 to 380 kilograms.

In the second half of the twentieth century, techniques for producing bovine super-breeds accelerated even more dramatically. Thanks to the practice of artificial insemination, which is used almost exclusively in modern herds, a single pedigree bull can now father thousands of cows without need of physical contact. Farmers no longer have to keep bulls, they just import sperm that is guaranteed to ensure the herd's thoroughbred credentials.

Farming for profit in a world where demand for food is ever-increasing tends towards the breeding of fewer, high-yielding species. The same trend that has happened to wheat (see page 199) is now also occurring in domesticated cattle. Hundreds of varieties, selected by humans over thousands of years to suit particular habits and lifestyles, are now on the edge of extinction yet the overall number of cattle on the planet, represented by a few super-breeds, rises exponentially with the increase in the human population.

Some of the environmental consequences of planet Earth playing host to 1.3 billion cattle, such as deforestation, have been obvious for many years. However, it is only in the last forty years that experts have become aware of the giant contribution that cattle make to global warming. A report published by the United Nations in November 2006 revealed that 18 per cent of greenhouse gas emissions come directly from the bellies of domesticated livestock – that's 5 per cent more than total worldwide emissions from global transport systems (road, rail, sea and air). Cows are by far the biggest culprits because of their huge numbers and their penchant for producing nitrous oxide and methane, far more powerful greenhouses gases than the carbon dioxide produced by human industry.

To keep up with global food demand, meat and milk production will have to more than double by 2050. The 230 million tonnes of meat produced in 2000 will have to rise to 465 million tonnes by 2050. The amount of milk produced is also projected to more than double over a similar timescale. Quite how emissions can be reduced while total quantities of food and drink from livestock double is a quandary about which scientists and governments have so far had very little to say. Who is prepared to stand up and persuade people to eat less meat or drink less milk? What government is bold enough to reward vegetarian eaters by cutting taxes on fruit and vegetables but increasing those on ecologically unsustainable meat?

There is a modern moral to be read into the old story of Jack, the boy who swapped the family cow for a handful of beans. Turning vegetarian is probably the biggest single personal contribution any one person can make in today's battle against global warming. Man's 12,000-year-old obsession with the wealth, taste, company and labour derived from domesticated cows is now a big factor in what is fast becoming a more vulnerable, volatile and significantly hotter world.

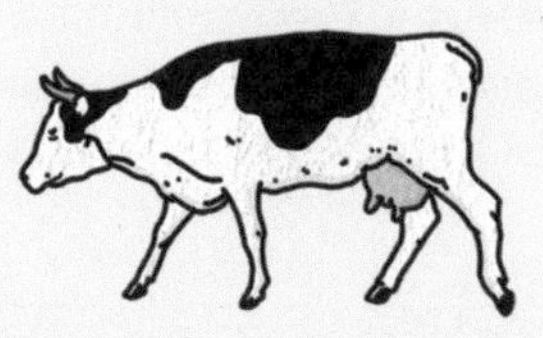

Chicken

FAMILY: PHASIANIDAE
SPECIES: GALLUS GALLUS
RANK: 61

How cocks and hens originating in the Far East were ideally suited to the machinations of modern fast food.

BIRDS HAVE NOT featured highly in our list of one hundred species that have changed the world. *Archaeopteryx* (see page 122), widely regarded as the ancestor of feathered flight, had a major evolutionary impact as the creature whose descendants came to dominate the skies. Birds were also remarkable survivors of the catastrophic meteorite collision 65.5 million years ago that smashed the rest of the dinosaurs into extinction (see page 126). Today only one avian class survives (out of an estimated eight that evolved following the emergence of *Archaeopteryx*). They are called the Neornithes. All birds – from ostriches to ducks and eagles to swallows – belong to this single class, but none have had as big an impact on the planet, life and people as *Gallus gallus*, the domesticated chicken.

Recent genetic research indicates that today's chickens are descended from at least two species of wild jungle fowl (red and grey), and not from a single wild ancestor as once thought. When two different animal species breed (horse and donkey, for example) their offspring are usually sterile (as with the mule). However, as the genetic ancestry of chickens suggests, cross-species mating can *sometimes* result in the generation of a new fertile species, especially, it seems, when different wild species are kept in captivity by humans. This fresh insight into the process of ancient domestication supports the concept of descent through 'acquired' genomes (see page 64).

The chicken's rise to dominance began somewhere in the Far East about 8,000 years ago. They were originally bred in places such as India as much for cock-fighting as for their meat and eggs. Indeed, the most popular meat breeds today have their origins in varieties such as the British Game Cock, bred primarily for the broad-set breast and strong legs that made it an excellent fighter. This is where the boxing terms 'bantamweight' and 'featherweight' originate. As early as the fourth century BC, Alexander the Great's troops brought fighting cocks back to Europe from their campaigns in India, although the first evidence of chickens becoming an important part of the Western diet goes back to the Romans, who fattened their chickens on a diet of milk and bread.

Roman legions then brought chickens to Britain. They soon became the bird of choice in Anglo-Saxon settlements, primarily as an egg-laying farmyard favourite. Meanwhile, Polynesian explorers – descended from Chinese farmers of the Yangtze valley – spread the bird farther eastwards across the Pacific, as far as Easter Island and even to the coast of the Americas. The recent discovery of chicken bones that date back to the pre-Columbian Americas is shedding new light on pan-Pacific expansion. Canoe-paddling Polynesians from Hawaii and

Easter Island, it seems, must have made the extraordinary journey to the west coast of the Americas decades before the arrival of Spanish adventurers. DNA analysis of ancient chicken bones in El Arenal, Chile, dates them back to the fourteenth century, before contact with Europeans was established. Vikings to the east and Polynesians to the west were, it seems, the real discoverers of the Americas, long before Columbus set sail.

Chickens were traditionally bred for cock-fighting in the UK at least until the mid nineteenth century. At one time the sport was as popular as football is today. Cockerels were consigned to a pit where they would, literally, fight to the death. As in ratting (see page 174), human observers would take bets on which bird would become champion. With the opening up of China to international trade from the 1840s onwards, chickens became a more popular source of food and nutrition. European 'Hen Fever' was fuelled by Queen Victoria's fancy for chickens. She even had a poultry shed built to house a flock of exotic Chinese Cochins. The keeping and breeding of exotic chickens for their looks and the curious brown eggs they sometimes laid had the cachet of royal approval. The age of the chicken show had arrived.

But it was not until the twentieth century that chickens swooped up the scale of those species that have most changed the world to become what is by far the world's most numerous bird. Today there are an estimated *twenty-four billion* chickens worldwide – that's more than three birds for every person on the planet. Most are bred for their meat, although egg production is also big business. Chicken is now the cheapest source of meat in the world and has dominated the cuisine of all human cultures from China to the Caribbean.

It is not a happy tale. According to a report on animal welfare written in 1992 by John Webster, a professor at Bristol University, modern chicken meat farming is '*in both magnitude and severity, the single most severe, systemic example of man's inhumanity to another sentient animal*'. The problem is that deep within the

Cock-fighting, the spectator sport of the nineteenth century.

domesticated chicken's gene pool lies a set of attributes that has proved perfect for the mass-production requirements of modern farmers.

The first breakthrough came by accident. In 1923 small-time Delaware chicken dame, Cecile Steele, ordered fifty chickens from a local seller, but received 500 owing to a mistake. She found that she could successfully farm the hens indoors inside purpose-built sheds, selling them off for meat when they were only a few weeks old. The trend was infectious. By 1929 around three million chickens were being produced in Delaware every year.

Mass-production techniques, pioneered by American Eli Whitney (1765–1825) in his water-powered armaments factory, proved ideal in the rapidly growing business of intensive chicken farming. Broiler belts automated the killing, plucking and butchering of birds so they could be sold in faraway markets, ready to cook. The discovery that, with a suitable vitamin D supplement, chickens could be kept indoors, twenty-four hours a day, cooped up in purpose-built factory sheds without natural sunlight, was a second breakthrough that came in the late 1920s. The less the birds moved around, the less feed they needed and the faster they grew. Techniques like these, begun in the US, were soon transported to Britain and beyond. With rationing of beef and pork during World War II, home-grown chicken meat and eggs became the West's favourite and cheapest foods.

But it was in the post-war era of the twentieth century, when American mass-production techniques first arrived in the UK, that the chicken really came into its own. Selective breeding has focused the domestic chicken's gene pool to provide exactly the right conditions to make mass-production of chicken meat cheap and efficient. Since the 1970s, the time it takes for a chick to grow into an adult bird so large that it can't even stand up has gone down from twelve weeks to just six, massively reducing the costs of feeding it. Antibiotics, originally administered to prevent disease in overcrowded chicken sheds, have accelerated maturation rates by suppressing the 'bad' bacteria that inhibit growth. The six-week life-cycle of a battery chicken, which never sees the light of day, compares with the twelve-year lifespan of a natural chicken allowed to feed in the open and not killed prematurely for its meat.

The transformation of egg-layers has been no less spectacular. Naturally chickens would lay a clutch of about twelve eggs, then sit on them to incubate the embryos over a period of about twenty-one days before they hatch out. Modern egg-layers have been selectively bred to produce an average of *300 eggs a year*, often helped by artificial lighting to stimulate relentless egg production. These birds are the most reproductive living creatures on Earth, laying up to eight times their body weight in eggs per year.

The extraordinary capacity of domestic chickens to produce meat and lay eggs in the highly cost-effective environment of intensive farming provided essential fodder for the fast-food industries of the developed world. In 1954, 'Colonel' H Sanders (it was a self-styled title, he was never in the military) opened his first Kentucky Fried Chicken outlet in Salt Lake City. After conquering the Wild West, cheap chicken fast-food outlets made their first complete circuit around the USA before arriving in the UK in 1964. So successful was the instant-chicken-eating revolution that by 1983 McDonald's burger-chain had launched its own range of finger food in the form of Chicken McNuggets. Since then, McDonald's has upped its retail presence from 6,000 to 30,000 restaurants worldwide.

KFC – the self-styled colonel who claimed to sell a million chicken dinners a day.

Increased meat consumption means increased demand for chicken feed. Soya beans (*Glycine max*) are one of today's most widely grown crops, which is largely thanks to the need to feed the world's twenty-four billion chickens (see page 401). Livestock feed now accounts for up to *half* of the world's entire agricultural produce – of which chickenfeed represents the lion's share.

Even though chickens are one of life's most efficient converters of plant sugars into animal protein, for every gram of protein they consume, their flesh yields only 0.33 grams of edible protein in return. This is the problem of the ongoing and growing human love affair with eating large quantities of meat. People in less developed countries, many of them in Africa, consume thirty kilograms of meat per head, per year on average. In more affluent societies, where eating meat has turned from being a luxury and a status symbol into a mass consumer habit, the average is closer to eighty kilograms per year. Most of this rise is down to cheap-to-produce chicken.

The impact on the planet concerns the scale of agricultural support needed to maintain this meat-eating trend. A sustainable response to the ever-rising human population would be for each person to reduce their average meat consumption and for governments to incentivize such behaviour. The reality today is the reverse. The wealthier people become, the greater their lust for meat – life's least environmentally efficient source of essential protein.

Without the highly pliable genes of domestic chickens, modern intensive farming could never have materialized since chickens would not have yielded to high-volume, low-cost industrial production methods. Without humans, chickens would never have had to put up with living on a mat no larger than an A4 piece of paper, in a shed with 20,000 others, for a six-week life that should naturally last about a hundred times as long.

Regardless of animal welfare (there are still no legal regulations regarding the welfare of farmed chickens in the USA) and whatever environmental issues are being stored up for future generations, there is no question that *Gallus gallus* represents the apex of human/animal domestication. This once exotic wild Asian guinea fowl is now the most populous domesticated species on Earth.

Rice

FAMILY: POACEAE
SPECIES: ORYZA SATIVA
RANK: 27

An agricultural miracle crop that sustains billions but seems to sow as many problems as it solves.

HIGH UP in a remote, land-locked, mountainous region of the Philippines, ten hours by bus from the islands' capital Manila, lies what many regard as the eighth wonder of the world. Ancient tribal people constructed this marvel – a vast series of stepped agricultural terraces that jut out sideways from the steep mountain slopes. They hand-crafted each one to make level fields for growing crops, using nothing more than gravel and rocks to flatten the ground and dry-stone walls to control the flow of water.

The terraces, built about 2,000 years ago, range across an area totalling more than 10,000 square kilometres. Artificial caverns high up in the mountains store rainwater. At the appropriate moment in the crop-growing cycle, the fields are flooded using an intricate network of dams, sluices, channels and pipes made from bamboo (see page 150). This is human construction on a massive scale, a feat of ancient engineering designed to harness and control the power of nature surely as remarkable as the fabled Hanging Gardens of Babylon or the Lighthouse of Alexandria. In 1995, the Terraces of the Cordilleras were officially given the status of a UNESCO World Heritage Site, putting them on a par with ancient Egypt's Pyramids of Giza.

The reason these people went to the enormous effort of building fields out of mountains can be summarized in a single word: rice. Island environments are precarious at the best of times. When these ancient human settlers arrived from Taiwan they knew their welfare depended entirely on cultivating the crops that they had brought with them. Rice, first domesticated along the Yangtze river valley in southern China, was, and still is, the most nutritious foodstuff ever known to man. Today more than 2.3 billion Asians – nearly half the world's human population – depend on this single crop to provide as much as 80 per cent of their daily intake of food.

Chinese people first began to domesticate rice between 8,000 and 5,000 years ago, at about the same time that Neolithic farmers in Mesopotamia were taming their first sheep (see page 218) and cultivating crops of wheat (see page 199) and barley. Like emmer wheat, wild rice is not genetically suited to agriculture. Its seeds shatter when ripe, scattering them as far as possible in the wind, making it hopeless for humans to gather as food to eat. The grains are also hard and gritty; nature designed them to survive long periods of dormancy on the ground, equipping them with a tough outer coating that protects them until the arrival of damp conditions when germination can begin.

But, like a few other select species, wild rice had a sufficiently diverse pool of genes that it could adapt in ways that eventually pleased people. Careful selection

The eighth wonder of the world? Rice growing on the Terraces of the Cordilleras in the Philippines, originally hewn out of the mountainside by ancient settlers from China.

and cultivation by early Chinese farmers produced varieties with seeds that stuck to the stalk, and soft grains that were permeable to water. These could be cooked in boiling water, quickly releasing a plentiful source of essential edible carbohydrates and protein.

So rich was this nutritional prize that Asians soon learned how to grow the crop for best results. They developed elaborate irrigation systems. They found that waterlogged paddy fields kept pests and weeds at bay. They sowed each plant in regimented rows so they could manage and harvest their crops more easily, increasing the yield. Advanced agricultural skills in growing rice largely account for why, until about the eighteenth century, the people of China led the world in their knowledge of agricultural science and innovation. The blast furnace, improved in AD 31 by court scholar and inventor Du-Shi, with its water-powered bellows, led to the manufacture of a new range of agricultural tools including hoes and adjustable ploughshares that could make heavy-clay soils suitable for growing rice.

Rumours of iron-making and rice cultivation caused people living in northern China to migrated southwards in a bid to unlock the region's fertile wealth. Conflicts provoked by the pursuit of riches in the form of rice ultimately united China's warring states into a single, mighty empire. From *c.* 221 BC, a steady, firm, centralized grip on power has been a more or less constant Chinese tradition, except when punctured by occasional but vicious bouts of civil war and rural revolt.

Human populations that live on rice tend to be more numerous than those that feed on wheat or olives because rice cultivation is highly labour-intensive. Like sugarcane, each plant has to be sown individually every year and fields have to be properly ploughed and irrigated at the optimum time to ensure a good harvest. Until new varieties of 'miracle rice' were bred by scientists in the 1960s (see overleaf) one hectare of rice could sustain 5.63 people for a year. This compared with just 3.67 people per hectare for the Western world's rival, wheat. The effects on population growth between rice and non-rice cultures were apparent by 1750 when the total population of Asia reached 502 million, more than three times that of Europe.

Big populations meant big governments and big armies, quickening the conquest of land by Asian rice. For the last 5,000 years forests have been uprooted for paddy fields and new rice varieties have been cultivated that require less water and can grow at higher altitudes. By 2,000 years ago, some rice-growing people had migrated from mainland Asia to Pacific islands such as Taiwan and the Philippines, taking their precious grains with them.

With the rise of Islam from the eighth century onwards, Asian rice spread throughout Europe and North Africa, along with a clutch of other significant Chinese inventions such as gunpowder, paper, wheelbarrows, compasses, umbrellas, stirrups, matches, playing cards, kites and chess. Human tastes in the Middle East were re-educated to accommodate this new, rich source of nutrition. Spanish paella, Italian risotto and the paddy fields of the French Camargue are all traditions that continue to this day, but were begun in ancient China and later transplanted to Europe. It happened thanks to the universal appeal of rice, the relentless zeal of Islamic traders and their enduring conflicts with Christian crusaders, who took rice-growing even farther west.

Meanwhile, in West Africa a related but different species of rice, *Oryza glaberrima*, was domesticated by natives living around the Gambia and Niger rivers. Africans developed their own expertise in rice cultivation, mirroring the irrigation techniques found in ancient China. Portuguese sailors who first explored the West African coast in the fifteenth century were astonished at the agricultural riches they found. Specialist long-handled shovels called 'kajandu' were used by the natives to turn over the earth and prepare their paddy fields. European explorers brought with them the Asian rice species, *Oryza sativa*, which has now largely replaced the African type.

Planting out paddy fields in the Far East, from a nineteenth-century Japanese print.

Expertise in rice cultivation travelled with African slaves that were shipped across the Atlantic to Charleston to work on American sugar, tobacco and cotton plantations from the seventeenth century onwards. Those with knowledge of how to farm rice fetched the highest prices at auction. Until the abolition of slavery, after the American Civil War (1861–5), Asian rice was one of the most profitable and widespread crops grown in the southern United States. It is still grown today along the banks of the Mississippi river. Rice cultivation in California caught on when Chinese labourers flocked there during the years of the gold rush (1848–55) taking their rice-growing skills with them. By then this crop had completed its circumnavigation of the globe.

Following World War II, when it seemed that many parts of the world were on the brink of starvation, it was to rice (along with wheat, see page 199) that scientists turned for a way of avoiding a great Malthusian famine. Painstaking cross-breeding techniques pioneered by experts at the International Rice Research Institute (IRRI), appropriately located in the Philippines, led to the development of a new variety – IR8 – with dwarf-like characteristics that could tolerate the higher yields generated by artificial fertilizers. When this variety was made available to the world in the early 1960s it was dubbed – for good reason – 'miracle rice'. Yields shot up in some places to as much as 10.3 tonnes per hectare, up from the average one tonne per hectare produced by traditional varieties. What's more, IR8 matured in just 130 days instead of the normal 170 days. Later improvements (breed IR36) have cut the growing time to just 105 days, allowing many farms to cultivate two crops per year – more than doubling their output.

The result is that instead of famine and declining populations, the last fifty years have witnessed a rice binge that has boosted the number of people from 2.5 billion in 1950 to nearly seven billion today. By far the biggest rise is accounted for in Asia – where rice eating predominates – up from 1.4 billion to 3.9 billion over the period.

All this leaves today's world with an almighty conundrum. Feeding ever-rising human populations will require a step-change in agriculture – as predicted by Norman Borlaug (see page 201). Since there is little uncultivated land left to press into new production, increases in supply must come from greater yields. Unfortunately, the yield of modern rice varieties is fast approaching biological limits, regardless of how much nitrogen fertilizer is dolloped on to fields.

There is, however, at least one genie left in the genetic magician's lamp. The type of photosynthesis used by *Oryza sativa* is the more primitive C_3 technique. If scientists could re-engineer the rice plant to operate on the more efficient C_4 method, as used by crops such as maize and sugarcane (see page 204), substantial new gains in yields could be achieved. So significant is this pursuit among experts that it has been described as the 'next frontier' in crop science, predicted to take at least another decade, perhaps two. Its successful accomplishment would be, according to some, *'the most audacious feat of genetic engineering yet attempted …'*

Genetically 'improved' plants may indeed be needed to safeguard future populations from the threat of starvation. But without the very careful and deliberate political management of birth rates, these varieties may also follow the rice legacy of previous generations, successfully sowing the seeds of human populations that carry on growing by the billions.

Maize

FAMILY: POACEAE
SPECIES: ZEA MAYS
RANK: 35

A highly modified, super-photosynthetic grass that has sustained human societies but which has often had sinister side-effects.

HUMAN SACRIFICE, cannibalism, vampires and obesity are horrors best left on the top shelf of Hollywood's library of 'video nasties'. Unfortunately, for people and animals alike, these terrors do not exclusively belong to the realms of fiction. They are also linked by a type of grass that dominated much of the Americas until about 500 years ago, when it suddenly spread across most of the rest of the world. Since then, this crop has become one of the three most widely cultivated, and potentially most lethal, species on Earth. American farmers call it corn, although to the rest of the world *Zea mays* is better known as 'maize'.

Quite how natives living in Central and South America bred wild grass of the *Zea* family into a species with an enormous ear (a cob) that contains a cluster of up to 300 highly nutritious, edible seeds (corn) is baffling. Despite extensive research, genetic sequencing and intense scrutiny by countless archaeological experts, there is still no single, convincing theory of how fast-growing teosinte, a wild spiky grass that hosts between five and ten inedible seeds, morphed into maize with its remarkably succulent, highly digestible cob. Certainly the process took thousands of years. Evidence of early domestication reaches back at least 5,000 years, around the Balsas river valley in Central Mexico, where wild teosinte still grows. A mixture of human selection and whimsical nature best accounts for how hybrids fused into a new super-species, *Zea mays*, with its highly complex genetic structure. Maize has more than twice the number of genes as man.

From about 1500 BC maize cultivation began to transform the lives of people living in Central and South America. New civilizations emerged that developed their own systems of government, writing, crafts and exchange. Since these regions lacked the domesticated animals of Africa and Eurasia (see sheep, page 218; cows, page 222; dogs, page 323; and pigs, page 215) they became almost totally dependent on crops such as maize. A creation story, contained in an ancient text called the *Popol Vuh*, copied by the Spanish missionaries in the sixteenth century, tells how the gods tried to create humans to keep them company. On the first occasion they used mud, but that failed. On the second occasion they used wood. Only on the third attempt were true people born after the gods managed to create them out of maize:

> *'They [the gods] came together in darkness to think and reflect. This is how they came to decide on the right material for the creation of man ... Their flesh was made out of white and yellow corn. The arms and legs ... were made out of corn meal.'*

Dependency on any one source of food puts a civilization at the mercy of nature – or the gods, depending on cultural beliefs. Olmecs, Mayans, Incas and Aztecs were civilizations that depended mostly on maize for their nutritional welfare. Currying divine favour to ensure good harvests and plentiful supplies of rain to water their crops became central to their way of life. Following three consecutive years of drought, any remaining stockpiles of maize would have gone rotten. With no animals as a back-up food supply, placating the gods to bring more rain through the practice of offering humans as a sacrifice turned from an occasional ritual into a religious routine. Despite the patchy records of the Aztec era (1248–1521) estimates of human sacrifices during the consecration of the Great Pyramid of Tenochtitlan in 1487 range from 4,000 to 40,000 individuals. The Flower Wars, waged during the fifteenth century, were campaigns that Aztec warriors fought against neighbouring states in a bid to secure a steady supply of sacrificial victims. Such practices had major political ramifications. Many of the Aztecs' enemies sought revenge by siding with Spanish conquistadors led by Hernán Cortés in the 1520s, hastening the demise of the indigenous Native American culture.

In times of drought, sacrifices were most often made to Tlaloc, the Aztec god of water and rain. Women and children were primary victims, their tears symbolizing the plentiful rains that Tlaloc would surely bring to water these desperate people's fields of maize.

Dependency on maize may also explain why cannibalism appeared on the menu of maize-dependent Native Americans living in the southern states of North America. Maize does not provide the full range of proteins, vitamins and sugars required for a healthy human lifestyle (unlike potatoes, see pages 206–7); meat is an essential addition. Anasazi people spread the practice of maize cultivation throughout the southern states of North America from *c.* AD 600 until *c.* 1150. Over-zealous deforestation (in order to grow maize crops) and a prolonged drought starting about AD 1130 led to their demise as they abandoned their holy city of Chaco Canyon soon after 1170. In a land devoid of domesticated animals and left with precious little game, the decision to eat human meat may have been a nutritional necessity. But it was not sustainable in the long term since, generally speaking, humans are reluctant to allow themselves to be domesticated as farm animals for consumption.

The arrival of European colonists in the early sixteenth century following the voyages of Christopher Columbus helped spread maize globally. As early as 1550 the hillsides of China were being cleared of trees to make way for this super-fast-growing crop that thrived on the sunny slopes of drier, higher terrain that were unsuitable for cultivating traditional varieties of rice. The population boomed. By the eighteenth century, European farmers were increasingly turning to maize as a source of feed for their livestock and as a staple crop for human consumption. By the early twentieth century *polenta*, a dish made from boiled cornmeal, was firmly established as a premium food in the north Italian diet.

The conquest of maize in Europe had further political ramifications. Balkan peasants learned to grow the crop at

Storing Mexican maize for a non-rainy day, as depicted in the sixteenth-century Florentine Codex, which was written and illustrated by early Spanish settlers.

high altitudes, sustaining clandestine alpine communities whose agricultural riches could be kept hidden from the grasp of Ottoman invaders. According to at least one authority, such affairs *'weaned the future political independence of Greece, Serbia and Romania from mountain cradles. In this corner of Europe an American product really did nurture freedom.'*

Nineteenth-century Europeans and their Spanish colonists who grew too dependent on this wonder crop became its victims just as much as those who were sacrificed by the customs of the ancient Americas. They suffered from pellagra, a vitamin deficiency disease that causes chronic insomnia, skin disorders, dementia and an extreme sensitivity to light. The ancients of America added lime to their crops to avoid this condition but Europeans knew of no such fix. People in Spain, Italy and the southern United States whose diets were almost totally dependent on maize were especially badly affected. Many people suffered and thousands died. Victims were kept in the dark, their disfigured faces inculcating a fear of vampires. The disease lasted until the mid 1920s when, finally, Western scientists re-discovered the cure – a simple dietary supplement to boost levels of vitamin B_3 (niacin).

Today maize is at the centre of yet another controversy. The USA cultivates more 'corn' than anywhere else in the world – about 270 million tonnes annually – over a quarter of the world's total. While much of it goes into feeding livestock, with all the consequences for global carbon emissions (see page 225), a great deal also goes into making cheap sweeteners for use in popular modern foods. High fructose corn syrup (HFCS), first synthesized in 1970, is produced using processed maize and is now America's most common sweetener, found in everything from fizzy drinks to cookies. Since its introduction, obesity in children aged between four and nineteen – the major consumers of these goods – has risen from 4 per cent to 15.3 per cent. With more than six million US children now classified as clinically overweight, obesity, which leads to a greater risk of type-two diabetes, is now America's greatest health problem. Recent reports suggest that corn syrup is a *'big part of the problem'*.

The race to wean modern economies off their reliance on fossil fuels is now propelling the human/maize relationship in yet another direction. The production of 'carbon-neutral' biofuels, in a process that often involves the intensive cultivation of maize, is being promoted as a possible solution to global warming (see also algae, page 56; and yeast, page 295). This is because the process releases only as much carbon as it removes from the atmosphere. Unlike the burning of fossil fuels, there is no net increase. But adapting maize for this new purpose comes at a considerable political and economic price. In 2007, global food costs soared partly as a result of subsidies given to farmers to grow maize not as food but as a feed stock for fermentation to produce ethanol for fuel. The ancient and historic association between politics and *Zea mays* looks set to continue well into the unforeseeable future.

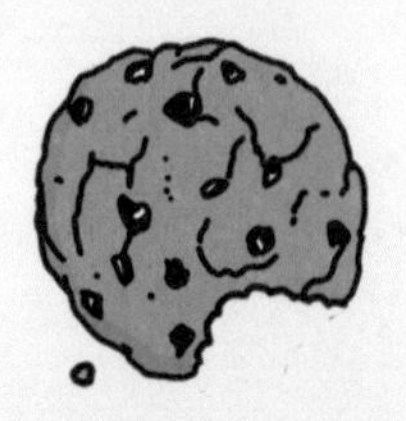

On Material Wealth

How certain species were perfectly suited to providing the wealth necessary to make life comfortable for civilized man.

PEOPLE GENERALLY want to live their lives as comfortably as they can. They also hope to live for as long as possible. These two desires have been common to every society since civilizations first arose about 6,000 years ago. While some people have been more successful than others at securing wealth for themselves, personal fortunes, like the rise and fall of nations, have always come and gone as ages pass. Jean-Jacques Rousseau (1712–78), in his *On the Origin of Inequality, Part 2,* described what distinguishes the rise of human civilizations from prehistoric human culture:

> *'The first man who, having enclosed a piece of ground, bethought himself of saying "This is mine" and found people simple enough to believe him, was the real founder of civil society ...'*

Material possessions, property ownership and the freedom to try to earn as much money as possible are central to the modern way of life. Private ownership provides an incentive to work, and a reward, usually in the form of money, is given in return for labour. Houses, cars and fashionable clothes are a few of the most sought-after material possessions in the Western world. Jewellery, pottery and cows are just as precious in cultures elsewhere.

The motivational power of private ownership has been evident for thousands of years. Aristotle, probably the most highly regarded philosopher of all time, said as much, writing more than 2,000 years ago:

> *'That which is common to the greatest number has the least care disposed upon it. Everyone*

thinks chiefly of his own, hardly at all of the common interest; and only when he is himself concerned as an individual . . .'

The seas and the air – often called the 'global commons' – are parts of the world that have never generally been owned either by individual people or nation states. They are therefore the least cared for parts of the planet. With the devastation of marine life and the ever-increasing levels of pollution and greenhouse gases, the oceans and sky are a living testament to Aristotle's ancient wisdom.

Private property

The concept of individual property ownership goes back to the beginnings of recorded history. In ancient Sumeria, where the first civilization can be traced back as far as 6,000 years, systems of writing were established to support commerce. Wheat, barley and domesticated animals in the form of goats, sheep and cows were such valuable commodities that making inscriptions in stone to record accounts of exchange led to the rise of a written language called cuneiform. Laws setting out rules for the inheritance of private wealth were innovations that quickly followed and have become ubiquitous throughout human societies ever since.

Laws, property and ownership rights are even more sacrosanct today. The United States' constitution is founded on the principle, derived from the philosophy of Englishman John Locke, that earning a profit from the cultivation of land is the cornerstone of natural liberty. Even the state cannot legitimately interfere with a person's private property. Such beliefs paved the way for American resistance to British colonial rule in the mid eighteenth century. Today, Article 17 of the United Nations Universal Declaration of Human Rights (1948) states the current global position unequivocally: '*Everyone has the right to own property.*'

The concept of property ownership does not simply spring from having access to easy-to-store consumable foodstuffs and compliant farm animals (see On Agriculture, page 191). Private property also depends on having physical power over other people – the ability to protect, transport or acquire possessions either by trade, threat or force.

Animal power

Horses (see page 242) and **camels** (see page 248) were domesticated not primarily for human consumption (meat or milk) but because of the *power* they gave those people who owned them over those who did not.

Horses, and their hybrid offspring mules and hinnies, provided an essential means of transport for carrying and trading wealth between different ancient, medieval and early-modern civilizations. Gradually they were bred into larger creatures that could pull chariots for waging war or for carrying people into battle. Horsemen have enormous advantages over people on foot thanks to the combined strength, speed and height of the animal and its human rider. People who own horses have traditionally been wealthier than those who do not. Horses are still status symbols today, a mark of authority in human societies that, like ants' nests (see page 154), have become more hierarchical as populations have grown.

For desert people, the equivalent power is provided by camels, creatures better adapted to arid environments. Species of similar significance in Arctic climes are reindeer. In each case these animals have satisfied human material wants and desires. In return for food, water and protection, they provide services in the form of transport and military strength. As a result, these creatures have become enormously successful, their numbers growing prodigiously during the few-thousand-year-history of human civilization.

Initially, power and inequality based on the ownership of animals like these could manifest itself only in places where such creatures dwelt. In Australia and many parts of the Americas there were no large animals capable of being domesticated (see pages 184–5). When contact was eventually established between these indigenous cultures and horse- and camel-breeding Eurasian societies, inequalities between people turned into inequalities between cultures and nations. This partly explains why it took just a few early-sixteenth-century Spanish conquistadors less than a decade to subdue tens of thousands of native Indians. Power came from the military advantages of horses and the introduction of diseases that were also derived from domesticated animals (see page 26).

Camels and horses are species whose ancestors originally evolved in the Americas but became extinct with the arrival of human hunters 13,000 years ago. Thanks to earlier migrations into Asia, via the Alaskan land bridge, these species survived. One wonders how differently the course of human history would have turned out had these creatures never made the trek.

The rise of human civilizations and the success of horses and camels are inextricably linked. After the successful conquest of the New World by Europeans, horses re-established their presence across the Americas. They are animals that now live alongside humans on every habitable continent of the world. In contrast, their close relatives, zebras, are an endangered species that live in a small pocket of the African bush. This equine cousin of the horse has never lent itself to domestication, showing what a difference the overly territorial instincts of zebras have made to their recent reproductive success.

A Spanish conquistador spears an Inca, capitalizing on the military benefits of being on horseback.

Creature comforts

Other non-consumable species have also been big winners with the rise of human civilizations. Wealth, as defined by the Scottish founder of modern economics Adam Smith (1723–90) is the *'produce of the land and labour that satisfies human needs and wants'.*

Humans living in large societies have always had material needs and wants. Species that have provided these most effectively include flax and **cotton** (see page 252) – crops not generally grown for consumption but for their hairy parts. The art of spinning fibres from the fruit of a cotton plant into thread for making clothes was first established by people living in the Indus Valley beginning *c.* 5,000 years ago, centred in what is now northern India and Pakistan.

But nothing grown by species in the plant kingdom has matched the allure and magic of fibre spun by invertebrate animals. When Leizu, wife of the legendary Chinese Yellow Emperor, accidentally dropped a moth's cocoon into her cup of tea, little did she realize what an impact her slip would have for the long-term prospects of the genes of *Bombyx mori,* the **silkworm** (see page 260).

Chinese merchants generated vast wealth from their exclusive knowledge of how to spin the fibre of this cocoon, designed by nature to protect the growing moth larva. When Asian people learned the art of sericulture and wove the fibre into a rich, shining cloth, they created the most valued ancient non-consumable commodity of all time. Silk-making was imperial China's most closely guarded secret until two sixth-century Byzantine monks smuggled some silkworm eggs in bamboo rods back to Constantinople (now Istanbul).

Silk's shiny appearance and exotic, unknown origins only increased its appeal to consumers living in Europe, North Africa and the Middle East. The adornment of human social elites with materials that glimmer (like gold) or shimmer (like silk) has proved to be a sure recipe for widespread success. Similar ingredients lay behind the extraordinary, if brief, bloom of the potyvirus that produced exotically patterned tulips in seventeenth-century Holland (see page 22).

Raw materials extracted from trees have been other sources of profit and power. **Rubber** (see page 256) was first extracted from palm trees in Central and South America by Olmec (meaning 'rubber') people for the purpose of making sports equipment. It gained a new lease of life as a sealant designed to increase the power of steam engines in the nineteenth century and it then took off as a material for making transport by bicycle, car and bus considerably more comfortable in the early twentieth century. European rulers, like Leopold II of Belgium (1835–1909) had massive private African estates planted with latex-producing vines, turning him into one of the richest individuals in the world.

To tree or not to tree

Trees have been critical to the growth of human civilizations. As the natural world's chief material for construction and fuel, what commodity could be more precious than wood? Despite this, careful forest management to ensure a secure supply of timber for future generations has been a constant failure of human societies from ancient times to today.

Many tree species have suffered at the hands of man. Early farmers felled forests to make way for pasture land. Today's destruction of the Amazon rainforest increases annually as demand for hardwoods for making durable furniture grows ever greater. Increasing urbanization, road-building and massively growing human populations are all stacking the odds against the future security of rainforests.

In contrast, fast-growing species such as Pinus – with their soft trunks that easily pulp – have thrived since Europeans began to make paper out of wood in the nineteenth century. Other fast-growing, climate-tolerant species such as **eucalyptus** (see page 264) have been transported by humans far from their continent of origin (Australia) to provide everything from fuel for steam trains in Africa, to furniture in the Californian gold rush (1848–55).

Human traffic

The generation of wealth through the acquisition and trade of private property has what is, to modern sympathies, a pernicious sting in the tail. One of the most commonly traded species throughout human history is man himself: *Homo sapiens*.

Slavery – the enforced exploitation of human labour and the trafficking of people against their will for profit – probably originated at about the same time as the domestication of the horse and camel. Certainly by 1800 BC Hammurabi, a ruler of ancient Babylon, made it clear that the institution of slavery was a widely accepted form of profit and trade. It is mentioned, etched in stone, on a tablet that specifies his famous code of 282 civilian laws.

In the ancient world people were usually enslaved either for being in debt or because they had been captured as prisoners of war. Some societies were addicted to slavery as a means to provide wealth to a ruling class. As many as 40 per cent of all humans living in Italy during the Roman Empire were slaves. Almost every human civilization has had its own thriving trade in trafficking slaves, from the Vikings who enslaved the people of Eastern Europe for sale to Islamic cultures, to Italian traders from Venice and Genoa who regularly bought and transported slaves captured by the Islamic Golden Horde, which they then sold on to Arabic markets in the Middle East.

Camels (see page 248) were the main instruments for the Islamic transportation of tens of millions of enslaved Africans eastwards via Mombassa. The fashion was adopted by Europeans from the sixteenth century onwards for exporting similar numbers of slaves westwards to the Americas. Even there, long before the European conquest, slavery was nothing new. Many of the Aztec victims of human sacrifice (see page 235) were enslaved prisoners of war.

Despite the abolition movements of the nineteenth century, trade in human slavery has continued. Globally as many as twenty-seven million people, most of them children, are still living in conditions of slavery today, according to a report by the British Anti-Slavery Society in 2005.

Ecological consequences

Species torn out of their natural environments and introduced elsewhere for human convenience have frequently had devastating environmental consequences. Examples range from the deliberate introduction of eucalyptus plantations (see page 264) to the inadvertent spreading of European earthworms (see page 99).

The predominance of a few such widely cultivated species has also increased the risk of mass extinction not normally present in the wild. The evolution of countless billions of species in nature is a biological insurance policy so that, whatever conditions on Earth may be in the future, at least some forms of life are highly likely to survive – the catastrophic Great Permian Extinction being a case in point (see *Lystrosaurus*, pages 118–19). Sexual reproduction is a microcosm of the same system, conferring variety within a wide species gene pool as protection against unexpected threats.

But the rise of human civilization and its few wealth-creating species subverts this prehistoric formula. The recent growth of large-scale, sexless single-variety crops has increased the need for pesticides to ensure their survival. It also makes the chances of a devastating virulent attack more likely, not less. An infection that successfully evolves a way of avoiding conventional chemical controls could sweep through an entire population of cloned crops, never encountering that element of natural resistance of species that reproduce via sexual reproduction in the wild (see pages 47–8).

This is why scientists today are engaged in a race to transfer genes from naturally occurring blight-resistant wild potato plants in South America into their latest cultivars for mass production (see page 53). The rise of super-crops from cotton to wheat is today forcing humans to further explore the science of genetic engineering in a desperate attempt to stay one step ahead of nature in a global effort to grow enough food to feed seven billion people.

Private property, social inequality, population growth and decreasing biodiversity stem from the symbiotic relationship between humans and a few compliant species that began several thousand years ago. To put it into perspective, the deliberate cultivation of such species is an innovation that has taken place in the last 1 per cent of the entire 2.5-million-year history of humankind.

Jean-Jacques Rousseau believed that the origin of human suffering, envy, injustice and inequality could be traced back to the first man who, having fenced in a piece of land, bethought himself of saying, '*This is mine*':

> *'From how many crimes, wars and murders, from how many horrors or misfortunes might not anyone have saved mankind, by pulling up the stakes, or filling the ditch, and crying to his fellows: Beware of listening to this impostor; you are undone if you once forget that the fruits of the Earth belong to us all, and the Earth itself to nobody.'*

Horse

FAMILY: EQUIDAE
SPECIES: EQUUS CABALLUS
RANK: 58

The loyal, hard-working, quick-learning servant of humanity that has seen more war and violence than any other creature on Earth.

THE PROPHET MOHAMMED is said to have ordered a herd of horses to be left without water for seven days. When they were released they raced towards an oasis, but before they got there the prophet sounded his war horn, calling the horses to battle. Five mares ignored the water and answered the call. Legend has it that all pure-blood Arabian horses descend from these five obedient mares.

The history of humanity is inextricably tied to the domestication of the horse. Today's pony clubs, flat-racing derbys or even Olympic eventing tournaments do not in the least reflect the historic impact of horses on man – and vice versa. Horses rose to pre-eminence among animals that learned to live with people because they provided a means of rapid transport and a source of labour in fields. It is only in the last half of the twentieth century, since the motorcar became a commonplace alternative, that horses have been sidelined to the more whimsical pursuits of leisure and competitive sport.

The family Equidae, to which modern horses, zebras and donkeys belong, arose about forty million years ago in what is now North America. They evolved from a fox-sized woodland-dwelling ancestor called *Hyracotherium.* As the Earth's climate cooled and grasslands replaced woodland, the need to flee from predators such as lions and sabre-toothed cats in the more exposed landscape meant that the fastest runners with the longest legs had a better chance of survival. These characteristics were passed down through the generations. By about five million years ago, after a series of evolutionary twists and turns, a wild animal had evolved called *Plesippus,* which looked much like a small modern-day horse. At some point during the ice ages (which began *c.* 2.5 million years ago) this species crossed the Alaskan land bridge into Asia where it made its home on the extensive Eurasian steppes.

Here, above and around the Black Sea, the fate of wild horses and people began to merge with dramatic effect from about 6,000 years ago. Until then horses were probably killed by human hunter-gatherers for their meat (they were depicted on wall-paintings in the 16,000-year-old Lascaux caves in southern France thought to have been created as part of a shamanic ritual to help locate wild beasts). Taming the instinct of fear out of these creatures may have come about as an incidental effect of rearing foals as a source of meat. By about 3500 BC, however, there is evidence of nomadic people experimenting with domesticating mature horses, still mostly the size of ponies, and discovering that with skill and training these animals were more useful alive than dead.

The Botai culture, Neolithic people living in the region now called Kazakhstan, may have been among the first to tame wild horses. Evidence from the remains of horse teeth studied by researchers at Exeter University, England, suggest riding bits and harnesses were in use as long as 5,500 years ago. They have also found the remains of food and drink in pottery vessels that suggest horse milk may have been consumed and even fermented into an alcoholic drink known as *koumiss*.

It is known for certain that by 2600 BC a hybrid breed of domesticated horse was being deployed in the Sumerian civilization of the Middle East because it is shown on the Royal Standard of Ur, a painted box unearthed by British archaeologist Leonard Woolley in the 1920s. The animals are depicted pulling chariots in which soldiers stand on a platform ready to fire arrows at their enemies.

The period between 2100 and 700 BC saw wave upon wave of men on chariots pulled by horses invading civilizations from Egypt in the south (Hyksos invasions) to India in the east. The epic Battle of Kadesh (1274 BC), fought between the Egyptian Pharaoh Ramesses II and the Hittite Empire from Turkey, involved an estimated 6,000 horse-drawn chariots. Whilst the Hittites specialized in close combat to knock out the enemy, the Egyptians would stand off at a distance, firing relentless volleys of arrows. The battle was a draw, and led to a compromise agreement in which both empires defined their spheres of influence over land later claimed by the Jews as theirs – promised by God.

The arrival of military power, initially through chariot warfare, brought huge cultural changes throughout Europe, Africa and Asia. Societies that made a living through trade (such as the Minoan and Indus Valley civilizations) and which worshipped a feminine deity of agricultural fertility (the Mother Goddess) were, from about 2000 BC, displaced by violent horse-using cultures that originated from the grassland steppes near the Black Sea. The Minoans, with their love of animals, as depicted on their frescoes (showing images such as leaping dolphins), were replaced by Mycenaean invaders whose tombs were full of objects made for war. The egalitarian Indus Valley society of trades, crafts and commerce disappeared,

Egyptian Pharaoh Ramesses II takes aim in the Battle of Kadesh, from an engraving by Achille Prisse d'Avennes, a mid-nineteenth-century French Egyptologist.

only to be replaced by chariot-riding invaders as described in the Vedic texts – probably dating from around 1700 BC. Class distinctions, initially into four castes, put soldiers and priests at the top levels of society with farmers and 'untouchables' at the bottom.

In Egypt the shock of Hyksos charioteers sweeping down from the North and invading the lower Nile (*c.* 1674 BC) caused a major rethink of imperial strategy when it came to burying their human gods, the pharaohs. Ostentatious pyramids that advertised the presence of buried treasure were replaced by underground chambers hidden to protect them from looting and theft. In Europe, a fashion for communal burial in barrows was usurped by a new trend in which individual graves were built for victorious invaders who were frequently interred alongside their weapons of war (including horses). The birth of a new social hierarchy also meant settlements were built at the top of hills for best protection and defence, not in lush valleys where the soil was most fertile.

All this happened because of the military power provided by the horse. By 700 BC the union of human and horse became even more closely melded. Selective breeding had increased the size of animals so that it was no longer necessary to wage battle on the back of an unstable, hard-to-manoeuvre chariot with its highly vulnerable charioteer. Nomads who herded flocks of animals (sheep, cows, goats) used their horses as a complete life-support system. They could literally live (even sleep) in the saddle. Horses also carried their rations, and sometimes these nomads even managed to ferment milk when on the move using leather pouches where it would curdle into cheese (to avoid the ill effects of milk intolerance, see page 195). Horsemeat from old animals was carried under the saddle as jerky. In extreme conditions they would even make a cut in their horse's neck and drink its blood as a source of protein.

Steppe nomads like these could choose to live peacefully, transporting goods between settled communities, or aggressively, raiding vulnerable towns and settlements. Sometimes they did both.

Horse-dependent Scythians were nomads from the area around the Black Sea whose rise to power began in about 700 BC. Their mounted warriors used a composite bow to unleash as many as six arrows before the first one had struck home – all the while riding a horse without using stirrups or reins for control. Skills such as these made the lethal combination of horse, man and bow into the world's most powerful weapons system until the invention of the machine gun in the late nineteenth century. As the armies of ancient Persia charged northwards in a bid to rid themselves of the Scythian menace once and for all, the capricious nomads simply disappeared – riding into the distance only to return to the fray at a time and place that suited them best.

When Idanthyrus, king of the Scythians, explained to the Persian Emperor Darius why he would not put to battle, he revealed the inner strength that a life shared fully with horses gave his people. *'If you want to know why I will not fight, I will tell you. In our country there are no towns and no cultivated land: fear of losing which or seeing it ravaged might indeed provoke us to a hasty battle'.*

After two years of fruitless efforts to confront the nomads, Darius withdrew.

The history of Asia and Europe would have been utterly different were it not for a series of other mounted nomadic invasions – probably provoked by climate change and population growth. The Huns, led by their famous leader Attila, made mincemeat of the declining Roman Empire in the fifth century AD, following which central European political control collapsed completely for several hundred years.

Mounted cavalry remained the mainstay of human conflict until well into the twentieth century. Muslim horsemen, experts in the art of selecting and breeding light, manoeuvrable and highly obedient breeds (following on from the example set by Mohammed) charged out of the Middle East in the seventh century, conquering settled civilizations across Asia, Africa and into Europe. Speed, reconnaissance and the precious element of surprise were theirs thanks to warfare committed on mounted horses.

Mongol chieftain Genghis Khan, born *c.* 1160, marshalled the most lethal nomadic force of all time, and his successors built the world's largest contiguous empire from China to Hungary. Mongol children were taught (by their mothers) to ride on horseback by the age of three, and by the age of five they could shoot with a bow while riding. Such complete symbiosis with the horse was the secret to their military success and, at least until the widespread use of gunpowder, settled societies had little or no defence.

Mongol tactics for subduing human towns and villages were the same as for hunting wild game. As many as 100,000 Mongol horsemen would stretch out in a line for sixty miles and, with precise discipline, riders at each end would gradually move further ahead until the army linked up into an ever-decreasing circle, capturing every living thing inside its impenetrable circumference. Once encircled, there was no escape – surrender or die.

Even after the arrival of infantry defences in the form of muskets, rifles and revolvers (such as during the American Civil War, 1863–5), the cavalry charge remained the most potent military force. Meanwhile it was the pulling power of horses that made possible the deployment of large military cannon, used extensively by European armies from the seventeenth century onwards. Barbed wire, trenches and machine guns eventually began to break the horse's grip on the killing fields of World War I (where an estimated eight million horses died) although by this time conditions were no less treacherous for infantry than for mounted cavalry. It was the emergence of motorized armoured transport – tanks and Jeeps – in World War II that finally took horses off the battlefield, although they remained much in evidence for hauling ammunition and supplies. Since then horses have continued to play a part in police forces for crowd control and remain at the heart of ceremonial military traditions in many countries.

Compare this to the history of the Native American empires (such as the Mayan, Incan and Aztec) which had no horses until European invaders of the sixteenth century reintroduced them to the Americas after an absence of perhaps 12,000 years. These people had no fortified settlements, no wheels, no carts, no bronze or iron weapons and no wars between nomadic and settled societies. Not that they didn't have violent histories of their own (see page 235) but without the horse, or any other large domesticated animals, their outcomes turned out very differently indeed.

Horses came to have such a profound impact on the history of human civilization for two main reasons. The first concerns their genetic versatility.

Scythian golden comb showing mounted armed combat, *c.* fourth century BC.

Mohammed's instincts about the benefit of breeding steeds to suit particular human needs have been consistently and successfully pursued at every stage of the human–horse relationship. Small feral ponies were 'domesticated' into species that could support warriors in battle. Those that survived the rigours of life with nomadic tribes were inevitably those that were best suited to the task. By the time of the Mongol invasions, steppe nomads had the most vigorous, robust, obedient and flexible breeds in the world.

Ninth-century Bedouin knights from Islamic Persia were probably the first people to take horse-breeding to a new, more precise level of scientific expertise by breeding only from the offspring of a single line of mares. The lack of grazing in the Middle East required strict control of herds, so they were especially fussy about looking after only those horses that served their masters best. They selected agile, 'hot-blooded' breeds that suited their style of warfare, which was based on attack by speed and surprise.

Following the Christian conquest of Islamic Granada in the fifteenth century, studs from the Carthusian monastery at Cartuja seeded the cavalry requirements of several European nations, such as the Habsburgs in Austria, with their highly skilled Lipizzaner ponies, creatures descended from high-spirited Andalusian stock. Other breeds such as today's British thoroughbred were established following the restoration of Charles II (1660), a keen horse-breeder himself. Their stock came from three Persian stallions: one was captured in war in Turkey, another was purchased in Syria and the last was a gift from the king of Tunis. Subsequent breeding from this stock proved vital for the effective supply of British cavalry forces from the eighteenth century onwards. They were also instrumental in establishing sports such as fox-hunting, polo, horse-racing and eventing that were conceived to keep the equestrian skills of an officer cavalry class well honed in peacetime, far cheaper – and a lot more fun – than maintaining large standing forces when they were not needed for war.

At the other end of the genetic spectrum were large, thick-set, calm-tempered 'cold' breeds that proved perfect for pulling ploughs in the heavy clay soils of northern Europe. The collar harness, an adaptation of a baggage-carrying device used by Asian camel traders, substantially increased the amount of land able to be pressed into agricultural use, boosting wealth for the beleaguered economy of early medieval Europe. These 'shire' breeds were later used for heavy transport work both in war (hauling artillery) and peace (pulling trams).

The second reason why horses have had such a major impact on human history is their highly sensitive nature. Horses are not traditionally much rated for their intelligence. Victorian training methods involved 'breaking in' a horse until it submitted to its owner – surely only stupid animals would succumb to such treatment? Recently more ancient training methods, as originally spelled out by Greek horse-guru Xenothon (*c.* 400 BC), have come back into vogue. Horses, said Xenothon, learn in just the same way as humans do – not through fear but by gentle communication and reward: *'for what the horse does under compulsion ... is done without understanding, and there is no beauty in it either, any more than if one should whip and spur a dancer ...'*

Horses are, in fact, among the most sensitive creatures on Earth. They can instantly tell when a fly lands on their back. Their ability to feel even the slightest movement of a rider is what enabled Mongol horsemen to control their horses hands-free. Without such sensitivity it would have been impossible for these raiders to fire off volleys of arrows whist charging past their enemies at thirty miles per hour. The myth that horses are just simple creatures that follow a traditional 'dominance hierarchy', in which the lead female is replaced by a human, is also in the process of being debunked. Recent research indicates that horses are naturally highly social creatures with complex group dynamics that moderate the aggressive characteristics of any one individual, providing social cohesion. It is this rational flexibility that has allowed horses to master their instincts and to be able to tolerate the most unnatural and terrifying environments, from battlefields to noisy urban streets, all while trying to follow the commands of a highly demanding bipedal ape: man. Given that horses are prey animals, whose instinct is to flee from threats and danger, their six-thousand-year relationship with man shows an extraordinary level of adaptability and an exceptional capacity for learning and understanding.

The upshot is that today horses are another of the modern world's most numerous species. China has an estimated eight million, Mexico 6.2 million, Brazil 5.9 million and the US 5.3 million. Compare this to its closest wild relative, the zebra, whose worldwide population numbers only a few hundred thousand with several varieties on the verge of extinction. There can be little doubt that the powerful partnership between people and horses has dramatically changed the course of human history.

Barefoot, ridden by Dick Goodison, winner of the Great St Leger Stakes in 1823 – a prestigious flatrace held at Doncaster racecourse every year since 1776.

Camel

FAMILY: CAMELIDAE
SPECIES: CAMELUS DROMEDARIUS
RANK: 87

A superbly adapted pack animal that made merchants rich through the trade of gold, spices, salt and slaves.

MANSA MUSA was a medieval West African emperor who ventured across the Sahara Desert on a once-in-a–lifetime pilgrimage to Mecca. He set out in 1324, accompanied by a retinue of 60,000 subjects and 12,000 slaves, stopping off at Cairo for a spot of retail therapy. So memorable was his visit that it was still being discussed twelve years later, according to contemporary Arab historian, Al-Umari, who wrote:

> *'This man flooded Cairo with his benefactions. He left no court emir nor holder of a royal office without the gift of a load of gold. The Cairenes made incalculable profits out of him and his suite in buying and selling and giving and taking. They exchanged gold until they depressed its value in Egypt and caused its price to fall.'*

Musa's wealth came from a mountainous region of West Africa called Bambouk, situated in the heart of his Empire of Mali. At that time, the region's three gold mines supplied more than half of all the bullion traded in North Africa, Europe and Asia. Such riches were entirely dependent on being transported overland across the inhospitable desert to the Mediterranean markets of the Arab and Christian world. Without Musa's caravan of eight camels, which between them carried more than two tonnes of gold, medieval history's most celebrated shopping expedition could never have taken place.

Camels, like horses (see page 242), evolved in North America about forty million years ago. The arrival of human hunters quickly led to their extinction *c.* 12,000 years ago (see pages 184–5). Fortunately by then, through various migrations across the Alaskan land bridge, stocks had already spread throughout Asia, reaching as far as the Middle East. Two-humped varieties (called Bactrian camels) thrived in the wetter climate of central Asia, whereas one-humped dromedaries, now by far the most successful species, made a living in dry, hot Arabian scrublands – habitats that became increasingly common throughout the Middle East as the climate dried out from *c.* 5000 BC.

Thanks to a range of brilliant evolutionary adaptations, camels can survive in places other mammals fear to tread. Contrary to popular belief, they do not store water in their humps. These are fat deposits that help camels survive long periods without feeding. Such humps also help conserve water by consolidating the heat-insulating properties of fat supplies away from the bulk of their bodies. Camels can also tolerate a wide range of body temperatures (from thirty-four to forty-one degrees centigrade) without breaking into a sweat. Other water-saving survival

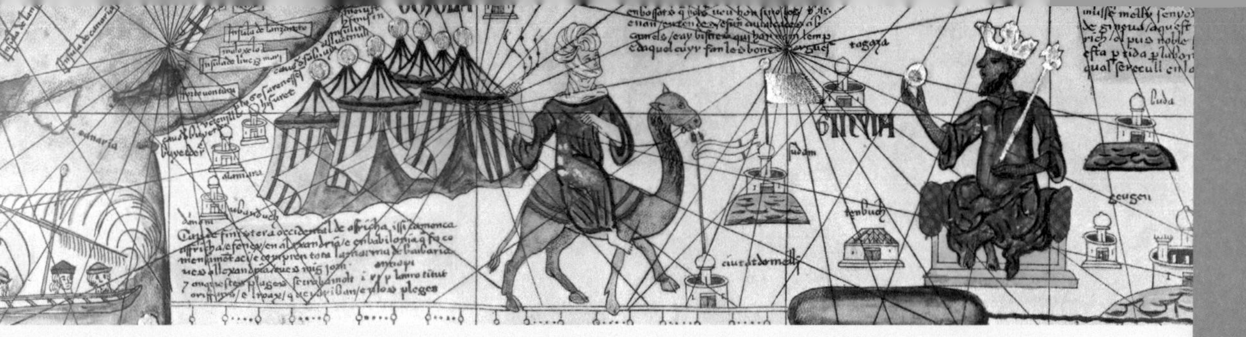

Golden wonder: Mali King Mansa Musa sits on a golden throne and is approached by a traveller on a camel. Part of a Catalan Atlas dated to 1375.

techniques include long legs that keep their bodies as far as possible from the scorching hot desert ground, oval-shaped blood cells that maintain a regular blood flow even when dehydrated, and nostrils that collect and return water into their bloodstream from vapour exhaled in their breath.

It is one of human history's most fortunate accidents that the only creatures well enough adapted to tolerate trans-desert migrations also happened to tolerate living in close proximity to people. Tribes living on the coasts of Yemen about 4,000 years ago are thought to have been the first people to domesticate the wild camels that roamed in the deserts that began just a few miles inland.

These tribes lived on seafood (sea cows, in particular, were a delicacy) and regularly sailed across the Arabian Sea to the east coast of Africa (present-day Somalia). The local manufacturing of incense gave them a reason to travel overland to markets in the Near East. Crossing the Arabian Desert with goods to take to market would have proved impossible without camels as pack-animals for carrying water supplies. Domestication was probably assisted by the lack of natural camel predators, making these animals naturally less fearful of humans. Salty seafood, such as sardines, proved to be a perfect source of bait to lure the camels close.

By 2,000 years ago the incense route had given birth to a number of wealthy trading cities, including Petra (now in Jordan) and Mecca (in Saudi Arabia). Camel-breeding and the development of saddles enabled large loads of merchandise (including humans) to be carried on camel-back. Camels became the chief means of transport across the Middle East, establishing overland trade routes connecting the Mediterranean to southern and eastern markets across the Syrian and Arabian deserts. They were also exported from Arabia by sea to Somalia where camel-breeding thrived on the arid East African terrain, eventually providing Berber traders in North Africa with a powerful pack-animal with which to transport their own commodities in the form of salt, slaves and gold across the expansive Sahara Desert.

It is an extraordinary testament to the impact of camels that wheels, invented by humans in *c.* 4000 BC, became almost completely redundant in places where domesticated camels were established as an alternative means of transport. The benefits of using this highly versatile pack-animal, as opposed to wheeled carts pulled by horses (or even camels), were pithily summarized by Major Arthur Glyn Leonard, a British military transport officer who proposed the introduction of the camel into South Africa in 1894.

> *'Let us sum up the special advantages of the camel over ox transport:*
> *1. Can carry or draw twice as much*
> *2. Faster and able to cover more ground daily*
> *3. Can do from 20–25 miles in one stretch*

4. Will make many more journeys in a year
5. Able to traverse ground that a wagon will sink in
6. No trouble fording rivers where wagons would have to be unloaded
7. Live and work four times as long
8. Greater powers of abstinence from food and water
9. Greater tenacity and endurance
10. Wagon liable to break down, upset or stick …
11. The additional dead weight of the wagon is at least a tonne, I should say.'

Following the seventh- and eighth-century Islamic conquests of the Middle East and North Africa, Muslim traders had a huge tactical advantage over Christian Europe with their control of overland trade routes made possible thanks to the triumph of legs over wheels. Such networks connected the Mediterranean with the gold hub of Sub-Saharan Africa and the silk and spice markets of eastern Asia.

Between *c.* AD 800 and 1450 the world's richest trading cities were the Muslim capitals of Cordoba, Cairo, Baghdad and Damascus, all of which thrived on camel-power. The old quarters of these cities provide a reminder of the advantages of four legs over wheels – their narrow, winding streets were not designed to accommodate a horse and cart.

Crowded Islamic cities like these provided much of the wealth that poured into fourteenth-century Renaissance Italy. Such riches mostly came from the selling of spices, silk, salt, slaves and gold transported by camel caravans across the deserts of Africa and Asia in exchange for European wool, timber and cloth.

Mansa Musa's ostentatious expedition through Cairo didn't go unnoticed by European merchants. Indeed it was as a result of his visit that map-makers began to include his empire on their charts, which were later used by Portuguese maritime navigators to find ways of circumventing the camel traders' overland monopoly. As a result, from *c.* 1450 onwards, they established their own overseas trade routes, cutting out the desert caravans of the Sahara and their Muslim middlemen.

The camel remained vital for maintaining cross-desert links throughout the world right up until the advent of motorized vehicles, such as Jeeps and Land Rovers in the 1930s. Camels, along with their Afghan handlers, were even transported and deployed in the outback of Australia from the 1860s, amply proving their value to European settlers eager to trade gold and other raw materials from mines buried deep in that continent's arid interior.

A dromedary camel, native to Arabia, but exported to Africa and Australia.

There was even an experiment to introduce camels to America at about the time of the Californian gold rush (*c.* 1848–55). This practice was restricted, however, for fear of equipping Native Americans with such a versatile, economically useful animal.

Yet camels, despite their immense value as a wheel-beating transport system, were never quite as globally successful as the domesticated horse (see page 242). After all, the most highly prized domesticated animals in history are those that humans find equally effective in war as well as peace.

The first recorded appearance of camels on the battlefield was by all accounts a spectacular success. Persian Emperor Cyrus the Great took great advantage of the fact that horses generally find the sight and smell of camels repellent. So, when Cyrus faced his arch-enemy Croesus of Lydia at the Battle of Thymbra in 547 BC, knowing that his army was outnumbered three to one, he placed the camels from his baggage train at the front of his forces. Greek historian and equine expert Xenophon relates how the tactic caused the Lydian cavalry to flee, turning the battle into a rout.

Cyrus's victory led to the Persian occupation of Anatolia, adding the Greek region of Ionia to its growing empire. The historical consequences of this camel cavalry victory were immense. Cyrus was now able to turn his attention southwards where he captured Babylon and freed the enslaved Jewish people, returning them to their former home in Jerusalem. Once re-established in their old capital, Hebrew scribes penned the Torah or 'Old Testament', part of their bid never again to be evicted from their Promised Land.

Bactrian camels, distinguished by their double humps, are better suited to wetter climates and higher altitudes than their one-humped cousins.

Despite such a promising start, camels could never truly compete with the horse as a weapons system. Camels were far too unstable to be reliable platforms from which to unleash a volley of arrows. Although the creatures could run at a constant twenty-five miles per hour, camel cavalry delivered nothing like the psychological terror of the sound and sight of horses charging at full gallop. Camels, for all their brilliant adaptations to desert conditions, simply aren't manoeuvrable enough to compete with horses in battle.

A second limitation is that one-humped camels thrive best only in dry, desert areas. Two-humped varieties have prospered in wetter, higher climes and the related llama of South America can even make its home on the sides of a mountain. Horses, however, are ultimately more versatile although, of course, they are nothing like as effective when trying to cross a desert.

The internal combustion engine finally revived the wheel's fortunes in places where, for as long as 3,000 years, domestic camels had kept it in the shade. Camel-herders still survive, however, especially in Somalia and Ethiopia, which between them host as many as half of the world's fourteen million population. While two-humped varieties are in decline (1.4 million), in Australia wild populations of one-humped camels – descended from ones that escaped their Afghan handlers – are thought to number as many as 700,000 in the central outback. With their populations growing at an impressive rate of 11 per cent per year, it seems that, like pigs (see page 217), these are creatures that can readily re-adapt to life in the wild.

Perhaps one day, if and when the era of motorized transport comes to an end, populations of camels like these will resume their former role as essential pack-animals. Just think of all the advantages: no need to build or maintain roads, breakdowns virtually unheard of and 'vehicles' automatically programmed to find their own fuel even in the harshest environments ... that takes some beating!

Cotton

FAMILY: MALVACEAE
SPECIES: GOSSYPIUM HIRSUTUM
RANK: 39

An environmentally demanding natural fibre that provoked a modern contest for industrial and capital supremacy.

GLOBALIZATION is an economic system founded on the worldwide trade of raw materials, supported by international finance, in which the industrial manufacturing of products on one side of the planet relies on their consumption as finished goods on the other. This contemporary (and, as I write, deeply stressed) model of human behaviour was pioneered by a powerful but ancient relationship between humans and a plant called cotton.

The human obsession with *Gossypium*, which envelops its fruit in a thick ball of twisted white fibres in the hope they may get caught in the fur of a passing animal, goes back as far as recorded history itself. There is still no scientific consensus on the plant's geographical origins. It was cultivated by the people of the ancient Indus Valley more than 4,500 years ago. They learned how to harness the fibre, spinning it into yarn and weaving it into cloth. Trade from growing, harvesting and processing fine cloth made of cotton formed the backbone of their economic welfare. Not much later, Native Americans on the other side of the world were busily doing the same thing with a different type of cotton. Although *Gossypium* plants native to Central and South America are of a different species from those that grow in India and Africa, genetically they are so closely interrelated that they must sometime, somewhere have evolved in proximity.

Was this genetic fusion a prehistoric freak of nature borne on the wind before the evolution of mankind? Or is the worldwide distribution of cotton itself evidence of prehistoric contact between the people of South America and Asia before recorded human history began? A similar debate still rages regarding the diffusion of human culture. Were the sacrificial pyramids of Chichen Itza, that so closely resemble the ziggurats of Babylon, examples of cultural convergence – in the same way that nature independently engineered systems of flight at least four times in evolutionary history (see pages 103, 120, 122 and 163)? Or does an unrecorded physical contact between the people of ancient Asia and America explain such architectural mimicry as well as the appearance of genetically related cotton plants on both sides of the world? It is an additional curiosity that people in Asia and Central and South Americas spun expertly with the same kind of spindles and wove their material on the same sort of two-barred looms. If such prehistoric contact did once exist, the story of cotton unearths globalization's most ancient roots.

The mysterious origins of cotton also flexed the minds of people in medieval Europe who lived far away from the native tropical heartlands of *Gossypium*. At that time it was widely thought that cotton came from the fruit of an exotic tree in India called the Vegetable Lamb of Tartary which, according to popular fourteenth-century travel writer Sir John Mandeville, *'bore tiny lambs on the end of its branches ...'* The creatures were thought to be connected by an umbilical cord from the tree to the ground where they grazed on the surrounding grass.

Cotton, silk (see page 260), wool (see page 218) and flax (see page 401), are the four global natural fibres that have traditionally clothed human civilizations. But in the last few hundred years something extraordinary happened that propelled cotton to the very top of the list of species that have had the biggest impact on human history.

Cotton's modern conquest began in the early seventeenth century with the triumphant colonization of the Americas by smallpox-bearing Europeans (see page 26). Desperate to find a crop that could be traded to make European colonization economically viable, New World pioneers experimented with planting cotton – along with tobacco (see page 289) and sugarcane (see page 202). In 1607 the first cotton seeds were sown by colonists along the James river in Virginia. This crop soon became the biggest generator of economic wealth for what were to become the southern United States.

But turning raw cotton cloth into finished products was strictly banned in the British colonies. Imperial officials preferred to keep profits to be made from manufacturing finished goods (such as clothes) close to home where additional supplies of raw cotton were also beginning to arrive from trade with India.

Potential fortunes were on offer at this British crossroads of international trade if only a way could be found of cheaply converting raw cotton into material wealth – bales of fibrous cellulose into mass-produced sheets, shirts and trousers. The business of automating the processes of spinning, weaving and dyeing was therefore reliant on British entrepreneurs whose rational problem-solving skills spun into overdrive at the prospect of turning cotton's short white fibres into yarns of strong, long material that could easily and cheaply be woven into cloth.

Cotton bolls on a branch of *Gossypium*. Nature designed them to stick to animal fur but they ended up woven into the fabric of human history with the rise of civilizations.

Such commercial incentives meant that between 1738 and 1800 a slew of ingenious human inventions transformed man's ability to harness the wealth of cotton. John Kay's flying shuttle (1733), Lewis Paul's spinning machine (1738), James Hargreaves' spinning jenny (1764) and Richard Arkwright's water frame (1769) led to the establishment of mass-production systems for processing cloth. By 1788 more than 210 cotton mills straddled the fast-flowing rivers of Lancashire, whose sources lay in the nearby Pennine hills. The city of Manchester – dubbed 'Cottonopolis' – lay at the heart of this industrial revolution, leading to the creation of enormous wealth for British companies that converted cotton grown in colonies such as America and India into clothes for sale (often back to the very same colonies). Cotton became the economic bobbin around which the systems of today's globalized trade were originally wound.

Cotton's detonation of the industrial innovation in Britain was aided by that country's conversion from energy supplied by wood to energy supplied by coal, and from water power to steam. It has been estimated that by 1889 nearly five million people were employed in Britain's cotton-processing and related industries, covering an area of 500 square miles in East Lancashire, more than a quarter of the county. Mill operatives accounted for only a fraction of the total employed within this mechanized mecca. Engineers, carpenters, coal workers, bleachers, dyers, candle makers and steam engine suppliers were just some of the other trades that fed on the burgeoning business of converting raw cotton into finished cloth.

Japan's industrial transformation was similarly triggered by cotton. Following the enforced opening of its economy to global trade from 1853, the industrialization of Lancashire was transplanted almost piece by piece to the prefecture of Osaka in Japan. Bales of raw cotton imported from India were processed in Japan, initially using machines supplied by Britain, to produce finished cotton goods for export to China and back to India.

Between 1885 and 1955 this new cotton industry revolutionized Japan's self-sufficient craft-based economy into an international factory model. By 1919, cotton processing represented more than 13 per cent of Japan's gross domestic product, which involved, according to one estimate, more than half a million factories. New management techniques and templates for corporate finance and capital liquidity through stock exchanges developed rapidly as a result, reorienting Japan into a major economic force in an increasingly globalized world.

Like the exploitation of other crops that appeal to humans (such as sugarcane, see page 202, and rice, see page 230), harvesting cotton was an extremely labour-intensive affair. In the profit-hungry environment of early-modern European colonization, its rise to economic predominance made sense only if the crop could be gathered cost-effectively. As a result, cotton-picking led to a dramatic increase in the use of slavery. The grim harvesting of cotton's material wealth was paid for by millions of kidnapped African men and women who were traded by European merchants in exchange for finished cotton goods, ammunition and arms.

Inventions such as Eli Whitney's cotton gin (1793) made it possible to clean cotton fifty times more quickly by using horse power. The human cost of such scientific progress was that demand for extra raw cotton fields soared. By 1803 more than 20,000 African slaves were being imported annually and were sold to work on American cotton plantations. By 1850 the United States was producing as many as three million bales of raw cotton, hand-picked by African slaves, for export to Britain each year.

Cotton's role as protagonist for the globalization of trade didn't just result in misery for African people. When supplies of raw American cotton were curtailed during the American Civil War (1863–5), British and French traders turned to Egypt as a source for growing and harvesting the precious fibre. As a result the Egyptian government took out substantial loans to invest in new plantations. When the Civil War ended, however, European traders abandoned Egyptian cotton, returning to cheaper American sources, and leading to Egypt's bankruptcy by 1876.

Injustices in India were no less profound. The effect of pitiful wages earned by cotton-picking labourers eventually led to a groundswell of nationalistic indignation, personified by their independence leader, Mahatma Gandhi. It was, said Gandhi, a scandal that Indian raw cotton was exported for processing in the UK and then sold as finished goods to Indian nobility, supported by the British Raj, who could afford to pay for it. Meanwhile, profits from the transportation, manufacturing and sale of cotton goods grown in India were bagged by British-owned enterprises. The 'free-trade' dynamics of such inequality, rooted in the growing and processing of cotton, were instrumental in India finally gaining its independence from Britain in 1947.

Cotton's most recent legacy, at least in terms of its relationship with modern human civilization, is ecological. This shrub doesn't yield its magic fibre for nothing. Unlike legumes, it has no self-fertilizing symbiosis with bacteria such as *Rhizobia* that nourishes the soil (see page 40). Instead, cotton-producing *Gossypium* is heavily dependent on plentiful supplies of fertilizer and water while its susceptibility to infestations and diseases such as boll weevil are notorious. To satisfy human demand, therefore, cotton has become the most human-dependent crop of all time. Pesticides, artificial fertilizers and massive irrigation projects have poisoned ecosystems and emptied seas. Shortly before the collapse of the Soviet Union in 1989, the full horror of Russia's urge to become self-sufficient in growing and processing cotton (dubbed 'White Gold'), was revealed to the world in a film called *Psy* ('Dogs'), by Soviet director, Dmitry Svetozarov.

Russia's cotton industry began in the 1870s but reached peak capacity in 1976. Its cotton plantations were irrigated by the diversion of two rivers that previously poured into the Aral Sea, once the world's fourth largest lake (68,000 square kilometres). Between 1960 and 1998 the amount of water taken to supply Russian cotton fields was doubled, causing the volume of water in the lake to shrink by a staggering 80 per cent. What was once a vibrant marine ecosystem that supported 40,000 Russian fishermen is now little more than a dead brackish puddle. As much as 90 per cent of the former lake has become desert.

Cotton is therefore a crop synonymous with the industrialization and globalization of the modern world. Despite the growth of artificial materials since the 1950s (such as polyester, nylon, acrylic), cotton is still the most widely cultivated, ecologically demanding and environmentally damaging, mass-produced natural fibre grown and processed all over the world today.

The Aral Sea, once a verdant blue inland ecosystem, has now almost totally disappeared as a result of over-zealous irrigation.

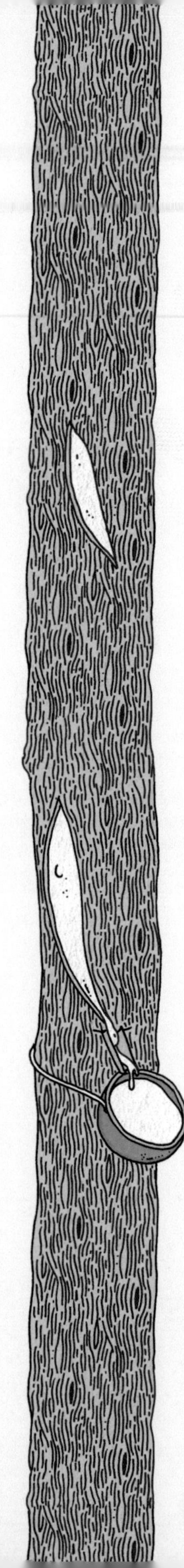

Rubber Tree

FAMILY: EUPHORBIACEAE
SPECIES: HEVEA BRASILIENSIS
RANK: 76

A relative late-comer to the crops that have most influenced humanity but one that had a colossal impact that made up for lost time.

KEW GARDENS is internationally renowned as one of the world's leading centres for horticultural excellence. Founded in the mid eighteenth century by the British royal family as a nursery park for their children, the gardens went on to house, from about 1840, a national botanical collection. Ornate glasshouses were populated with exotic species from all over the world. To fulfil its new purpose, the estate was extended to cover an area of 120 acres in a prime location alongside the river Thames in west London. It is something of a surprise to find that this innocent-sounding, plant-loving enterprise was once at the centre of an international conspiracy which turned out to be one of the most lucrative heists in all history.

The plot did not involve much money. At least not at the stage when Kew Gardens was involved. Rather, it concerned something far more valuable to the avaricious instincts of various nineteenth-century British imperial officials. Their sights were set on a new kind of treasure that grew on the other side of the world. They were after seeds. Not just any old seeds, but those that belonged to a particular type of Peruvian tree that thrived along the south bank of the mighty Amazon river.

Beginning in about 1740, when European exploration of the New World was still at its height, two Frenchmen set out on an expedition down the Amazon. What most impressed Charles Marie de la Condamine and François Fresneau as they explored the interior of Peru was how the natives living on the banks of the Amazon seemed to 'milk' a certain type of tree. The thick white substance, which the Frenchmen called 'latex' (derived form the Latin word for 'milk') was processed in makeshift smokehouses, turning gooey gum into elastic sheets that we would recognize today as a type of rubber. The Frenchmen sent samples back to their home country where they set about conducting a series of experiments to see what possible commercial uses this mysterious substance might have.

The native Amazonians were in fact pursuing a most ancient tradition – one that was first established more than a thousand years before. A people called the Olmecs (meaning 'rubber people'), who lived in Central America, were the first civilization known to have used sap from a tree called *Castilla elastica* to make rubber. They mixed the latex with the juice of a Morning Glory plant (*Ipomoea alba*), dried it and cut it into strips to construct bouncy rubber balls around a solid core. The balls were used for playing a sacred game called *ulama*. Teams of players would try to propel the ball through a hoop, built in a high stone wall, without using their hands or feet. Losers were often sacrificed to the gods, and their heads were sometimes used to form the core of the next generation of sacred rubber balls.

The latex tapped by the Amazon natives that the eighteenth-century French explorers had observed came from a different species – the Para rubber tree, or *Hevea brasiliensis*. This material, the Frenchmen found, could be used rather successfully for waterproofing clothes by dipping fabrics in a solution of latex and turpentine. For the next hundred years small quantities of rubber were exported, mostly as a curiosity, from the Amazon to Europe and North America, where it was used, among other things, for making waterproof shoe coverings, more commonly known as galoshes.

Rubber was not an instant hit in the rapidly commercializing and industrializing nations of Europe because it suffered two serious flaws. When it was too cold it turned brittle, losing all its elasticity. When it was too warm it became soft and gooey.

Attempts to cultivate a variety that didn't go brittle when cold and sticky when hot were handicapped by the fact that a rubber tree starts producing milky sap only after six years of growth. It was therefore left to human curiosity and ingenuity to find ways to compensate for what natural genetics could not achieve in an acceptable human timescale. As luck would have it it was found that the problems could be solved by heating latex laced with sulphur to 170 degrees centigrade for ten minutes. Two quite different types of entrepreneurs discovered this process – called vulcanization – between 1835 and 1843: the usually bankrupt maverick Charles Goodyear in America and a shrewd Protestant businessman called Thomas Hancock in Great Britain.

Vulcanization transformed global demand for rubber. The timing of this discovery was impeccable. High-pressure steam power, invented by Cornishman Richard Trevithick in 1801, was by now being used for self-propelled ships and railway trains to revolutionize transport all over the colonial world. The elastic properties of vulcanized rubber could be applied to make seals for steam engines. The old method of sealing the steam cylinders, involving oil-soaked leather, was inefficient when subjected to high pressure, allowing steam to escape, reducing the engine's power and efficiency. Not so with vulcanized latex, which preserved its elasticity even at high temperatures, allowing steam cylinders to be fitted with rubber seals that didn't leak. Vulcanized rubber seals led to ever more powerful engines, used in everything from cotton mills to electricity generating stations.

Charles Goodyear shows colleagues how he learned to vulcanize rubber after accidentally experimenting with latex on his stove. From an engraving published in 1871.

Rubber springs were used to make the suspension for railway carriages, horse-drawn carriages and trams, making journeys more comfortable. With the invention of the bicycle and later the internal combustion engine in the nineteenth century, rubber found a new use in the manufacture of pneumatic tyres (commercialized by John Dunlop in 1885). By this time demand for rubber became truly enormous and the amount that could be supplied from wild trees tapped by natives in the Amazon basin simply wasn't great enough, even though South American exports of rubber had rocketed

DUNLOP TYRES
LEAD THE WORLD
FIRST IN 1888
FOREMOST EVER SINCE

Easy rider: by the late nineteenth century global demand for rubber far outstripped supply.

from 200 tonnes a year in the 1830s to more than 20,000 tonnes by 1890.

Opportunities to cash in on this commodity boom in rubber were not missed by imperial wannabes such as Leopold II, King of Belgium (ruled 1865–1909). Leopold established his own rubber-growing empire in what is now the Democratic Republic of Congo. Latex milked from this region had traditionally been used by natives to attach arrowheads to their hunting spears, but unlike latex from the Amazon it came from a wild vine called *Landolphia*.

Leopold's private army used forced labour to extract rubber worth an estimated £14 million between 1898 and 1905. The price paid by the natives was horrific: many had their hands and penises cut off if they failed to deliver sufficient quantities of rubber. The population of the Belgian Congo dropped by ten million during Leopold's years of tyranny. Some died from a concurrent sleeping sickness epidemic (see tsetse fly, page 140) but most deaths were due to unimaginable levels of brutality. British sea-clerk Edward Morel and civil servant Sir Roger Casement finally exposed the genocide, which Leopold had disguised under the auspices of a humanitarian project, in 1906.

Growing international demand for vulcanized rubber in the last quarter of the nineteenth century propelled Kew botanical gardens to employ the services of a struggling freelance British explorer called Henry Wickham, who visited South America when he was twenty. In 1875 Sir Joseph Hooker, then director of Kew, offered Wickham a reward of £10 for every 1,000 rubber seeds that he could collect from the Amazon basin and have transported back to London.

With unbridled enthusiasm (and plenty of help from bribed natives) Wickham extracted as many as 70,000 seeds and shipped them back to London via Liverpool, earning himself a handsome £700 (and later a knighthood) for his troubles. When the Brazilian authorities discovered what had happened they branded the Londoner a despicable thief for carrying out an 'exploit hardly defensible in international law'. But, strictly speaking, at that time there was no official prohibition on the export of seeds.

Wickham's *Hevea brasiliensis* seeds arrived at Kew in June 1876. About 10 per cent germinated in the botanical garden's greenhouses, of which almost 2,000 were immediately exported to Ceylon, now Sri Lanka. The plan was to establish rival rubber plantations under British control some 10,000 miles away from their native country of origin. After several unsuccessful cultivation attempts (and further shipments of seedlings from Kew), twenty-two healthy plants were eventually forwarded from Ceylon to the botanical gardens in the British colony of Singapore. Sir Henry Ridley, director of the gardens, was under no illusion as to the potential value of rubber plantations. Under his stewardship the seedlings were sent to Malaya where an initial 2,000-acre plantation was established in 1898. By 1920 colonial Malaya's newly introduced rubber trees covered more than two million acres of what was once tropical rainforest and supplied the bulk of the British Empire's ever-increasing demand for vulcanized rubber.

Demand for rubber continued to rise, mostly as a result of the popularity of the motorcar from the 1930s onwards. Between the two world wars scientists set out to find a way of making rubber artificially, using oil as a raw material. Synthetic rubber gradually began to supplement natural rubber, especially when sources of natural latex from south-east Asia were interrupted during World War II (1939–45). Today about 40 per cent of the world's annual output of sixteen million tonnes of rubber comes from milking nature's prime latex-producing species, *H. brasiliensis*. The rest is synthesized from oil.

It is natural not synthetic latex that looks set to be the most viable long-term solution to mankind's requirements for rubber, however. Synthetic substitutes use at least three times more energy in their manufacture. What's more, rubber trees absorb carbon dioxide as they grow, so rubber produced naturally is more environmentally sustainable than artificial alternatives based on finite supplies of fossil fuels. Also, once *H. brasiliensis* comes to the end of its useful latex-producing lifespan (about twenty-five years) its wood provides an excellent resource for making furniture – a trend that has become increasing fashionable in the early twenty-first century.

Henry Wickham employing Amazonian women to weave him enough baskets to transport rubber seeds back to Kew.

But there is a downside even to using rubber from natural sources. Once fixed by man into its vulcanized form no type of rubber can easily be disposed of. When a rubber product comes to the end of its useful life (eventually it oxidizes and becomes brittle or wears down from excessive use) there is no known large-scale economically viable recycling process. As a result, more than 90 per cent of all rubber used today is incinerated or discarded in landfill sites, both of which cause serious environmental damage (see also *Pseudomonas*, page 38). It is estimated that of the more than one billion waste vehicle tyres discarded annually only 10 per cent are recycled in any form – accounting for a waste-pile of *three billion* tyres in Europe and *six billion* tyres in America. What's more, consumers often discard their vehicle tyres when they are only halfway through their viable life, because frequently they replace all four wheels even when just one or two are worn down.

Materials that evolve through organic processes can be broken down by nature's systemic recyclers such as fungi (see page 82), earthworms (see page 97), beetles (see page 100) and bacteria. Not so with materials that humans produce purely for their own needs. As modern uses for rubber become ever more various (from bathtub ducks and condoms to airport baggage conveyor belts), people have yet to prove that they can intervene with the same ecological sympathy as nature.

Silkworm

FAMILY: BOMBYCIDAE
SPECIES: BOMBYX MORI
RANK: 86

A highly domesticated moth that has enriched civilizations for thousands of years but which seldom gets the chance to grow up.

SEXUAL REPRODUCTION has generally ensured that no single human being is exactly the same as another, in either looks or personal taste. Nevertheless, some naturally occurring materials, especially those that are shiny, sparkly and rare, do seem to have universal appeal. Elements such as gold and silver, or precious stones (such as diamonds) have been coveted by humans since the beginning of recorded history. Their ubiquitous appeal has underpinned the wealth of civilizations. Not all such riches result from inert geological processes. High up on the list of the world's most illustrious materials is the unlikely-sounding excretion of a certain caterpillar, which seeks to protect itself by spinning a salivary mixture into an impenetrable cocoon. This silk has been prized by humans all over the world as one of the most luxurious materials on Earth.

Quite a few invertebrate animals produce silk as a natural product. It forms a vital part of their survival strategies in the wild. Spiders spin webs of silk to catch flies, and bees sometimes produce a silk-like substance to help construct their nests. But the sort of silk that has captured the imagination of men and women for at least the last 6,000 years belongs to a type of moth, *Bombyx mori*, whose caterpillar larvae feed on the leaves of mulberry trees. After feasting for as long as a month and shedding their skins four times, these six-legged caterpillars begin swinging their heads from side to side in the shape of a figure of eight. Two glands near the caterpillar's lower jaw give off a fluid that hardens into fine silk threads as it comes into contact with the air. At the same time the caterpillar excretes a gum called sericin that cements the two threads of silk together. In this way the worm-like creature constructs a highly reinforced cocoon in which it lives for three weeks while it transforms itself from a lava into a moth. Almost all the silk that has ever been traded in the world has come from the cultivation of this single species.

Legend has it that Leizu, wife of an ancient Chinese emperor, was the first person to discover the potential wealth locked up in the cocoon of *B. mori.* She had been charged by her husband to find out why his grove of mulberry trees was looking so sick. As she took a break from her investigations a cocoon from a caterpillar dropped into her cup of tea. As she looked down she saw it unravel. Entrepreneurial curiosity took over and, it is said, Leizu worked out how to weave the material into fabric.

As Leizu discovered, when dipped in hot water, fibre from the cocoon of *B. mori* unravels into a single thread, sometimes stretching up to *one kilometre* long. This makes silk quite unlike wool (see page 218) or cotton (see page 252), which yield only short fibres that stretch to a few centimetres at most. They have to go through a complicated process of spinning to turn them into long threads of useable yarn. Making fabric out of silk is simpler. First, kill the caterpillar by dipping the cocoon in boiling water (to avoid it hatching and therefore severing the thread) and second, twist and wind the long fibres from several cocoons on to a single bobbin to make a robust, highly versatile material, ideal for weaving into cloth.

Silk is not just convenient because of its uninterrupted length. It is also one of the toughest naturally occurring substances in the world, a legacy of its original evolutionary purpose as a sturdy changing-room. It didn't take long, once the secret of silk-making had been discovered, for human reason to put such a valuable material to work. Some of the earliest battle garments in history were tightly woven Chinese silk vests, which offered a degree of protection against incoming arrows. Genghis Khan (1162–1227) issued them to his Mongolian horsemen. Enemy arrows that reached their intended targets still became embedded in human flesh but were surrounded by silk, rather than shattering. This made them easier to extract and caused much less injury. Even in the late 1800s American bullet-proof vests were made out of silk and provided effective protection against shot from black powder handguns.

Bombyx mori seldom gets the chance to reach adulthood, but when it does, it looks like this.

But silk's biggest appeal was to the human senses. Thanks to the triangular arrangement of its molecules, silk reflects the light in a way that gives an iridescent sheen. This, together with the fact that it is lightweight, warm and elastic (which stops it creasing), means silk is an ideal fabric for clothing. Finally, when dyed, cloth made from silk has a deep, rich appearance that makes those who wear it look and feel really rather special.

This is why kings, emperors, popes and aristocrats have traditionally chosen silk as their preferred material for clothing. Archaeological evidence suggests that people living along the Yellow river valley in China between *c.* 6000 BC and *c.* 3000 BC were the first to practise sericulture. Silk was so highly revered by ancient Chinese society that it passed sumptuary laws to prevent peasants from wearing the fabric. This prohibition continued until the Ming dynasty (1368–1644). Raw silk was exported from China to India until about AD 100 when the Indians learned the art of sericulture themselves. Expert craftspeople turned the material into vibrant coloured saris as garments for adorning the well-off higher castes. Poor low-caste people stuck to wearing cotton or wool instead.

Silk was used as a symbol of wealth, status and privilege in human societies – from Beijing to Cairo – right from its very earliest manufacture. Fragments of a silk thread recovered from the hair of an Egyptian mummy (*c.* 1070 BC) indicate that some of the earliest-known trade contacts between east Asia and the Mediterranean were based on Chinese silk. Silk garments became so popular in Roman high society that a decree was passed by Emperor Tiberius (42 BC–AD 37) in an attempt to prevent so much gold passing into Chinese hands. Clearly it was a material that appealed greatly to Roman ladies because it made other people take more notice of them, much to the chagrin of writer and historian Seneca the Elder (54 BC–AD 39) who wrote:

'I can see clothes of silk, if materials [that] *do not hide the body nor even one's decency can be called clothes ... Wretched flocks of maids labour so that the adulteress may be visible through her thin dress, so that her husband has no more acquaintance than any outsider or foreigner with his wife's body ...'*

When the Byzantine Greeks took over the rump of the eastern part of the Roman Empire beginning in the fifth century AD, silk had already become an essential part of the imperial arsenal for re-establishing law and order. Justinian I (emperor of the Eastern Roman Empire from AD 527–65) had a three-pronged approach to restoring imperial authority: military re-conquest, legal renaissance and the establishment of indigenous silk production.

To Justinian, the priority was to set up his own private silk-making industry. He sent two Nestorian monks as spies to China where they successfully managed to smuggle a batch of *B. mori* eggs back with them to Constantinople (safely tucked inside a bamboo cane). Until then, the art of silk-making had been a closely guarded secret and production was limited to the Far East. Japan had been producing silk since *c.* AD 300, when they learned the secret after kidnapping four Chinese silk-maids. Justinian established mulberry groves and imperial silk factories in Greek cities such as Corinth and Thebes. Tribal leaders and ambassadors from other cultures were treated to displays of royal opulence inside the capital Constantinople, and given sumptuous gifts of silk in exchange for political alliances. This proved that societies can survive – at least for a while – by diplomatic cunning as much as military might.

However, such opulence eventually contributed to the empire's demise at the hands of Christian crusaders. They sacked the imperial silk-making city in 1204 when they were supposed to be ousting Muslim 'infidels' in Jerusalem. As a result of the raid, *B. mori* and its human cultivators spread farther eastwards into Christian Europe via Norman Sicily. By the time of the Italian Renaissance (*c.* 1450) sericulture had became a source of wealth for the rich ruling families of Venice, Genoa, Lucca and Florence. So many artisans had relocated from the Byzantine empire to Italy that by 1472 there were eighty-four workshops and as many as 7,000 silk craftsmen in the Italian city-state of Florence alone.

European lust for the richly coloured, shimmering fabric reached its zenith in France in the fifteenth century. King Louis XI (reigned 1461–83) applied the same tactics as Justinian, although nearly 1,000 years later. In 1480 he relocated an entire silk-growing and processing industry to the French city of Tours in an effort to avoid paying Italian merchants exorbitant prices and to reduce his country's spiralling trade deficit. The policy was reinforced under King Francis I (reigned 1515–47) who granted the city of Lyon a monopoly on silk production. By 1560 more than 12,000 people were employed in this single city's silk-weaving industry.

It was here that Joseph-Marie Jacquard (1752–1834) invented a new type of loom that could weave portraits out of silk cloth as if they were painted pictures. It worked by employing a series of punched cards (rather like a pianola scroll) that could control every individual line in the weaving of a single piece of cloth. Like woodblock printing, once the cards had been designed, they could be used again and again. Five years after the death of Jacquard this mechanized form of silk weaving inspired British inventor Charles Babbage to use punch cards to program his famous 'Calculating Machine' – the world's first programmable computer.

Another country that profited from wealth generated from silk was Iran. Located at the heart of the ancient overland trade routes that connected Asia to Europe, a succession of Persian regimes had long been successful at adding value to raw silk from China by turning the shimmering fabric into luxurious silk textiles. The Safavid dynasty (1501–1722) was able to use the wealth generated by selling silk rugs to European powers, via merchants in Christian Armenia, to re-equip and retrain its military forces in weaponry based on gunpowder, which was beginning to replace traditional forms of medieval warfare. The Shi'a form of Islam is still predominant in that region of the Middle East today, largely as a result of the Safavids' silken success.

But *B. mori* caterpillars never quite reached the status of a globally dominant species, even if the silk products of their labours did. British monarch James I tried to grow 100,000 mulberry trees near his palace at Hampton Court, London, with a view to copying the success of the Byzantines and French. But for some reason his caterpillars never took to their newly adopted home, although limited production was successful at a castle in Kent called Lullingstone, and continued there until the middle of the twentieth century.

The caterpillars weren't a big hit with colonialists in North America either. Small-scale production succeeded in New Jersey in the 1800s but by then cheap Japanese silk imports, which flourished alongside the growth of that country's cotton industry (see page 254), had forestalled the economics of mass production in the United States. By the middle of the twentieth century the break in supply of raw silk as a result of World War II led to the creation of artificial alternatives. Nylon, first synthesized in 1935, caught on as a replacement for silk and quickly came to be used in everything from parachutes to ladies' stockings. The same plastic fibre has since replaced other natural fibres such as wool and cotton, especially for use in sheets, clothes and other consumer fabrics.

But the human attraction to the secretions of silkworms lives on, even if their silk is no longer a mainstay of twenty-first-century material wealth. These caterpillars are so easy to breed and their silk production so prolific that geneticists have recently experimented with using them to manufacture other chemicals and fibres that may be useful to man. One Japanese team has successfully introduced genes responsible for manufacturing the protein collagen (see page 59) into the DNA of silkworm eggs. Potentially the silk produced by these genetically modified varieties of *B. mori* could be used to make a form of semi-artificial skin.

B. mori is probably the least fortunate of all the main domesticated species. After thousands of years of human selection these moths are now genetically so far removed from life in the wild that even as adults (when they are allowed to hatch out of their cocoons in order to lay fresh eggs) they cannot fly or feed without human assistance. Meanwhile, the vast majority are killed before they are allowed to mature in order to protect their 1,000-metre-long thread.

Eucalyptus

FAMILY: MYRTACEAE
SPECIES: EUCALYPTUS GLOBULUS
RANK: 16

A highly invasive species uncorked by European colonists who discovered its treasures down under.

SCHOOLS and charities are among the many organizations that today promote the idea of planting a tree for life. The thought appeals to those, perhaps living urban lifestyles, who want to feel they can still do something to improve the global environment. Here's the blurb from a woodland charity that promotes the idea of 'giving a tree as a gift' on its website:

> *'Our dedicated tree-planting gift is probably one of the most eco-friendly and unique presents you can buy! It's a gift that doesn't "cost the Earth" and it's suitable for all occasions, a very special way of giving something back to the planet. It truly is the gift of a lifetime!'*

Appealing as it sounds, planting trees willy-nilly without asking probing questions such as *What type is it? Where is it to be planted?* and *What are its likely ecological side-effects?* can sometimes lead to more environmental harm than good. The need for caution is well exemplified by the story of eucalyptus.

There are about 700 species in this highly successful family of flowering plants (see angiosperms, page 129). Almost all of them evolved in Australia, where they were isolated from the rest of the world until the Pacific voyages of English explorer Captain James Cook (sailed 1768–1779), which trail-blazed European settlement to the Australian mainland. Among the exotic species first encountered by Cook's on-board botanist Joseph Banks (1743–1820) were gum trees, species of eucalyptus never before seen by non-aboriginal people.

Eucalyptus trees evolved between fifty million and thirty-five million years ago. Climate change and the geology of Australia largely explain why these angiosperms behave as they do today. Australia was once a part of Gondwana, the southern part of the super-continent Pangaea that began to break up *c.* 250 million years ago. Having avoided colliding with any other landmasses since, this place has some of the oldest, least-disrupted terrain on Earth. In the absence of a good geological stir, typically provided by volcanic activity from colliding tectonic plates, life-supporting minerals and nutrients have been sucked dry. Species that have survived are those that are most intensely competitive for access to the scarce raw materials necessary for life.

Over the last thirty-five million years, Australia's climate has also become increasingly arid. Divining for water is therefore another essential survival skill that has, in this place, made the critical difference between survival and extinction. Just as when mammals developed their own bag of essential survival adaptations in the shadow of the dinosaurs (see pages 157–9), so eucalyptus learnt how to thrive in the

dry, nutrient-poor landscape of what is now modern Australia.

Their range of survival skills is breathtaking. First, they grow extremely fast – reaching full maturity in just six years. In a competitive environment, the quicker a tree can grow and establish its canopy to snatch as much sunlight as possible, the better. Second, eucalyptus trees have superb skills for finding water even when sources are buried deep in the ground. Long roots and rich, intricate connections with mycorrhizal fungi (see *Prototaxites*, page 86) combine to provide a powerful solution. But it is eucalyptus oil – famous today for its disinfectant, antibacterial, antifungal and anti-insect properties – that comprises this tree family's ultimate weapon. Rotting leaves and sticky gum (exuded from the roots and bark of the eucalyptus tree) kill off microbes and poison the surrounding soil, thereby warding off other species that might threaten supplies of vital nearby nutrients.

If a lightning strike happens to trigger a forest fire, dense-growing eucalyptus gives the arboreal equivalent of a victory whoop. The whole forest goes up in flames all the more readily thanks to its highly volatile, combustible oil. Then, just as a phoenix rises from the ashes, these die-hards grow back from charred stumps thanks to tubers and buds that lie buried under layers of protective bark. With all remaining competition removed at a sweep, the smouldering ash provides a perfect carpet of nutrients from which forests of eucalyptus re-grow.

Aboriginal Australians co-existed with forests of eucalyptus for tens of thousands of years before the first European colonists arrived. Infusions made from eucalyptus oil were a traditional medicine used for treating bodily pain, sinus congestion, fever and colds. Oil from one of the most widely grown species today, *Eucalyptus globulus*, is still used to make cough sweets and as an antiseptic.

Potential for material gain was not lost on Captain Cook and his crew when, at the end of the eighteenth century, they began to familiarize themselves with these highly successful species. Banks, also botanical advisor to King George III, was instrumental in ensuring eucalyptus seeds were brought back to Britain where they were cultivated in botanical gardens such as Kew (see page 256). Within fifty years the prodigiously fast-growing properties of this tree family helped turn it into one of the most widely cultivated and globally transported groups of trees in human history. With demand for wood soaring as a result of high-pressure steam power, the building of railways and the construction of mines, species of eucalyptus (especially the red gum *Eucalyptus camaldulensis* and the blue gum *Eucalyptus globulus*) were, for ambitious European industrialists, like manna from heaven.

Wherever nineteenth-century European colonists went in their global pursuit of material wealth, eucalyptus trees followed in their wake. In 1850 the trees were introduced to California at the time of the gold rush to support the massively increased energy requirements of an immigrant population that soared by as many as 300,000 people in just three years. Eucalyptus trees respond well to human coppicing. They quickly recover after being hacked down thanks to their well-buried buds, techniques developed in response to decimation by fire.

Edmundo Navarro de Andrade (1881–1941) was a Brazilian agronomist who has been credited with establishing more than *twenty-four million* eucalyptus trees in plantations that stretched alongside his country's rapidly expanding railway network. As a fast-growing hardwood it provided copious supplies of fuel for steam engines, sleepers for railways tracks, telegraph poles for communications systems and fencing to keep animals off the tracks. By 1925, de Andrade was exporting eucalyptus to the USA as experimental material for woodpulp in order to make high-quality brilliant white paper. Such pulp is among the most popular today, thanks to the advent of modern laser printers and photocopiers.

British authorities introduced eucalyptus trees into the African country of Kenya in 1902 to supply fuel for the Kenya-to-Uganda railway. In South Africa these same trees were introduced to provide props to support shafts in gold mines. Meanwhile, Spanish and Portuguese authorities authorized the growing of vast eucalyptus plantations as a source of revenue from pulpwood farming. They still dominate the forests of Extremadura on the western Iberian peninsula. The Asian conquest of eucalyptus was no less spectacular. China, Thailand and India are all host nations to enormous plantations that have been established over the last 200 years to support the export of pulp, the need for firewood and the worldwide demand for the therapeutic properties of its oils, 70 per cent of which come from China.

But in the last twenty-five years, some countries have begun to realize that their Victorian ancestors' headlong rush into planting fast-growing alien species without first investigating their ecological effect on native species may not have been so wise after all. Since eucalyptus evolved in nature's toughest environments, when given the gift of transplantation into wetter, more fertile, temperate zones, their ability to dominate all other life-forms turns them into agents of ecological destruction. Species of fast-growing evergreen eucalyptus make huge demands on local water resources. As a result, nearby wetlands are frequently reduced to desiccated scrubland or desert.

In South Africa species of Australian eucalyptus suck up more than 70 per cent more water than drought-resistant native African acacias (see page 146). The result has been as catastrophic as that from cotton-growing around the Aral Sea (see page 255). Water flowing from river systems has reduced by three quarters. Such dry conditions also turn the habitat into a tinder box in which frequent forest fires benefit the well-adapted eucalyptus even more.

The pros and cons of eucalyptus plantations stirred up a great debate in India during the mid 1980s with landowners pitched against locals in a battle between those who wanted more and those who wanted less. Landowners claimed they could create wealth for communities (themselves included, of course) by exporting the fast-growing timber for pulp. Locals said they could no longer grow their vegetables or graze their animals because the plantations had made the ground dry and toxic. The fact that money for new plantations was coming from a social forestry programme designed to support local communities only embittered the debate.

Thailand's plantations of *Eucalyptus camaldulensis* were first established in the 1950s. During the 1990s Finnish investors were behind the development of further eucalyptus crops for exporting as pulpwood, supported by generous tax-subsidies from the Thai government. Meanwhile local environmentalists lobbied furiously, claiming that local livelihoods were being destroyed by sour soil and that rice cultivation had become impractical owing to a chronic shortage of water.

So acute have concerns been in some places that authorities have begun to turn the clock back by literally ripping these highly invasive trees out of the ground. Working for Water is a South African project launched in 1995 that is clearing as many as 200,000 hectares of eucalyptus plantations a year in a desperate bid to restore badly needed water supplies. Meanwhile, concern for wildlife has driven Spanish authorities to declare the forests around Monfragüe, near Trujillo, as a natural wildlife park in an attempt to protect some of Europe's rarest species of storks, vultures and eagles as well as the almost extinct Iberian lynx. The park has recently established a programme of systematically removing eucalyptus plantations and replacing them with cork and holm oaks to revive the local ecosystem and raise the water table. Oaks, unlike eucalyptus, are indigenous species whose life strategies have co-evolved with local wildlife including black feral pigs that feast on their prodigious quantities of acorns (see pages 143–5).

The great eucalyptus debate still rumbles on. Countries with the highest levels of economic growth tend to be those that top the list of places where eucalyptus plantations predominate. Geneticists have recently found ways of further helping the success of these species by cloning those that grow fastest and give the best cuts for timber. Such projects thrive in Brazil, India and Kenya. Today as many as twenty million hectares of eucalyptus plantations fan out across all the inhabitable continents of the globe. So impressive is the success of this tree family over the last 200 years that it has come to cover an area of cultivation almost the size of the United Kingdom in an age when worldwide deforestation is at an historic high. It is an extraordinary conquest by a single tree genus, dominated by just a few species such as *Eucalyptus globulus*.

How are we to deal with such a prolific success story in the environmentally conscious twenty-first century? Some see this family of trees as best placed to help absorb high atmospheric levels of CO_2 and to form sustainable coppices for providing timber, paper and energy supplies. Others say these species are at the sharp end of an ecocide spread by humanity's increasing population and lust for material wealth.

Whichever way the eucalyptus debate eventually turns, remember that there is more to consider than meets the eye when giving someone the gift of a tree for life.

11 On Drugs

How certain plants and fungi diverted the development of human culture by bestowing a bevy of addictive habits.

I AM A DRUG ADDICT. It's not something I have ever discussed before. In fact I haven't thought about it that much until now because the types of drugs I am referring to are not illegal. They are so embedded into the everyday fabric of human culture, past and present, that any attempt to outlaw them today would either sound absurd or has already been tried and failed.

Drugs are chemicals that provoke a specific physical or mental change in behaviour. They are also substances that, to varying degrees, induce someone to want to maintain or increase their use. The drugs I use are the commonplace fixes of **tea** (see page 282), **coffee** (see page 278) and chocolate (see **cacao**, page 274). Tea and coffee contain caffeine and chocolate also contains *theobromine* and *phenethylamine*, drugs known to trigger neuronal pathways in the brain that affect human behaviour. Most people find these sensations – be they stimulants, sedatives or mild aphrodisiacs – highly attractive, if not actually necessary to their everyday lives.

I have a cup of coffee here beside me. It is now 7.45 a.m. and it would be quite extraordinary for me if I did not. Does that mean I could not *live* without coffee? I don't think so. Does it mean I am an addict? Well ... maybe it does. If addiction involves the establishment and feeding of a regular habit then, while I am sure I *could* give up coffee, the fact that I do not is sufficient, I suppose, for me to have to own up to a certain degree of dependence.

Any substance not necessary for survival but which people nevertheless find desirable or irresistible and which provokes some change in behaviour tends to be classed as a drug. Wine, beer and spirits contain alcohol, cigarettes contain nicotine. Today these are more legally borderline drugs than caffeine-based tea, coffee and chocolate. Liquor is sometimes legal, sometimes not (depending on which country you live in and your age). Stronger drugs such as **cannabis** (see page 286), cocaine (see **coca**, page 300) and opium (see **poppy**, page 308) are mostly outlawed because modern societies deem their addictive qualities and behavioural changes too threatening. All these substances – legal or otherwise – create some level of craving or desire in the human brain. They are also all derived from certain species of plants and fungi that have become hugely successful as a result of their mind-warping, habit-forming qualities.

Drug culture

Human addiction to chemicals found in plants and fungi is as old as humanity itself. Not that much is known about the rituals of holy men or shamans in the pre-agricultural hunter-gatherer world because systems for recording history, such as writing, were not established until the first civilizations appeared in the Middle East, North Africa and Asia *c.* 6,000 years ago. Nevertheless, archaeological evidence in favour of a prehistoric symbiosis between humans and mind-bending drugs is unequivocal.

Certain plants and fungi are renowned today for their *hallucinogenic* qualities, especially those found in flowering plant families such as acacia, mimosa, virola and echinopsis. Fungi ranging from powdery **ergot** (see page 304), which parasitizes rye grass, to the garish red-capped *Amanita muscaria* toadstool, depicted in fairy-stories from *Alice in Wonderland* to the tales of the Brothers Grimm, were ingredients variously used in spiritual rites since prehistoric times. Ergot, which also infects barley crops, is a hallucinogenic fungus from which the modern drug LSD is derived. Chemicals in such psychoactive species interfere with the neurochemical pathways in the brain, affecting an individual's moods, perception, cognition and behaviour. By using such drugs, shamans believed they were able to contact the spirit world and travel on astral journeys to commune with the dead.

Thanks to the scarcity of such plants and fungi, and the specialist knowledge required to brew them into sufficiently intoxicating potions, the shaman became accepted as an exclusive intermediary between the human and divine world. It is from such ancient customs that the idea of a priesthood – a select group of people who have contact with the spirit world – has evolved.

Such holy men were also regarded by their communities as healers. Ceremonies designed to help fight diseases included chanting, music and the administration of certain herbs, flowers and fungi (such as cannabis or opium), which were thought by many to have therapeutic or pain-relieving effects. Some Eastern cultures have traditionally administered natural stimulants (such as ginseng and herbal teas) to promote feelings of wellbeing. A legendary emperor of ancient China, called Shennong, is said to have concocted 365 different medicines out of various minerals, plants and animals, and is regarded as the founder of Chinese herbalism, still widely practised throughout the world.

Modern equivalents, at least in developed countries, are professional medical doctors and pharmacologists. Although many drugs are now synthetic, some of the most significant are still derived from natural products such as the antibiotic fungus **Penicillium** (see page 314); the anti-malarial drug quinine (see **cinchona**, page 316) and painkilling aspirin (derived from the bark of the willow tree and the leaves of the meadowsweet plant).

Little Alice gazes in wonder as a puffing caterpillar, seated on a toadstool, offers her some advice.

Ancient addictions

Once humans began to settle in towns and cities and develop complex societies, their appetite for drugs was no less intense. Every civilization on Earth had its own unique cocktail of plants and fungi that it relied on to make sense of the world. Indeed, the relationships between humans, religion, art and culture are inextricably interwoven thanks to the sometimes mind-blowing effects of relatively few species of plants and fungi.

Rituals in ancient Egypt revolved around the consumption of blue and white lotus flowers (*Nymphaea caerulea* and *Nymphaea lotus*), varieties of water lily that grow along the banks of the River Nile. Recent research shows that these plants contain quantities of a drug called apomorphine, which stimulates the production of dopamine in the brain, leading to feelings of euphoria and sexual excitement. The famous Egyptian *Book of the Dead* refers to the use of these flowers in sacred rites. In the tomb of Tutankhamen, discovered by Howard Carter in 1922, a gold-plated shrine depicts a pharaoh holding a huge *N. lotus* in his hand. Other frescoes found on tombs in Luxor show a ritualistic funeral dance in which women dance with garlands of *N. caerulea.* They are entreating men to drink from vases that contain a mystical potion.

Mystic rituals in ancient Greece were formalized from *c.* 1600 BC onwards in an annual religious festival known as the Eleusinian Mysteries. It revolved around Demeter, the goddess of agriculture, and her daughter Persephone, who was imprisoned in the underworld, Hades, for several months a year. When she returned to the Earth (after the rains had come in spring to restore plants to life) a ten-day celebration would follow in which people were initiated into contact with the spirit world. A special potion called *kykeon*, made from fermented barley, was prepared just before the climax of the festival. It was taken by the initiates to break a fast, following which they took part in a day of top secret events inside a great hall called the Telesterion. Anyone who broke the secret of what went on inside faced death.

The Ninnion Tablet, kept in the Archaeological Museum of Athens, depicts the feast and shows the initiates being offered *kykeon*, which was administered from a vessel by a priestess. Experts believe the psychoactive potion may have contained ergot. Although the exact contents of the potion are still debated, the use of psychedelic drugs was almost certainly what transported these people into contact with what must have seemed like a truly different world.

Further to the east, psilocybin mushrooms and a plant called hemp (cannabis) are thought to have been the likely candidates for a potent brew that was sacred to central Asian peoples who invaded northern India from *c.* 1700 BC onwards. A whole chapter of the highly venerated Hindu text the *Rigveda* is dedicated to the glorification of a drink called *soma*, as are many verses in the Zoroastrian *Avesta*, sacred to people in pre-Islamic Persia. So powerful was the drink that it was believed to be consumed by the gods themselves, which explains why, when imbibed by humans, it was thought to put them in direct touch with divinity. *'We have drunk Soma and become immortal, we have attained the light, the gods discovered ... '*

Experimentation with mind-changing plants and fungi was also endemic on the other side of the world. Native Americans had a penchant for a cactus called peyote (they still do) that is native to southern Texas and throughout Mexico (its botanical name is *Lophophora williamsii*). Peyote contains a highly hallucinogenic chemical called mescaline that triggers a rush of rich visual and auditory 'other-worldly' experiences.

Evidence of peyote use has been dated back to at least 3600 BC (found in an archaeological site in Texas). Seeds from this cactus were often mixed with other intoxicating plants such as morning glory in the later Aztec civilizations (1440–1521). At least five sacred plants and fungi have been identified in the eleventh book of the 'Florentine Codex', which depicts priests visiting ambassadors and nobility drinking a sacred brew. Psychedelic drugs were also administered during religious festivals and to human victims before they were sacrificed to the gods.

Everyday people were just as frequently using intoxicating plants and fungi. According to sixteenth-century Spanish invaders, Aztec ceremonies involved mixing psilocybin mushrooms with chocolate and honey, after which the natives 'danced and wept all night long'.

Chewing the leaves of **tobacco** (see page 289) and the psychoactive herb *Salvia divinorum* was also popular among indigenous Americans. In the Inca Empire, centred on what is now Peru, young women from the Emperor's 'House of the Sun Virgins' chewed maize to help ferment a sacred brew called *chichi* made out of maize and coca leaves (the same plant from which modern cocaine is derived). During their religious festivals it was a *state requirement* for the population to get inebriated so that the gods could look down on their people and see they were happy.

Modern habits

With the emergence of the monotheistic religions of Judaism, Christianity and Islam, the natural, ancient tradition of intoxicating oneself with hallucinogenic mushrooms and plants became almost universally taboo. To Christians, God was no longer directly accessible via visions and trances, instead He came down in the form of a human, Jesus Christ, who was best consulted, said this holy priesthood, via prayer and confession or a simple hymn or chant.

From the fifteenth century AD, people who continued to indulge in making potions out of plants (a common combination in Europe was *henbane*, *datura* and *belladonna*) were cast out as witches. The hallucinogenic trips induced by these drugs brought about the idea that witches brewed some form of flying ointment, administered vaginally using a pole – the origin, apparently, of travel by broomstick. Following the Christian equivalent of a *fatwa* imposed by Pope Innocent VIII (1432–92) witches were sentenced to death by burning at the stake on the grounds that they were possessed by the devil.

Yet even the Christian Church, which went to great pains to stamp out pagan and Celtic rituals, could never quite resist the other-worldly effects of intoxicating substances such as alcohol to celebrate its sacrament. The fungus that turns innocent grape juice into inebriating alcohol, **yeast** (see page 293) has also played its part in Christian worship, in the miraculous transformation of wine into the blood of Christ (as in the Catholic belief of transubstantiation) or more simply as a symbolic representation (Protestant consubstantiation). Nor was Jesus himself a party-pooper when it came to drinking alcohol – as attested by his first recorded miracle in which he turned pitchers of plain water into fine wine at a friend's wedding feast.

In the Islamic prophet Mohammed's holy book, the Koran, the practice of drinking alcohol, which remained acceptable in many (but not all) forms of Christianity, was condemned outright:

> *'They ask you (O Mohammed) concerning alcoholic drink and gambling. Say "In them is a great sin and (some) benefit for men. But the sin of them is greater than their benefit".'*

Despite the enormous popularity of Christianity and Islam over the last 2,000 years, efforts by church and state to control the popular use of intoxicating and stimulating substances have mostly failed.

Mohammed's protestations notwithstanding, much of the Muslim world has long since succumbed to the craving for alcohol. Only the strictest Islamic states, such as Saudi Arabia, still prohibit its sale.

European Methodists and Puritans from the seventeenth century onwards put a big effort into banning alcohol. For a while America joined in the fight by prohibiting the sale and consumption of alcohol in 1919. In the end, the effort proved impossible to enforce, leading to alcohol's re-legalization fourteen years later.

Meanwhile, from the mid eighteenth century, a different source of transcendental escape began to grip Western culture. From the poetry of Samuel Coleridge (1772–1834) to the fantasies of *The Wizard of Oz* (when Dorothy falls asleep in a field of poppies), high society from Britain to China revived the use of a drug known at least since Neolithic times – and probably long back into

Selling, dealing and smoking dope in British India, the source of China's nineteenth-century trauma.

prehistory – which was traditionally administered for pain relief.

Opium is a highly addictive euphoric drug derived from a type of latex extracted from crushed immature poppy seeds. When mixed with tobacco, it gave rise to a new smoking habit. Trans-cultural enthusiasm for the opium den dramatically altered the political, economic and cultural history of the modern world (see pages 308–13). Attempts to control its use in China, beginning in 1769, proved an abject failure. Efforts to prevent the use of its modern derivative, heroin, have been equally unsuccessful in the Western world. Despite its occupation by British and American forces since 2001, Afghanistan is still the world's largest producer of opium poppies, a crop that dominates the country's economy.

Other booty from Europe's early-modern overseas conquests included coffee, tea and chocolate, which by the mid nineteenth century had become consumer 'must-haves' in the colonial capitals of Europe. While they are only mild stimulants compared to the highly addictive qualities of opiates, their extraordinary ability to permeate economies, societies and living rooms transcended all forms of social, racial and pecuniary divide. The high-tech, hectic pace of life in modern, post-industrial nations has only accelerated the relentless cultivation of these stimulating plants.

'Soft' but addictive substances were never enough to satisfy the cravings of some people, whose habits hark back to the shamans of the not-so-distant hunter-gatherer past. Heroin, LSD and cocaine have thrived, some say, thanks to their prohibition beginning in the early twentieth century. Another rather less addictive plant, cannabis, has leaves which produce what has arguably been the most popular psychoactive drug of all time, variously known as marijuana, ganga, hashish or pot. From ancient religious rituals to modern pop festivals, cannabis is now estimated to be the most lucrative cash crop grown in the United States today, beating corn, soya and maize. Not bad going for a plant that's supposed to be illegal to grow.

The dawn of conciousness?

The simplest explanation as to why some plant and fungal species have such profound behavioural effects when ingested, imbibed or inhaled by humans is that it is an evolutionary accident. What were originally natural survival responses by plants and fungi to threats from microbes, insects or grazing animals have, for some reason, come to have bizarre, psychedelic effects on the human brain ranging from increased awareness and feelings of euphoria to dizziness, nausea or depression. After all, what should people expect if they eat, drink or smoke what nature intended as insecticides, pesticides and antibiotics?

But some experts believe that the human appetite for stimulation by drugs and our susceptibility to addiction is so ancient and so deeply ingrained in our physiology and way of life that there must be a more logical explanation.

Hallucinogenic plants and fungi are well known for their ability to push the boundaries of human perception. One intriguing theory proposes that when early humans experimented with intoxicating plants and fungi they became increasingly aware of themselves and their surroundings (see cannabis, pages 286–7; and ergot, page 304). Rather than their brains being solely concerned with the biological business of functioning and surviving as a living being (homeostatic mechanisms), the effects of the drugs on their minds began to endow them with a sense of self – of their past, present and future – something we now call consciousness.

What power such qualities could bestow on those early hominids! Like eating from the fabled forbidden Tree of Knowledge in man's primeval Garden of Eden, people on drugs could now more easily outwit those with little or no sense of self. Significant survival advantages like these could have naturally selected humans that were more sensitive to mind-stretching survival-enhancing substances. Before too long, humanity may have become populated by individuals whose propensity towards addiction was far greater than previously. In the process these plants also benefited – initially by becoming items of value and veneration – but ultimately because they came to be counted among the very few select species to be cultivated by man.

Is the idea that human consciousness was itself derived from an ancient symbiosis between certain plants and fungi so absurd? Such a theory explains how humans have evolved to become so susceptible to addiction (perhaps even helping explain the evolutionary origin of other modern addictive human habits such as constantly checking for emails or stock-market prices ...). It also provides a neurological mechanism for the evolution of what we call consciousness. Finally, such mutual symbiosis is typical of the co-evolutionary forces that are the hallmarks of what makes flowering plants so successful (as exemplified in the development of their pollination and fertilization strategies – see On Biodiversity, page 128).

Whatever the merits of such an ingenious theory for the genesis of human consciousness, the macro-historical impact of highly addictive plants and fungi goes without question. These are species that have played a pivotal role in the development of all human history – across cultures, nations and civilizations – be it religious, medicinal or recreational. As a result many of them are truly among those species that have most changed the world.

Cacao

FAMILY: MALVACEAE
SPECIES: THEOBROMA CACAO
RANK: 81

How the therapeutic charms of a South American tree have been hijacked by the modern world's addiction to high-speed confection.

IT IS BIZARRE to think that until just a few hundred years ago no one outside the Americas had ever heard of chocolate, let alone tasted it. Only the tribes and civilizations of Central and South America, who first began to domesticate the fruits of the tree *Theobroma cacao* about 1,500 years ago, knew of its delicious aromatic flavour and therapeutic powers. Yet today, just 500 years after its discovery by invading Spanish conquistadors (the merest blip in the 2.5-million-year history of humanity) these seeds, commonly called cocoa beans, have come to dominate the hearts, minds and stomachs of millions, if not billions, of men, women and children all over the world.

Nowadays cocoa – and especially its refined form, chocolate, produced from the seeds of the cacao tree – is big business. Products worth more than £10 billion, ranging from confectionery bars to sweet-flavoured drinks, are sold annually worldwide. Most Western cultures are besotted with chocolate, whether presented by adults as a romantic gift on Valentine's Day or fantasized about by children in stories about winning the prize of a tour around a mysterious confectionery factory followed by a lifetime's supply of chocolate. The fates of those five 'lucky' children who happened to discover a golden ticket tucked inside a scrumdiddilyumptious chocolate Wonka bar have entranced readers and viewers ever since British novelist Roald Dahl wrote the story of *Charlie and the Chocolate Factory* in 1964.

So why are people crazy about chocolate? Have humans always been so susceptible to its delicious flavour? What has been the impact of our apparent addiction, past and present, which stems from the seeds of this otherwise unspectacular South American tree?

When Carolus Linnaeus, the famous eighteenth-century Dutch botanist, first described this tree species, he chose its name especially well. *Theobroma* is the Latin for 'food of the gods' and *cacao* derives from *cacahuatl*, the name used by Mayan and later Aztec people who cultivated these trees in what is now Mexico in the pre-Columbian era. Genetic evidence clearly shows, however, that the cacao tree did not originate in Mexico, which is where Europeans first encountered the hot, rich, aromatic drink made from its beans whilst staying as uninvited guests at the palace of Aztec Emperor Moctezuma in 1519.

It is native to lowlands around the Amazon river in what is now Brazil and Peru. This is where the most exclusive chocolate available today comes from – a variety of *T. cacao* called criollo. This is a variety that has the highest number of seeds per pod, the sweetest flavoured pulp and requires the least amount of time for the fermentation of its beans. Such characteristics are typical of careful and painstaking selection by humans leading to a domesticated crop that is of most use to man.

Once they have flowered, cacao trees produce fruits called pods. Each one typically contains thirty to fifty large almond-like seeds or beans, which are surrounded by a thick, sweet pulp. The beans and the pulp are cut out of the pod and left in the sun, where they ferment naturally. Once dry, they can be processed into chocolate, either for drinking or eating.

The first archaeological evidence of human cacao cultivation dates from around 1100 BC. Jars from this era containing traces of chemicals from the processing of chocolate beans have been uncovered by archaeologists working in modern-day Honduras in Central America. They show that widespread cultivation and trade must therefore have existed well before this time since Honduras is a considerable distance from the trees' centre of origin in the Amazon lowlands.

Indigenous peoples of South and Central America put a high value on the trade and consumption of chocolate. By the time of the late Aztec Empire (*c.* 1440–1521), tribute payments from neighbouring states were made in the form of sacks of cocoa beans. When Hernán Cortés and his retinue of Spanish invaders first encountered Aztec Emperor Moctezuma their curiosity was aroused by this mysterious hot drink. According to their accounts Moctezuma drank fifty whipped cupfuls of hot chocolate from a golden goblet every day. His attending nobles also consumed copious quantities, they said, as many as 200 daily helpings between them.

Inside each cacao pod there are between thirty and fifty almond-like seeds, immersed in a thick, sweet pulp.

Such high levels of individual consumption would, by most people's reckoning, amount to addiction. Additional evidence of chocolate's addictive qualities is borne out by the global effect cocoa beans have had, beginning with their cultivation, transportation and trade by Spanish-American colonists from the late sixteenth century onwards.

Naturally occurring active ingredients in chocolate include *theobromine* and *phenethylamine*. The former dilates the blood vessels, increasing the flow of blood throughout the body and brain, resulting in a feeling of general mental and physical wellbeing. The latter, which belongs to the amphetamine family of drugs, affects neural pathways in the brain, potentially leading to feelings of contentment and euphoria. It is the effect of this chemical that has given rise to cocoa's legendary aphrodisiac qualities. However, recent medical research suggests that chocolate's romantic power is probably derived more from tradition than from chemistry since in most people the liver metabolizes the chemical before it can have any significant impact on sexual arousal.

But the therapeutic qualities of cocoa, especially in its unprocessed, natural form, are well recognized. A recent study, conducted by American medical researchers, suggests that native people called the Kuna, from Panama, have a greatly reduced risk of heart disease, cancer and diabetes compared to the general population of developed societies. This is thought to be due to their prodigious intake of cocoa. Typical adults consume as many as forty cups of their own natural brew of drinking chocolate a week. Cocoa's unique combination of respiratory stimulant (*phenethylamine*) and blood-vessel relaxant (*theobromine*) is thought to lead to an overall reduction in blood pressure that can extend the human lifespan.

So when European explorers discovered cacao during their conquests of the Americas, its combined cocktail of delicious taste, 'feel-good' factor and health-giving benefits should have had a seriously positive effect on the global population. Unfortunately, that did not turn out to be the case.

How Cadbury's promoted the delights of its Dairy Milk chocolate as long as one hundred years ago.

The initial spread of chocolate beans throughout Europe began innocently enough. Like Moctezuma, Spanish royalty and their attending nobles adopted the fashion, enjoying the flavour of drinking chocolate, but often adding vanilla (rather than chilli, preferred by indigenous Americans) to suit the European palate (see page 346). When the Spanish-groomed Habsburg Princess Anne (of Austria) married French King Louis XIII in 1615, she took the chocolate-drinking habit to France where its aristocratic appeal spread further. For the following 200 years the therapeutic effects of chocolate-drinking were therefore made available to upper-class people in the ancient regimes of early-modern Europe.

But then, in the first half of the nineteenth century, all that changed. The impact chocolate should have had when it eventually became available to the mass market was nothing like what Aztec natives or even Spanish and French nobility had previously experienced. With the forces of industrialization and mass-production in water-powered mills, chocolate's health-giving destiny took a dramatic turn. When Dutch chocolate-maker Conrad Johannes van Houten (1801–87) pioneered a pressing process for milling cocoa he removed the centre of the bean – called the nib – in order to make the resulting paste more digestible. It also had the commercial benefit of making chocolate paste easier to combine with other substances, such as milk and sugar.

As a direct consequence of van Houten's hydraulic press, chocolate could now be made into solid bars for immediate consumption, as opposed to concentrated cubes useful only for dissolving into a hot, therapeutic drink. Milk-and-sugar-enriched solid chocolate bars had their natural bitterness removed (and with it most of their healthy chemicals), turning what was an exotic health drink into a sucrose-rich confection. Along with many other modern fast foods, mass-produced chocolate has substantially increased levels of tooth decay and played its own part in the acceleration of the modern pandemics of diabetes and obesity that plague the populations of developed countries today.

Once van Houten's patent expired in 1838, a clutch of enthused Victorian chocolate entrepreneurs leapt on to the chocolate-making bandwagon. Englishman

John Fry brought out his first chocolate bar in 1847, followed by the Cadbury brothers two years later. As steam-power began to take over from water-wheels, the cult of the modern chocolate-factory, Willy Wonka-style, was born.

Unfortunately, Roald Dahl's story about the man who wanted to find an honest heir to his chocolate factory reveals a dark side to the worldwide spread of cacao since the emergence of chocolate bars in the 1850s. Child-like 'Oompa-Loompas' – Willy Wonka's mythical midget ethnic-looking factory staff that some say resemble African pygmies – still reflect the terrible human price paid today for mass-produced cocoa. Few consumers are aware of the living and working conditions in the world's largest cacao plantations, located in West Africa, as they gaze longingly at their Valentine's gift of a box of dark chocolates.

The Côte d'Ivoire (formerly known as the Ivory Coast) exports nearly half (46 per cent) of the world's current production of cocoa beans. Its plantations were established between 1890 and 1960 by French colonialists who sought to profit from their newly acquired territories by growing cash crops such as coffee (see page 278) and cacao. As a result, this war-torn nation is still dependent on whatever income it can gain from the export of its beans to chocolate-processing corporations in Switzerland, England and Holland via the London commodity exchange, which runs the international trade in cocoa beans.

Families desperate to scrabble together a living from the harvesting of beans recruit every member who can help, however young. Other children are sold as slaves and trafficked into the Côte d'Ivoire from neighbouring countries such as Mali because the use of free labour is the only way of keeping prices for international markets sufficiently low. So little is the 'free market' willing to pay for its chocolate addiction that it requires as many as 200,000 child labourers toiling on cacao plantations in this poverty-stricken West African country to provide the raw product at an economical price.

A US Department of State study, published in 2005, makes dismal reading regarding the human impact of sustaining the modern global chocolate habit. In its report into human rights in the Côte d'Ivoire it said: *'Approximately 109,000 child labourers worked in hazardous conditions on cacao farms throughout the country in the worst forms of child labour.'*

International commitments made by worldwide chocolate companies in 2001 were supposed to have resolved the problem by 2005, removing the *'worst forms of child labour'* – although even today, according to human rights watchdog the International Labor Rights Forum, some commitments have not been met. However, in March 2009, UK confectionery company Cadbury pledged to source cacao for its most popular chocolate bars from Fairtrade plantations. More than a hundred million Cadbury's Dairy Milk bars are sold in the UK each year.

What was once a natural product cultivated and consumed as a therapeutic remedy and royal delicacy by the indigenous people of central and southern America is now a mass-produced health hazard mostly grown by poverty-stricken African peasants, many of whom are children. The fruits of African labour feed the addictive habits of post-industrial nations that are being made fat and diseased by chocolate.

Coffee

FAMILY: RUBIACEAE
SPECIES: COFFEA ARABICA
RANK: 85

A sub-tropical plant that helped commercialize modern society with its stimulating effect on the body and mind.

KALDI was a nomadic goatherd who lived in tenth-century Ethiopia. One day, it is said, his goats went crazy after munching the red berries of an unknown wild bush. Maybe out of curiosity, or perhaps boredom, Kaldi ate some of the fruits himself. Not long after, he was spotted dancing gaily among his goats by some passing monks who lived nearby. They took the berries, crushed them, seeds and all, into a thick paste, added hot water and were the first humans to sample the highly stimulating delights of what we now call coffee. They liked the brew because it helped them stay awake for all-night prayers.

Any plant that has so transformed the world in just a few hundred years deserves a colourful origin, even if the story of Kaldi, goats and monks is more legend than fact. What is certain is that until the tenth century there are no records of humans consuming the fruit or seeds of the coffee plant. But after the discovery of its stimulating nature, coffee went on to establish itself as one of humankind's most popular addictive substances.

The invigorating effect of coffee derives from caffeine, a naturally occurring chemical insecticide. *Coffea arabica* (and its other popularly grown cousin *Coffea canephora*) are plants that have especially high levels of caffeine in their seeds. The chemical also sours the soil, preventing other plant-life from growing too close and competing for resources. It is rather bizarre to think that the world's most popular human beverage is, in its natural state, a growth-stunting pesticide.

In humans the effects of caffeine are now known to inhibit the action of various receptors in the brain, resulting in a build-up of chemicals that induces a state of mental and physical stimulation that can last for several hours. When caffeine is taken regularly, the brain compensates by establishing more receptors, so that the stimulating effects of caffeine are reduced. However, if the drug is withdrawn, the excess number of receptors causes an imbalance in the behaviour-regulating systems of the mind, leading to headaches, nausea and irritability. These patterns of tolerance and withdrawal are classic symptoms associated with addictive drugs, to which the human brain seems particularly susceptible (see also page 273).

Aside from its addictive tendencies there is no medical evidence that caffeine is a particularly dangerous drug. Recent research shows various positive health benefits, including reducing the risk of Parkinson's disease, increasing short-term memory functions and lowering the chance of heart failure. Although such medicinal diagnoses could not have been known to African and Arabic people, coffee's global conquest since *c.* AD 1000 has a great deal to do with the stimulating effects of caffeine.

Unlike wine, beer or spirits, which typically make humans feel dizzy, disorientated and drowsy, coffee has the opposite effect. It is the classic 'pick-me-up', increasing alertness and mental precision. Such qualities made this drink the number one favourite among Islamic people, for whom drinking alcohol was prohibited by the Prophet Mohammed (see page 271).

From Ethiopia, news of coffee's energizing effects spread to Yemen, on the southern tip of the Arabian peninsula, where trans-Arabian/African trade was already well-established (see camel, page 249). Then it reached Muslims on their annual pilgrimage to Mecca where the drink, which became known as the 'wine of Araby', was found also to stimulate conversation, trade and commerce. Coffee houses sprang up in Islamic trading cities, such as Damascus, Cairo, Baghdad and Istanbul. By the fifteenth century such establishments became centres of trade and exchange between Islamic and European merchants. Coffee sharpened the wits of haggling traders to make sure each came away with the best possible deals.

European coffee houses brought the association between coffee and commerce even closer. Until *c.* 1650 European people depended on beer for liquid nourishment. In England the average family drank about three litres per person per day – children included – and businessmen traditionally consumed a pint during their mid-morning break.

But by the middle of the seventeenth century, as coffee houses first began to appear in London, a dramatic transformation was under way, as documented by British historian James Howell in 1660:

> *''Tis found already that this coffee drink hath caused a greater sobriety among the nations. Whereas formerly apprentices and clerks with others used to take their morning's draught of Ale, Beer or Wine, which, by the dizziness they cause in the brain made many unfit for business, they now ply the good fellows with this wakeful and civil drink.'*

The stimulating, sobering effects of coffee soon fused with the entrepreneurial spirit of overseas exploration that swept across early-modern Europe from the seventeenth century onwards. It was a powerful cocktail. By 1700, according to one estimate, there were a staggering 3,000 coffee houses in London alone – that's one café for every 200 people living in the city at that time. Drinking coffee became an addictive habit in bourgeois classes, a fashion few social climbers, or those wanting to further their careers, could resist.

When Edward Lloyd opened his coffee house on London's Tower Street in 1688, little did he imagine that it would evolve into a vibrant meeting place for mariners, ship's captains, merchants and insurance brokers. By 1774 what began in this

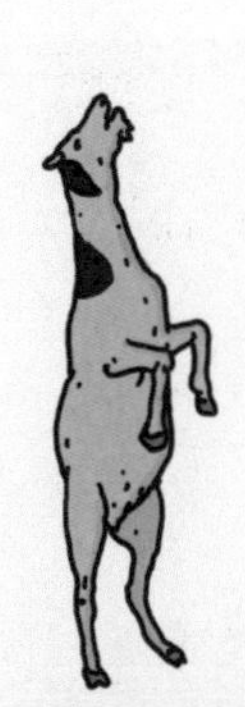

humble café had transformed into the world-famous institution Lloyd's of London that is still the largest insurance brokerage in the world. Meanwhile the posting of stock and commodity prices in Jonathan Miles' Coffee House, beginning in 1689, cemented the relationship between coffee and commerce even further. This café's stockbrokers eventually built their own premises in 1773, originally dubbed 'New Jonathan's', but which later became known as the London Stock Exchange.

Journalism was another trade born in the coffee house, thanks to the mental stimulation afforded by the habit of drinking coffee. Political and literary topics were discussed with feverish enthusiasm in the coffee houses of London. As a result weekly newspapers such as *Tatler* (started in 1709) were founded to publish the gossip they overheard (hence the name). The paper even used a coffee house as its editorial office. Meanwhile literary and philosophical giants of the French Enlightenment, such as Voltaire, Diderot and Rousseau, spent many of their waking hours talking and writing in the coffee houses of France.

By the beginning of the eighteenth century it was clear that the popular European appetite for coffee presented substantial commercial opportunities for overseas explorers and colonists – if only they could find places to cultivate the crop cheaply enough. Dutch traders were the first Europeans to establish coffee plantations overseas. Pieter van den Broeck (1586–1640), a Dutch cloth merchant, smuggled live coffee seedlings out of the Yemeni port of Mocha in 1616, despite the fact that their export was strictly prohibited by the Muslim powers. He used the seedlings to establish the first plantations on Dutch colonial territories on the island of Java. The popularity of coffee increased exponentially throughout Europe following the defeat of the Turkish forces outside the walls of Vienna in 1683. Polish soldiers found sacks filled with coffee beans in the camps abandoned by fleeing Turkish troops, seeding the start of a thriving coffee-house revolution among the liberated inhabitants of down-town Vienna.

Human deception, rivalry and cunning all played their part in turning coffee that originated in Arabia and Africa into a global mega cash crop which today boasts an estimated annual trading value of $33 billion. The crop's biggest break came when sneaky French naval officer Gabriel DeClieu transported coffee to the Americas in *c.* 1720 after taking an illicit cutting from a royal bush in the botanical garden belonging to French king Louis XIV. After a stormy Atlantic crossing (in which he is said to have sacrificed his last water rations to keep the plant alive), DeClieu landed in the French Caribbean colony of Martinique. Originating from his coffee plant cutting, plantations were soon established across the region, as well as on the South American mainland just north of Brazil in French-controlled Guyana.

Now Portuguese authorities in Brazil were desperate to get their hands on this lucrative crop. They sent a certain captain, Francisco de Mello Palheta, on a mission to French Guyana to procure coffee seeds from the island's governor. The story goes that after failing to secure his booty by negotiation, the captain so charmed the governor's wife that she presented him with a pretty bunch of flowers as a farewell gift. Tucked inside was a stash of precious coffee seeds from which the plantations of Brazil were established soon after the captain returned home. Towards the end of the eighteenth century, coffee from Brazil was re-introduced

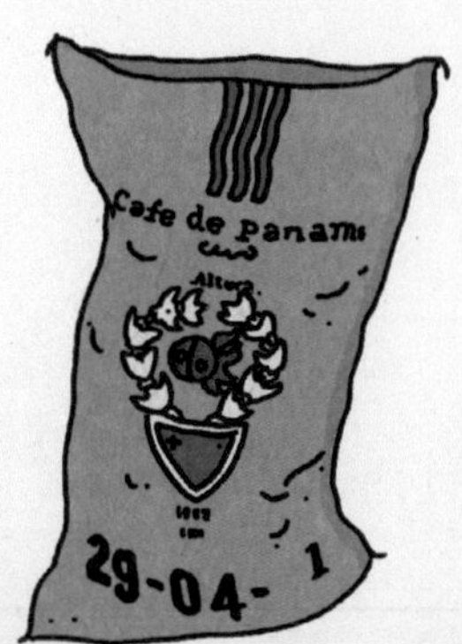

to Kenya and Tanzania, completing a global circumnavigation that had begun less than 1,000 years before.

Today coffee is by far the most popular chemical stimulant in the world, with an estimated *500 billion* cups drunk annually. To support such enormous consumption, 7.8 million tonnes of coffee beans are cultivated each year. Brazil accounts for about 40 per cent of worldwide supply. In that country alone the growing and harvesting of coffee employs as many as five million people who tend to three million individual coffee plants.

Scales like these have been possible only because historically the price of coffee has been low enough for consumers to afford their daily caffeine-fix. African slave labour, mostly on South American and Caribbean plantations, provided an essential commercial 'kick-start' to make this labour-intensive crop economically suited to European markets. More recently, increasing yields have been masterminded by changing the way coffee is grown. In the 1970s the US Agency for International Development granted $80 million to South American coffee growers on the condition that they plant their bushes in rows in the full sun instead of under the shade of a forest canopy (as they are traditionally grown). Whilst full sunlight causes coffee bushes to ripen faster and yield more berries, these growing techniques have had disastrous ecological side-effects, such as large-scale deforestation and habitat destruction. Together these have devastated indigenous bird life that depends on native forest canopies for its natural habitat.

Full sunlight coffee varieties are also highly demanding on pesticides and fertilizers (mostly supplied by US companies) and require vast quantities of water to generate maximum yield. According to one recent estimate, so much water is used in the cultivation and processing of coffee that its takes 20,000 litres of water to produce a one-kilogram jar of coffee – that's an astonishing 140 litres of water per cup. Underground aquifers that contain slow-to-replenish 'fossilized water' are being drained by hard-up coffee producers from India to Vietnam who, thanks to fierce international competition among growers, cannot grow their produce responsibly and still make a profit. As a result, the land they are cultivating is turning into desert and an ecological disaster is developing underground that is equivalent to the effect cotton-farming has already had on the surface with the shrinking of the Aral Sea (see page 255).

Coffee's grip on human culture was vividly captured during the historic Apollo XI space mission on the night of 20/21 July 1969. Shortly after the Eagle lunar module landed, command module pilot Michael Collins interrupted preparations for Neil Armstrong's first steps on the Moon's surface when he radioed the Johnson Space Center in Houston to say: *'Excuse me for a minute, I'm going to have a cup of coffee.'*

Stimulating berries from the coffee plant, which apparently sent Kaldi and his goats leaping for joy.

Tea

FAMILY: THEACEAE
SPECIES: CAMELLIA SINENSIS
RANK: 71

A plant with highly versatile leaves used for everything from beverage-making and fortune-telling to a possible remedy for HIV/AIDS.

IF A CONTEST were to be held to find the plant species that has its roots most deeply buried in the history of human civilizations, then tea would have to be a serious contender. While the globalization of the leaves of *Camellia sinensis* (now commonly transported in small perforated bags) is a relatively recent affair, the establishment of this plant as a medicinal, spiritual and convivial companion for humankind began several thousand years ago. Tea's place of origin, as the name *sinensis* suggests, is specific to China but its impact today, be it physical, mental, political or economic, is truly international.

All the tea in the world comes from a single plant species thought to have originated somewhere in south-west China not far from Tibet. Here the world's oldest tea plantations still thrive, such as Jing Mai, in Yunnan, where tea trees are still being harvested in a traditional organic fashion more than 1,000 years after they were first planted.

For Chinese people, drinking tea is a habit as venerable as their 5,000-year-old civilization itself. According to ancient Chinese legend, the trick of brewing the leaves of *C. sinensis* in boiling water was first discovered in *c.* 2737 BC, by fabled emperor Shennong. One day he was drinking from a bowl of hot water when a sudden freak blast of wind blew a few leaves from a nearby bush right into his cup. As he looked down, he saw the colour of the water begin to change. Being a curious fellow Shennong took a sip. That's when he discovered the delightful and restorative effects of drinking tea.

Although our modern tea-drinking habit is regarded as an everyday comfort, for thousands of years it was the most significant natural ingredient to Chinese herbal medicine and Asian spiritualism. For Laozi, a spiritual leader and philosopher who lived in the sixth century BC (at about the same time as the Buddha), tea was the elixir of life. On one of his journeys around China, during the violent Warring States period, Laozi is said to have been offered a cup of tea by a customs inspector named Yin Hsi. As they consumed the drink, conversation flowed. Captivated by what the philosopher said, Yin Hsi implored Laozi to write down his thoughts in a book so others could benefit from his wisdom. It is thanks to tea, therefore, that the *Dao De Jing* (The Way of Virtue), a collection of Laozi's most famous sayings, such as *'He who knows he has enough is rich'*, has been passed down through history to become among the most influential writings in all Chinese culture.

It is known that tea cultivation was widespread in China by *c.* AD 800 thanks to a treatise on tea-growing called *Cha Jing* (The Classic of Tea) written by a former circus clown called Lu Yu between AD 760 and 780. The book details all aspects of tea's origins, cultivation, processing, brewing, serving and drinking, showing that, by this time, the little leaves of *C. sinensis* had diffused deep into the heart of Chinese culture from where the plant spread even farther to the East.

Records suggest that tea plants reached Japan in the early ninth century, where they were used by priests during religious ceremonies. A Zen Buddhist monk called Eisai (AD 1141–1215) travelled extensively in China and brought green tea back to Japan. In 1211 he wrote a highly influential two-volume book called *Kissa Yojoki* (How to Stay Healthy by Drinking Tea) which promoted the tea-drinking habit enthusiastically. The book opens by saying: *'Tea is the ultimate mental and medical remedy which has the ability to make one's life more full and complete.'*

It goes on to detail tea's many medicinal benefits and offers practical advice on how to cultivate, brew and serve the precious drink.

When Western mariners began to unlock the markets of Asia, beginning in the sixteenth century, tea was one of the first products they shipped to Europe. Sleek sailing ships called 'clippers' formed the backbone of a maritime trade in which tea grown in China was transported back to markets in Amsterdam and London.

An advertisement from one London tea house in 1660 summarizes the benefits to prospective customers of Europe's newly discovered brew:

> *'It makes the body active and alert. It offers relief against violent headaches and vertigo ... it cleanses the vital fluids and the liver. It fortifies the stomach, improves digestion ... it is good against nightmares, it eases the brain and strengthens the memory ... one infusion is sufficient to allow one to work through the night, without doing injury to one's body.'*

But it wasn't just the medicinal effects of tea, drunk in its unadulterated traditional Chinese form, that turned this tonic into the everyday classic we are addicted to today. The habit of sweetening tea with milk and sugar not only made the drink more suited to the sensitive European palate, but also gave a commercial boost for products such as sugar grown on European colonies on the other side of the world in the Caribbean and Americas (see page 204).

Between 1700 and 1950 sweet, hot, milky tea became Britain's favourite drink, transcending all barriers of class and gender. Anna Maria Russell, Duchess of Bedford (1783–1857), invented the Victorian aristocratic custom of taking afternoon tea, and later Joseph Lyons and Co. (established 1894) built a nationwide network of teashops and corner houses that sustained the spirits of Britain's middle- and working-class men and women throughout two world wars. From *c.* 1700 onwards, the habit of drinking tea spread to the Americas, even surviving the symbolic stigma of crates being tossed into Boston Harbour (1773) in protest at the importing of tax-free tea by British traders keen to undercut local American fare.

So great was demand for tea in Britain's domestic and overseas colonial markets that paying Chinese cultivators in precious silver bullion began seriously to undermine the Empire's balance of payments. By the late eighteenth century directors of the Honourable British East India Company came up with two solutions to the problem.

The first involved the East India Company diverting some of its extensive farming resources in a region of India called Bengal into fields for growing poppies, from which they extracted opium. An illegal triangular trade was then established in which British tea traders would buy Chinese tea in Canton using credit notes that could be redeemed against opium smuggled across the Chinese border from India. In this way the British paid for tea using a drug grown by oppressed natives in India while China became addicted to opium (see poppy, pages 310–11).

The second concerned the smuggling of *C. sinensis* plants to establish rival tea plantations in British-controlled India. Robert Fortune (1812–80), a gardener and orchid specialist with the Royal Horticultural Society, was tasked by directors of the East India Company to locate the tea plant within China and send seedlings back to India (a crime punishable by death). In 1848, disguised as a Chinese mandarin with shaven head and pigtail, the 35-year-old Fortune travelled to China under the pseudonym Sing-Wa (meaning Bright Flower). After travelling extensively up the Yangtze river and into the tea-growing Wuyi Mountains of south-west China, he located and transported a number of tea plants. Using his own horticultural expertise, he packaged them up in layers of soil, like a trifle, which he then encased in a glass box and arranged to be clandestinely transported to India. Within a generation newly established Indian tea plantations, centred around Darjeeling, were outstripping China's supply both in quantity and quality.

Despite the global popularity of tea, it is only recently that a scientific explanation has begun to emerge as to why an infusion of the leaves of *Camellia sinensis*, one

Tea plantation in Darjeeling, India, pictured here in the 1860s shortly after Britain's 'Fortunate' quest.

of about eighty species of *Camellia*, has been so successful in a world dominated by humankind.

Like coffee (see page 278) tea contains the stimulant chemical caffeine. Therefore, all the stimulating, 'pick-me-up' addictive effects associated with coffee are also present in tea. But tea has more to offer, which may help explain why it was originally cultivated with such enthusiasm so many thousands of years ago in the Far East where it grew naturally in the wild.

An early Chinese text on herbal medicine claims drinking tea could help cure tumours, abscesses, bladder infections and lethargy. Over the last few years, scientists have found that tea does indeed provide significant protection against many types of cancer. A study by researchers at Kyushu University, in south-west Japan, has found an antioxidant called *epigallocatechin gallate* (EGCG) that binds to a receptor on malignant lung cells, slowing their growth significantly. Another study conducted on mice at the University of California suggests that drinking green tea in conjunction with standard drug treatments also inhibits the growth of breast cancer.

The very same antioxidant chemical, EGCG, is also being trialled with HIV/AIDS sufferers. Scientists believe its binding properties help block the virus from destroying human immune cells. According to Mike Williams, researcher at the University of Sheffield, drinking green tea could even *'reduce the risk of becoming infected with HIV'*.

Tea's health-enhancing secrets are also being revealed by research into another compound, theanine, found only in tea plants and certain species of mushroom. This chemical crosses from the bloodstream into the neural pathways of the brain where, according to medical trials conducted in 2007, it increases alpha brainwave activity, inducing a calmer yet more alert state of mind. When theanine is combined with caffeine, as it is in tea, the cognitive impact is much greater than with caffeine alone (as in coffee). Improved visual, numeric and verbal information processing and performance are just some of the observable benefits.

German scientists have recently discovered why Western tea drinkers may not have benefited from the medicinal effects of drinking tea in the same way as Asian people whose lifestyles and culture have evolved with tea over thousands of years in the Far East. When milk is added, many of tea's therapeutic properties are blocked. Compounds called caseins counteract tea's ability to increase blood flow around the body and inhibit the disease-busting ability of EGCG to bind to cancer cells.

The success of tea, taken with or without milk, is an extraordinary story. Tea takes its place today as one of the world's three most widely consumed beverages. At 3.5 million tonnes cultivated per year, the weight of the world's tea crop is about equivalent to cacao and half that of coffee (see page 278). Now add to that the ancient table-top art of tasseography, the practice of divining tea leaves, and its place in the future looks even more assured ...

Cannabis

FAMILY: CANNABACEAE
SPECIES: CANNABIS SATIVA
RANK: 54

A psychoactive weed that captivated the imagination of people long before the beginning of recorded history.

WHAT COULD be more innocent than young children enjoying the crazy sensation of spinning round in circles on the playground? A few years later they may seek the thrills of a high-speed roller-coaster ride that hurls them through the air on tracks twisted into the shape of a giant cork-screw. Sensations such as bungee-jumping off a suspension bridge or parachuting out of an aeroplane at 3,000 metres appeal to others in adulthood. People of all ages, it seems, are naturally attracted – and some addicted – to the feeling of having their bodies driven beyond whatever margins may have been set by nature.

The desire to push human consciousness beyond its normal operating limits also helps explain how a particularly fast-growing weed became the most popular, and is probably the most ancient, recreational drug of all time. Tetrahydrocannabinol (or THC) is a psychoactive chemical produced by the plant *Cannabis sativa* that has given human culture everything from pot to marijuana and hashish to bhang.

Angiosperms such as *C. sativa* (which belongs to the Rosid family) evolved in the Cretaceous Period (*c.* 145–65.5 million years ago) and have become some of nature's most brilliant protagonists in the development of chemical attack and defence (see page 132). Evolutionary experts believe that THC was originally synthesized to fend off bacteria, insects and over-zealous herbivores. Its tactics are far more subtle than simply producing fatal toxins. Instead of poisoning its adversaries outright, cannabis plants disorientate, disable or bamboozle potential predators so they simply forget where the plant they nibbled came from in the first place. This has the added advantage of preventing a herbivore that happened to have natural immunity from cannabis toxins (a possibility afforded by sexual reproduction, see pages 47–8) from decimating plant populations.

When Stone Age humans experimented by eating the female flowers of *C. sativa* they experienced first-hand the dazzling effects of THC on the human mind. Modern medical researchers were amazed to find that THC activates a vast number of receptors called CB_1 located in the human brain. In 1992 two Israeli scientists discovered that these receptors are used by a chemical called anandamide that is produced naturally in the brain. This chemical regulates short-term memory. Its job is to stimulate forgetfulness – especially important for liberating the mind from unpleasant recollections such as the pain of childbirth or severe physical trauma. Usually it achieves its objectives in a very short space of time, and is metabolized by the body in just a few seconds.

THC messes with this naturally occurring mental process. Not only does it invade neural pathways normally traversed by anandamide, but it lingers in the brain for

several hours, as opposed to just a few seconds. The result can be literally mind-boggling, significantly exaggerating an individual's awareness of what is happening in the present-tense – but temporarily blocking experiences from the past.

It explains why smoking cannabis is renowned for making its users feel as if time has stopped and as if every sensation is extra-special. Music is said to peel apart with every voice and instrument momentarily isolated, standing out in solo. What are normally ordinary, everyday objects become inherently more meaningful. Sensations such as taste and touch are similarly stopped in time – as if digitized in slow motion – appreciated and enjoyed frame by frame in a zone of temporal suspension. Such sensual exhilaration causes anxieties about the future and worries about the past to disappear into blissful irrelevance ... well, at least until the weed wears off.

This combination of increased psychological awareness of the present with the lack of toxic aftershock helps explain why cannabis extracts such as hashish and marijuana have been immensely popular drugs. THC in cannabis may have originally evolved as a protective mechanism to make herbivorous predators forget where the plant grew, but sometime during the emergence of modern humanity its highly pleasant sensual powers almost inevitably resulted in deliberate cultivation for its extra-powerful sensory side-effects.

Ancient shamans of the prehistoric hunter-gatherer world are known to have used cannabis in their herbal potions, concocted to induce a trance-like state for contacting the spirit world. Sacks of marijuana leaves discovered by archaeologists in China have been found buried next to mummified bodies dating back 3,000 years. From the earliest recorded religions such as Hinduism, to American churches founded in the twentieth century, cannabis has almost always played a ritualistic role in some form or other.

The worship of Shiva, most important of the Hindu gods, has been celebrated in northern India for more than 1,000 years by drinking bhang, a potent milky drink made from crushed cannabis flowers, also a favourite among Sikhs. Traces of hashish have been found in Nordic cultures, thought to have been used in erotic festivals dedicated to their fertility goddess Freya. Even the Jewish Torah (Old Testament) is now widely thought to have condoned the use of 'kaneh-bosm' (cannabis) as an essential component of the oil of anointment, prescribed by God as an initiation rite to be given to Hebrew priests and kings (see Exodus, chapter 30, verses 22–3).

Despite the Muslim ban on intoxicating substances (see page 271), the whirling dervish dances of Sufi Muslims have traditionally been associated with the higher states of awareness brought on by the effects of cannabis. For some Christian denominations, such as the Rastafarian movement founded in Jamaica in the 1930s, the taking of cannabis is a fundamental ingredient in the spiritual experience of uniting the physical world with dimensions of divinity.

Secularism has not in any way suppressed the global appeal of cannabis. Despite its prohibition in the early twentieth century (along with most other recreational drugs, see page 302), smoking pot became a potent expression of peace, love and freedom following World War II. The association of cannabis with popular culture from modern jazz to The Beatles climaxed in the 1960s. Despite a 'war on drugs' waged by the US government since the early 1980s, cannabis is one of America's most lucrative cash crops, estimated to be worth approximately $36 billion per year.

Cannabis is grown today in as many as 172 countries, mostly indoors under bright lamps to avoid detection. An unintended consequence of modern prohibition has been the cultivation of cloned varieties that yield up to 25 per cent THC instead of the traditional 2–3 per cent from plants traditionally grown outdoors. The market is now flooded with new strains of highly potent marijuana such as Skunk, Northern Lights and California Orange, which some medical researchers believe could lead to far more significant long-term health issues for users, especially for those with pre-existing mental health problems.

Governments today are therefore in a quandary. While the plant's universal appeal and ease of cultivation makes prohibition laws largely unenforceable and prohibitively expensive to put into effect, there are misgivings preventing the relaxation of the rules owing to new strains coming on to the market.

Cannabis succeeded in the ancient human world because its chemical properties helped people get closer to a spiritual nirvana that made sense of everyday life. It continues to succeed in the modern day for some of the same reasons, providing many with a temporary escape from the anxieties of consumerism and modern urban living. Cannabis is a drug that stretches human consciousness beyond its normal limits, a more potent form of that feeling all of us first tasted as youngsters when spinning around in circles on the playground.

Indian deity Shiva prepares to drink a cup of bhang, made from crushed cannabis flowers.

Tobacco

FAMILY: SOLANACEAE
SPECIES: NICOTIANA TABACUM
RANK: 67

A highly obnoxious plant that continues to decimate populations despite clear evidence that smoking is disastrous for human health.

ONE OF THE MOST celebrated moments in all human history was the accidental discovery of the Americas by a European mariner in 1492. Christopher Columbus collided with these unknown continents while attempting to find a new trade route to China. He was trying to complete his journey in a lot less time than Portuguese rivals who were on the same errand but had set sail eastwards around the southern tip of Africa.

Columbus's colossal miscalculation of the actual distance to China had monumental consequences for global history. European horses, gunpowder and diseases decimated Native American populations but led to the successful conquest and colonization of South, Central and North America by European opportunists closely followed by incumbent African slaves. As many as twenty million indigenous Americans are estimated to have died as a result, mostly from smallpox. Today, a very small number of their descendants cling on to America's traditional way of life in small specially administered reservations.

Most accounts of American history tend to overlook the fact that the natives unwittingly inflicted their own much bigger holocaust on the invading Europeans. When sixteenth-century colonists began to settle in North America, earning their living by cultivating the land and trading overseas, little did they know that they were, quite literally, sowing the seeds for the premature demise of hundreds of millions of people all over the world. In the last 500 years the true human cost of the Native American people's gift of a plant belonging to the same family as the common potato (see page 206) has come to haunt the modern world. The worldwide adoption of growing and smoking tobacco – a Native American habit – has since led to the death of billions of people.

Statistics about smoking today speak for themselves: 1.2 billion people smoke worldwide, that's 33 per cent of the global adult population; 5.4 million deaths result from tobacco-related causes every year, 70 per cent of which are in developing countries; 100,000 children take up the habit of smoking every day, half of whom live in Asia; 50 per cent of those that try to kick the habit can't manage for more than a week without suffering a relapse; and the amount spent *every week* on caring for those suffering from tobacco-related illness in the UK totals more than £100 million.

Nicotiana tabacum is one of the most influential plants in human history because of its highly addictive power and devastating effects on human health. Nicotine, its most important active ingredient, is a naturally occurring insecticide that is especially concentrated in the sixty-four species of plants in the genus *Nicotiana*.

Once inhaled, the compound takes just ten seconds to travel through the bloodstream and cross into the brain. Here it binds to various receptors making people feel relaxed, contented – even euphoric – while also stimulating the release of glucose, boosting energy levels, raising the body's metabolism, enhancing concentration and increasing overall levels of alertness.

However, it is the *absence* of these effects that causes people who smoke to feel a desperate urge to have another dose. Nicotine is metabolized by the body within two hours and its addictive properties are driven by the urge to restore its mental and physiological effects.

Studies on addiction suggest that a common feature of human consciousness is the habit of dividing up time into manageable segments. Lighting a pipe or smoking a cigarette satisfies this natural inclination, inserting a series of pleasurable commas into everyday life. A smoking addict finds that, in the absence of a nicotine fix, time drags on *ad infinitum*, sinking deprived individuals into a depressive abyss. The brain's response to a flood of nicotine is to build additional chemical receptors, so developing tolerance to the drug. But when the drug is absent, the sensitivity of all these neurons causes severe withdrawal symptoms. These extra neurons are more or less permanent. An ex-smoker is said to be like a person who has decided to give up having sex. Despite kicking the habit, it's impossible to return to a state of virginity.

But, addictive as it is, nicotine does not kill. Large doses, if administered through the skin, could be lethal but unlike users of cocaine (see coca, page 300) or LSD (see ergot, page 304), no cigarette smoker has ever died from an overdose of nicotine. Levels in tobacco are simply not high enough. Diseases such as cancer, emphysema, or cardiovascular ailments such as heart attack and stroke, are statistically much more likely to occur in smokers but they have nothing to do with nicotine. As many as nineteen different cancer-provoking chemicals have been isolated in cigarettes, the most lethal of which are radioactive isotopes of lead and polonium that rip through living cells, shredding them from the inside out.

The power of the tobacco plant to take command of people's lives is well demonstrated by the many hoops it has effortlessly leapt through to become a world-conquering species. *N. tabacum*, now grown in at least ninety-seven countries worldwide, is a hybrid of two or more other tobacco plants. It originated somewhere in the Americas where a slightly different variety, called *N. rustica*, had been used by natives for thousands of years as a way of trying to contact the spirit world. European accounts reveal how addicted the North American natives were to nicotine, as Jesuit missionary Paul le Jeune described in 1634:

> *'Their fondness for this herb is beyond all belief ... They go to sleep with their reed pipes in their mouths, they sometimes get up in the night to smoke ... I have often seen them gnaw the stems of their pipes when they had no more tobacco. I have seen them scrape and pulverize a wooden pipe to smoke it. Let us say with compassion that they pass their lives in smoke, and at death fall into the fire.'*

Native Americans were regarded by most Europeans as savages whose habits and culture included everything from cannibalism to devil-worship. Despite these prejudices, within 150 years of Columbus's arrival tobacco plants and the Native

American practice of smoking through a pipe had spread via European explorers throughout the world. Other cultures, it seems, were just as susceptible to the charms of tobacco. Spanish explorers transplanted the crop to the Philippines in 1575 from whence it quickly spread to China. At the same time Portuguese traders took tobacco to Africa, India, Java and Japan. By 1621 smoking tobacco was widespread in the streets and mosques of Persia, imported via traders in Baghdad. By 1700 the Persians had their own tobacco-growing industry, and were exporting as many as four thousand water-pipes to smokers in India each year. By the end of the seventeenth century tobacco was, by any definition, a global crop, which had spread, says one historian, *'like ripples from a handful of gravel tossed into a pond'*.

Its passage was unstoppable. Even the most influential powerbrokers such as King James I (ruled England 1603–25) were powerless to halt its advance, despite his famous *Counterblast to Tobacco* (written in 1604), denouncing the filthy habit of smoking as a custom that is *'lothsome to the eye, hateful to the nose, harmefull to the braine, dangerous to the lungs, and in the blacke stinking fume thereof, neerest resembling the horrible Stigian smoke of the pit that is bottomelesse'*.

James's Persian contemporary Abbas I (Shah of Iran from 1587–1629) was so angered by the spread of tobacco-smoking among his troops that he tried to outlaw the habit by threatening to crop and burn the noses and lips of offenders. But it was no use. The human predilection for tobacco proved then, as now, politically and economically inexorable.

Norfolk farmer John Rolfe was one of the plant's chief protagonists. He took the seeds of sweet-tasting *N. tabacum* from the Caribbean and planted them along the banks of the James river in Virginia in 1612. Virginian tobacco quickly became the world's favourite blend. At the beginning of the seventeenth century, annual UK imports of North American tobacco totalled just 114,000 kilograms. One hundred years later they had risen to seventeen million kilograms.

Tobacco's big break came about because of a contest. Until 1870 it took a skilled manual worker one minute on average to hand-roll four cigarettes. By this time the worldwide demand for American cigarettes was so great that Virginian cigarette-making company Allen & Ginter offered an enormous cash prize ($75,000) to anyone who could automate the process. James Bonsack, a humble American inventor, seldom makes it into history books, but surely his achievement was as monumental as that of any king or emperor.

When Bonsack's automatic cigarette-rolling machine cranked into action in September 1880 it could process a staggering 12,000 cigarettes an hour, as compared to the yield of a manual worker who averaged 240 in the same time. By 1895, North America's cigarette output had increased from sixteen million to *forty-two billion* a year. Two world wars, mass-production and a series of carefully targeted marketing campaigns that glamorized smoking for women all contributed to tobacco's exponential rise in popularity in the first half of the twentieth century.

Then, one day in 1950, a British medical expert threw a monkey wrench into the works. It took the British government seven years to digest Richard Doll's conclusive report that linked smoking tobacco with lung cancer. From 1957 onwards, government advice began to inform the public that smoking, however enjoyable, was potentially disastrous for human health.

But it then took another *fifty years* for smoking in public places to be banned in Britain, following similar restrictions across most other European countries. In tobacco's American birthplace, coming to terms with the knowledge that smoking causes ill-health (reducing life expectancy on average by at least ten years) was, like coming to terms with the ban on slavery, extremely hard for many people. Corporations addicted to commercial profits (not nicotine) fought battles with consumers and campaign groups to disprove the link, or at least make sure any proof was inconclusive. A conspiracy called Operation Berkshire involved the collusion of America's biggest tobacco companies in an effort to forestall any federal legislation limiting the extent of their business.

It was only when one company was caught genetically engineering a variety of *N. tabacum* to double its dose of nicotine but lower its incidence of cancer-producing tar that the federal authorities finally intervened. They found that cigarettes made with fields of tobacco called Y1 were being sold as ultra-lights (supposedly the least harmful) but, unbeknown to the public, thanks to their higher levels of nicotine they were doubly addictive. On 22 November 1998, the Tobacco Master Settlement Agreement forced America's cigarette companies to stump up $365 million to go towards healthcare programmes in return for immunity against most forms of retrospective prosecution.

Concerted government intervention and public health awareness programmes have now resulted in a decline in tobacco-smoking in many developed countries. In America today only 21 per cent of the US adult population are smokers compared with 42 per cent in 1965. However, in the less-developed world (especially Africa and Asia), where governments are less able to raise awareness among their people or more open to exploitation by giant tobacco corporations, the total number of smokers still increases daily. According to a report by the World Health Organization, published in 2002, tobacco consumption in developing countries is rising at an average rate of 3.4 per cent per year.

The same report also reveals that the impact of *N. tabacum* today, thanks to its highly addictive insecticide, is every bit as pronounced, despite most people knowing the consequences of smoking to their health. It includes the following chilling statistics: ten million cigarettes per minute are sold across the world every day; twelve times more British people have died from smoking than were killed in World War II; $150 billion a year is the cost of smoking-related healthcare in the United States alone.

'I'm sending Chesterfields to all my friends,' said the actor (and future president) Ronald Reagan in 1952, shortly after links between smoking and cancer were first established.

Yeast

FAMILY: SACCHAROMYCETACEAE
SPECIES: SACCHAROMYCES CEREVISIAE
RANK: 8

A microscopic fungus with a macro-economic impact throughout all human history.

A CURIOUS correlation seems to exist between the size of a living thing and its potential effect upon planet, life and people. Generally, it seems, the smaller a life-form the greater its impact. The Oomycete water mould (see page 51), cyanobacteria (see page 34) or smallpox (see page 24) are all good examples – as is the extraordinary microscopic fungus commonly called yeast. There are roughly *twenty billion* individual cells in just one gram of yeast. Each cell is capable of performing the equivalent of Biblical miracles – for example turning sugary fruit juice into wine – a chemical metamorphosis that has dramatically influenced the culture of human civilizations.

More than 1,500 yeast species have been identified by scientists to date, although by some people's reckoning these represent only about 1 per cent of those that exist in nature. These microbes feed on sugars and thrive in every environment – land and sea – spreading, when necessary, as spores in the air. But there is one species in particular, *Saccharomyces cerevisiae,* that has played a profound role in history. It is almost exclusively thanks to the action of this single-celled microscopic fungus that humanity has been able to enjoy everything from leavened bread to fine wine. What's more, some of our best prospects for fuelling sustainable industrialization and transportation systems in the future are derived from by-products associated with this quasi-miraculous yeast.

Yeasts make their living by digesting sugars (such as glucose, fructose or sucrose) to make energy. The waste products of this process, which is called fermentation, are carbon dioxide gas and a liquid compound called ethanol. It is these waste products that have been of such enormous value to humans.

Whoever was the first person to discover how to bake bread will forever remain a mystery, but painted hieroglyphs show that baking was widely practised by people living in the civilization of ancient Egypt (see also pages 199–200). Someone probably left out a mixture of flour and water on a warm day and this then became contaminated by *S. cerevisiae,* a naturally occurring yeast that frequently grows on fruit skins. This doughy mixture would then have fermented as the yeast began to break down sugars in the flour, causing it to rise with the output of carbon dioxide gas. Heat from a baker's oven encapsulates this sponge-like texture, turning soft dough into the light, fluffy bread we are so familiar with in our everyday lives. Meanwhile, the ethanol would have evaporated in the heat of the oven, but not without leaving behind the irresistible aroma of a freshly baked loaf.

Such smells have had ubiquitous appeal throughout human history and the art of baking leavened bread, begun in ancient Egypt, became that civilization's most precious export. Everyone around the Mediterranean soon wanted grain (cultivated alongside the river Nile) with which to bake their own bread. Each time, it seemed, a miracle was performed in the oven as uncooked but fermented dough metamorphosed into a wonderful, crispy loaf. Certainly it seems that the Israelites enjoyed leavened bread during their captivity in Egypt since the Bible tells us that God commanded them during their exodus to celebrate their release by the taking of unleavened bread. It was meant as a permanent reminder that they were in such a rush to leave that there was no time to wait for the dough to rise.

Yeast's other miracle has been no less dramatic. Ethanol (a type of alcohol) is a colourless liquid used to make different types of intoxicating drinks – the most famous of which are beer, wine and spirits (such as gin, whisky or rum). When mixed with acidic fruits, *S. cerevisiae* ferments sugars in a way that seems to have an especially pleasing effect on the human palate, producing fragrant chemicals called esters. Other yeasts, such as *Zygosaccharomyces* or *Brettanomyces* produce off-putting, antiseptic flavours. These are organisms that sometimes illicitly creep into the fermentation process, particularly through the use of old barrels, much to the dismay of wine-makers throughout the ages. *S. cerevisiae* (often called brewer's or baker's yeast) has no such nasty side-effects, which is why strains from this species of yeast have most commonly been used for producing almost all the alcoholic drinks ever consumed by man.

Like baking leavened bread, no one quite knows when it was that people sipped their first cup of wine or draught of beer. Some of the earliest evidence of wine production goes back 6,500 years to Georgia, in central Asia (see also grape, page 297). It then seems to have spread throughout Mesopotamia and northern Greece. Ancient Sumerian people (*c.* 5000 BC) are known to have set aside as much as 40 per cent of their annual grain harvest for the production of beer (hops began to be used only from medieval times). Before that no one really knows if hunter-gatherer humans ever enjoyed the dizzying by-products of *S. cerevisiae*.

What is certain is that the appeal of drinking alcohol caught on throughout human history, even more so than smoking nicotine – and has proved just as addictive. Ethanol, like THC (see pages 286–7), is a psychoactive drug. In small quantities it produces feelings of relaxation and cheerfulness, usually provoking behavioural changes such as the loss of inhibitions and an increase in self-confidence. As blood-alcohol levels rise, it depresses the central nervous system leading to feelings of lethargy. High doses cause confusion, dizziness, stupor, coma and sometimes death. Most often too much alcoholic drinking ends in unconsciousness, which has the advantage, unlike some other drugs (see coca, page 300), of breaking its immediate administrative cycle.

Most cultures have taken advantage of naturally occurring brewer's yeast to ferment alcoholic drinks – from rice wine in Japan and palm wine in Africa to malt beer, cider, conventional wines (see grape, page 296) and spirits. Such universal appeal has been driven, at least in part, by the requirement of human consciousness to schedule time into manageable, differentiated blocks (see also tobacco, page 290). For someone who is addicted to alcohol, this takes the form of oscillating between being drunk and sober. Modern drinking cultures reflect this instinct.

Certain times of day are culturally approved for alcoholic consumption (after 6 p.m., for example) while others are not (in most countries it is generally taboo to drink alcohol at breakfast). Routines like these, the nuts and bolts of temporal consciousness, are rhythms reinforced by sensors in the brain, their numbers augmented due to the toxic effects of alcohol. Once the effects of an alcoholic binge wane, and sufficient time passes, the body inevitably begins to itch for another slurp.

Alcoholism today does not account for as many deaths as nicotine, but its impact on society is no less devastating. The World Health Organization's *Global Status Report on Alcohol (2004)* provides a sobering snapshot. As many as sixty different diseases and injuries are caused by alcohol consumption of which the most pernicious are cancers, suicide and motor accidents. Nearly two million deaths a year are alcohol-related (3.2 per cent of all deaths) – a third of which are caused by the drug's stupefying effects – called 'unintentional injury'. Global patterns of consumption are uneven, however.

In some Muslim countries where the sale of alcohol is banned (see page 271) consumption is extremely low. More than 95 per cent of the adult population in Iran, Saudi Arabia and Egypt are teetotal. In other countries, however, despite attempts to outlaw alcohol (such as US Prohibition 1919–33) the picture is reversed. Teetotallers are especially rare in places like the UK (12 per cent), Norway (6 per cent) and Germany (5.1 per cent). Such patterns help explain why as many 55,000 young Europeans, aged between nineteen and twenty-nine, die each year as a direct consequence of alcohol abuse.

The by-products of the fermentation process of *S. cerevisiae* are not all alcohol-related, however. A yeast extraction process pioneered in the nineteenth century by German chemist Justus von Liebig (1803–73) makes a nutritious, vitamin-rich, edible paste from the left-over sediment of brewer's yeast. Marmite (love it or hate it) is especially popular in Britain, which has been manufacturing this vegan health food since the early 1900s.

A British favourite as advertised in the 1930s 'for health and good cooking'.

Other potentially positive uses for the by-products of ubiquitous *S. cerevisiae* are represented by the recent interest in biofuel as a potentially sustainable means of fuelling the modern economy. Ethanol, which can be used to power everything from cars to power stations, is usually made by fermenting crops of sugar or maize. More than three million hectares of sugarcane is already being farmed in Brazil specifically to fuel the 25 per cent ethanol quota mandated by its government in its efforts to reduce fossil fuel consumption. As many as 6.8 million cars in Brazil are now able to run on ethanol fuel, which is available in 35,000 of the country's petrol stations. To what extent such practices can be made genuinely sustainable, without destroying the world's rainforests, increasing food prices or using so much fuel and fertilizer in the farming and harvesting of crops that the environmental benefits of biofuel become purely political/academic achievements, remains to be seen.

Brazil's lead is now being rapidly followed by other nations such as the USA. Fuel from ethanol looks set to ensure that the highly significant story of how a microscopic, single-celled fungus transformed the eating and drinking habits of human civilizations continues to power on into the twenty-first century and beyond.

Grape

FAMILY: VITACEAE
SPECIES: VITIS VINIFERA
RANK: 73

A species that has benefited from both fungus and man, turning its innocent juice into a drink fit for the gods.

APPARENTLY it's a matter of complete biological accident that the wine I am drinking is white. (It's just after six in the evening, so past taboo time for a tipple – see page 294). Geneticists have discovered that an extraordinarily improbable double-mutation occurred sometime in the 150-million-year evolutionary history of grapevines that turned a few naturally occurring red fruits into green-coloured freaks of nature.

Vitaceae is a family of about sixty species of flowering trees. Among them are several that bear grapes – juicy, sweet fruits that are also home to naturally occurring yeasts (see page 293). But it is the grape from a single species – *Vitis vinifera* – that has had such a powerful impact on world history and human culture.

More than 70 per cent of all grapes grown today are pressed and fermented into wine – a smooth alcoholic drink enjoyed particularly by the inhabitants of Europe, the Americas, Australia, New Zealand and South Africa. Plantations of *V. vinifera* now inhabit every continent in the world (except Antarctica). Latest figures estimate that the worldwide area dedicated to cultivating grapes extends to about 76,000 square kilometres – nearly twice the size of Switzerland – and the rate of cultivation is increasing by 2 per cent per year.

The human love affair with grapes has a great deal to do with that naturally occurring yeast, *S. cerevisiae* (see page 293), which coats the skins of the fruit with what looks like a dusting of light white powder. It can only be supposed, rather like the discovery of how to bake bread, that someone left grape juice out in the sun and the process of fermentation began. Before the liquid turned into sour-tasting vinegar, they must have drunk it and discovered the pleasing effects of ethanol on

Picking, pressing and pouring grapes into juice for making into wine – a divine drink in ancient Egypt.

body, mind and spirit. Vineyards need full-time attention; processing grapes takes time and keeping wine from going sour requires technology in the shape of air-tight containers, something possible only with the invention of impermeable pottery jars capped with sealing wax. Innovations such as these emerged in central Asia and the Middle East alongside the first settled societies beginning *c.* 8,000 years ago.

The success of the grapevine *V. vinifera* is therefore synchronous with the rise of human civilization. A freak of nature may have given this vine a helping hand. About 5,600 years ago waters from the Mediterranean burst through the Bosporus, flooding thousands of square miles of land, raising the region that is now the Black Sea to sea-level. This was where it is thought the first vine-growing people may have lived, in the proximity of where *V. vinifera* grows in the wild.

Vitis vinifera: first cultivated around the Black Sea.

Survivors of the flooding of the Black Sea pushed their way to the east, south and west, taking with them expertise in how to grow vines and turn grape juice into wine. Such a sequence of events also helps explain why the word 'wine' is so strikingly similar across such a wide range of distantly related languages from the Hittite *wiyana* to the ancient Greek *oinos*, Arabic *wayn*, Georgian *gvino* and Latin *vinum*.

By *c.* 3000 BC seeds from *V. vinifera* had been carried, planted and cultivated by humans ranging in a wide arc from Egypt to Greece and deep into what is known as the 'Fertile Crescent' towards the border with India. The earliest archaeological evidence of winemaking traces its origins to Georgia, just south of the Caucasus (near Mount Ararat). Archaeologists have found microscopic remains of tartaric acid, a wine residue, left on ancient ceramic shards. Large pottery vessels used for storing wine were among the many precious objects discovered by Howard Carter when he uncovered the tomb of Tutankhamen in 1922. It shows that by 3,300 years ago this drink was popular with the ruling Egyptian elite – an absolute 'must-have' for any self-respecting pharaoh to have in his tomb to sustain him in the afterlife.

But the ancient world's premier vintners – and those most in love with the taste, culture and after-effects of drinking wine – were undoubtedly Greek. These people lined their wine jugs with tree resin as a natural preservative, antibiotic and insecticide (see page 95). Greek wine called *retsina* still retains this ancient flavour. Wine was the drink of the gods. The Greeks even had their own wine god, Dionysus. Drinking parties called 'symposiums' were held in his honour during which men discussed key issues of the day, hatched political plots or simply partied all night long. A large wine jar, called a 'krater', would stand in the middle of a room. It was flanked on three sides by sofas on which men only would recline. Drinking games (such as *kottabos*, which involved hurling empty goblets at targets on the walls) and a combination of homo- and heterosexual activities were all stimulated by copious cups of wine.

Although the concept of processing grapes into wine for the pursuit of politics and pleasure was first practised by the well-to-do of ancient Greece, it was the Romans who spread winemaking to all parts of their imperial domains and seeded the first vineyards in Spain, France and Germany. The Romans had their own copycat version of the symposium, which they renamed a 'convivium' – from which derives our modern word 'convivial' – used to describe a social occasion loosened by wine. They worshipped Dionysus, the god of wine, but renamed him Bacchus.

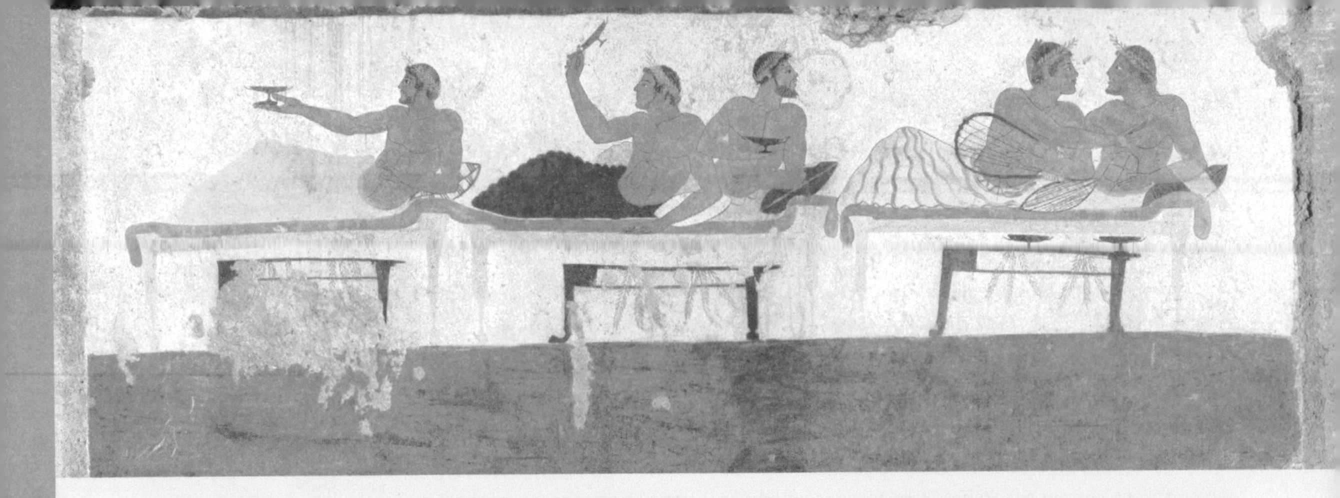

Punch drunk: a symposium of Greek men enjoying the fruits of fine vines.

Cultural fusion between divinity, fine wine and high society was reflected in the emerging religions of the Mediterranean world. Pagan worship of wine through the gods is not culturally so very different from the idea of wine turning into the Blood of Christ during the Catholic sacrament. The Holy Eucharist, still celebrated today by hundreds of millions of Christians across the globe, requires the drinking of wine to commemorate the death on the cross of the son of God, Jesus Christ, who is recorded by St Paul in his letter to the Greek Corinthians as saying: *'The cup of blessing which we bless, is it not the communion of the Blood of Christ?'*

Following the decline of the Roman Empire the market for wine collapsed although the link between wine and divinity lived on. From *c.* AD 500 to 1000 Europe's premier vintners were mostly monastic. Benedictine monks had to maintain healthy vines in order to supply the wine needed to celebrate their holy sacrament. Vineyards were also a handy way of earning extra cash from trade with European aristocracies, which gradually began to reassert themselves from the ninth century AD onwards under the auspices of the feudal system.

By the fourteenth century, the thirst for wine among the royalty and aristocracy of England had become insatiable. From 1305–9 nearly 300,000 barrels of wine were exported by sea from Gascony, in France, to England – that's *900 million litres* a year. When Europe's conquest of the Americas turned to colonization from the sixteenth century onwards, settlers took their vines with them. Jesuits from Spain established their own vineyards to secure sufficient local supplies of Eucharistic blood.

Artificial selection constantly runs the risk of rubbing up against nature's systems (see On Rivalry, page 369). A lack of genetic variety in *V. vinifera*, like the potatoes brought back by colonists from the New World (see page 208), meant that vineyards full of genetically identical grafted fruits were extremely vulnerable to attack. And when the predator finally arrived, the effects were devastating.

Monsieur Borty was a French wine merchant credited by history with importing a number of American vines, which he planted somewhere along the banks of the river Rhône in the summer of 1862. Little did he know that their roots were infested with tiny yellow venomous aphids that had survived the Atlantic crossing, probably because new-fangled transportation by steam ship had dramatically reduced the time it took to travel from America to Europe.

So prolific is the reproductive capacity of this aphid species, *Daktulosphaira vitifoliae*, that a single female can produce as many as 25.6 billion descendants in just eight months – all without the assistance of a male. These aphids suck out sap from the roots of *V. vinifera* and replace it with their own specific cocktail of venom. No one knew the cause of the mysterious vine-killing disease called *phylloxera*, until one day in 1868 when French biologist Jules-Emile Planchon came up with his aphid-infestation theory after some ramblers told him they had spotted the insects nonchalantly marching along some vine roots *'strolling along like good bourgeois going into a restaurant with walking sticks in their hands'*.

Despite Planchon's discovery no one could fathom how to rid the vines of this pest. So desperate was the French government that it offered a 320,000-franc prize to anyone who could find a way to kill the pest. While organic chemists tried every conceivable type of pesticide, French winemakers resorted to experimenting with everything from cow's urine to tobacco and elder leaves to crushed bones dissolved in sulphuric acid. Nothing seemed to kill the happy aphids and within fifteen years the French wine industry had collapsed to half its former size. By 1875 more than 40 per cent of all French vineyards had been destroyed – some 2.5 million acres – with another 1.5 million infested.

Without a cure it is hard to see how the French wine industry – and with it *V. vinifera* – could have escaped complete collapse, with the almost inevitable result that very few people would be drinking wine today. One of the only places that escaped infection was (and still is) Chile. With such scarcity of supply, prohibitive prices would have excluded all but the super-rich from the convivial drink. Other species of *Vitis* that grew in the Americas had far too 'foxy' a taste to suit the refined palates of Western wine drinkers.

Yet ironically, it was from these American species that a solution finally came that saved the French wine industry from annihilation – and with it revived the fortunes of *V. vinifera*. Two French wine growers, Leo Laliman and Gaston Bazille, suggested grafting the European winemakers' *V. vinifera* on to the roots of other American *Vitis* species that was immune to the aphid plague, thus restoring some sense of genetic diversity. Despite entrenched opposition from French vintners, who were paranoid that such dramatic surgery would impair the delicate flavours of their high-class wines, trials of this hybridization process proved successful.

From the mid 1870s on, the enormous task of uprooting, grafting and replanting every vine in France got under way, a process that came to be known as the 'reconstitution'. Today there is hardly a vineyard of *V. vinifera* grown anywhere in the world (except Chile) whose rootstock is not derived from a different Native American species. Since roots simply channel nutrients and water up from the ground, no difference in flavour has ever been detected in this now ubiquitous hybrid species since the chemistry of turning water into grape juice is still undertaken by the genes of *V. vinifera* that are able to thrive, aphid-free, safely above the ground.

There is as yet no known way of eradicating these aphids, so the French prize remains locked in a bank vault, still waiting to be claimed. It was judged that the solution pioneered by monsieurs Laliman and Bazille did not, strictly speaking, lead to the eradication of the pest. Despite its brush with catastrophe in the hands of nineteenth-century cultivators, *V. vinifera* is now a global success story. About 70 per cent of grapes are pressed into wine – and the rest are enjoyed simply for their refreshing juice or eaten as fresh fruit or dried, using a variety of processes, into raisins, currants and sultanas.

Coca

FAMILY: ERYTHROXYLACEAE
SPECIES: ERYTHROXYLUM COCA
RANK: 96

How highly addictive leaves from this South American plant flavour a soft drink that is popular throughout the modern world.

THE STEPAN CORPORATION is an American chemical company founded in 1932. The business, located in New Jersey, just twenty miles west of New York City, makes compounds used in everything from soap to fabric softeners and from paints to toothpaste. But it also lays claim to one curious fact, not listed on the company's website, which is this: it is the only corporation legally allowed to import the leaves of the coca plant into the US.

Each year Stepan imports more than one hundred tonnes of coca leaves derived from a flowering plant which grows under the forest canopy in Colombia, Bolivia and Peru. Usually the trade in coca leaves is an illicit one, thanks to the highly addictive compound that is often extracted to make a recreational drug called cocaine. But Stepan's importation is for an entirely different use. It processes the leaves in vast vats on its nineteen-acre industrial estate, during which it extracts a cocaine-free, non-narcotic flavouring that can be added to a recipe for the world's 'favourite' soft drink, named after the plant itself: Coca-Cola.

For at least 2,500 years before European explorers first discovered the Americas in the late fifteenth century, South American people had enjoyed chewing coca leaves. The Incas (1438–1533) considered the plant a gift from the gods thanks to its stimulating, energy-enhancing effects. In traditional native South American cultures a wad of coca leaves is tucked between the gums and the cheek and gently sucked and squeezed in the mouth for up to three hours at a time. This ancient technique allows a low but constant dose of cocaine, an alkaloid drug, to filter into the bloodstream where – like caffeine – it acts as a stimulant, restricting blood vessels, increasing the heart rate and readying the body for action. Consuming cocaine in this measured way avoids any sudden rush of the drug to the brain. Modern-day effects, such as mind-blowing euphoria, addiction, withdrawal and, occasionally, death, are largely due to the consumption of cocaine in a purified form via injection or inhalation.

When Spanish conquerors first came across South America's coca-chewing natives they banned the practice, dismissing it as primitive and un-Christian. But they changed their minds when locals refused to work in the silver mines – such as those at Potosí, which by the 1650s had begun to make many Europeans extremely rich. Payment in coca leaves provided sufficient incentive, it was found, for slaves to extract all that silver and gold on behalf of their European oppressors.

For more than 200 years people outside South America remained blissfully ignorant of the secrets of what coca could do to their minds. Then, in 1858, an Italian doctor called Paolo Mantegazza (who regularly corresponded with

Charles Darwin) returned from practising medicine in South America to take up the post of surgeon-general at a leading hospital in Milan. His paper entitled 'On the Hygienic and Medicinal Properties of Coca', published the following year, exposed the euphoric effects of chewing coca, which he had discovered for himself after spending time among the indigenous natives of South America: *'God is unjust because he made man incapable of sustaining the effect of coca all life long. I would rather have a life span of ten years with coca than one of 10,000,000,000,000,000,000,000 centuries without ...'*

His words triggered a race among European medics to find the secrets of these wonder-leaves. Within a year, Albert Niemann, a graduate student at Gottingen University in central Germany, had written a dissertation explaining how to isolate the compound cocaine from the leaves of coca. Three years later a Corsican pharmacist called Angelo Mariani patented a new drink, made by mixing wine from Bordeaux with cocaine from coca leaves. Over the next twenty years his cocktail became an international hit, eventually attracting the attention of morphine addict John Pemberton, the man who, in 1885, developed his own American cocawine alternative, which he later called Coca-Cola.

Swig back a mouthful of Pemberton's original French Wine Coca today and you'd get quite a kick; the original recipe contained up to six milligrams of cocaine per fluid ounce of wine. Pemberton also added caffeine from the kola nut as an additional stimulant, the combined effects of which would most certainly have lived up to his claims that his coca drink would benefit *'scientists, scholars, poets, divines, lawyers, physicians, and others devoted to extreme mental exertion'* as well as being *'a most wonderful invigorator of the sexual organs'*.

Some of the most famous pieces of late-nineteenth-century literature were written under the influence of the coca plant – such as Robert Louis Stevenson's schizophrenic novella *The Strange Case of Dr Jekyll and Mr Hyde*, a 60,000-word narrative written during a six-day cocaine binge in 1889. At about the same time Sigmund Freud (1856–1939), the nineteenth-century father of sexual psychology and psychoanalysis, extolled the benefits of cocaine use in a book called *On Coca*. He wrote of its ability to support *'long, intensive, physical work without any fatigue'*. Even the Holy Pontiff himself, Pope Leo XIII (1810–1903), took fondly to the convivial effects of cocawine, endorsing the drink by awarding a gold medal to Angelo Mariani for *'services to the benefit of humanity'*.

Papal bull: Leo XIII endorsing the therapeutic effects of cocaine-laced wine.

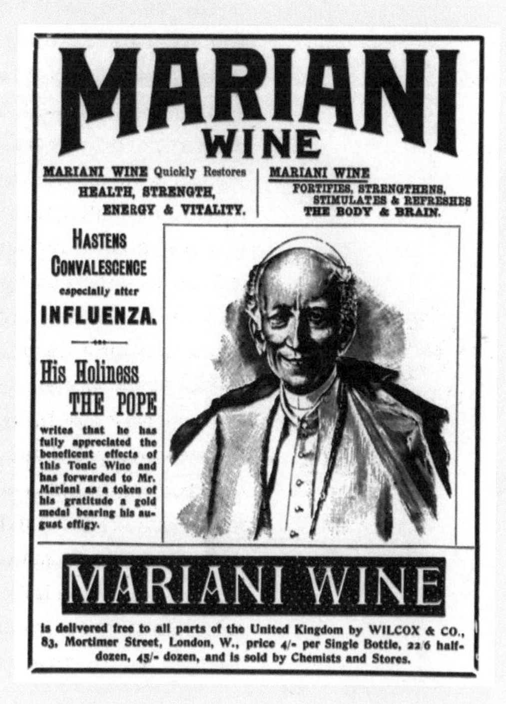

The nineteenth-century love affair with the coca plant continued with cocaine tooth drops, marketed in the US as a family-friendly solution to toothache. Meanwhile a new technique for removing cataracts was pioneered by German surgeon Carl Koller in 1884 who showed how an application of cocaine could numb the cornea, providing blessed relief in an age when cataract removal was likened to having an eye punched through with a red-hot needle. Until 1916, syringe kits, like those used by

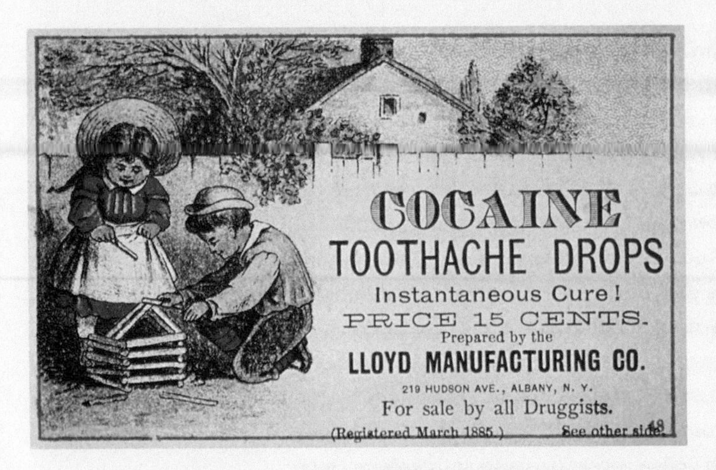

Teething troubles: early users of cocaine were ignorant of possible consequences.

fictional super-sleuth Sherlock Holmes in Arthur Conan Doyle's detective stories, were sold as lifestyle accessories in upper-crust shops such as Harrods of London.

But by the early twentieth century, Western governments, paranoid about the idea of their troops getting high on drugs, eventually began to wake up to the epidemic of cocaine (and heroin, see page 310) addiction that was sweeping through American and European societies. The International Opium Convention of 1912 led to a general prohibition on drugs, with countries such as the US following it up with their own legislation, such as the Harrison Act of 1914, which banned the sale of heroin or cocaine except under licence.

The change of mood had begun as early as 1903 when the secret recipe for Coca-Cola was surreptitiously downgraded to remove coca's psychoactive ingredients. From this time on only spent coca leaves, with most of their active compounds removed, were used to flavour the drink, although its caffeine content and some molecular traces of cocaine remain. Despite the removal of its main psychoactive ingredients, this energy-giving, sweet-tasting drink had worldwide appeal, furthered by some exceptionally smart marketing tactics by the company's chief Robert Woodruff (1889–1985), who, among other things, created the image of Santa Claus still familiar in many parts of the world today in his red and white costume.

The efforts by Western governments to purge the popular use of cocaine as a recreational drug were undone during the 1970s. Since then the plant has undergone a dramatic renaissance to become one of the world's most widely used addictive substances. Today more than 150,000 hectares of coca are cultivated deep in the forests of Colombia, Peru and Bolivia supplying an estimated fourteen million addicts worldwide. Most of it is grown illegally and turned into a cocaine powder in small makeshift factories that are re-established as fast as the authorities locate and destroy them, although 30,000 hectares of coca are legally cultivated in Peru for medicinal purposes. The government is trying to revive coca tea as a safe, traditional way of harnessing the plant's natural medicinal qualities and as an antidote to altitude sickness for visiting tourists.

Cocaine's modern revival began among America's upper-middle classes. In 1974 the *New York Times* magazine called it 'the champagne of drugs'. Its popularity surged a decade later after someone discovered that by mixing cocaine powder with sodium bicarbonate the drug could be inhaled in a smokable form giving a more instant, addictive high. The spread of this variant, called 'crack' after the sound it made during its manufacture, was boosted by a combination of severe economic dislocation in South America, giving birth to cartels of drug-growers, and a large social underclass (most of them black) in America's sprawling inner cities, which became the drug's urban sales force.

In its new form, the drug then spread to Europe, where its largest communities reside in Spain, Italy and the UK. New markets for the drug are growing in other European countries and West Africa.

There is an intriguing postscript to the story of the coca plants' success, both through its use as a highly addictive recreational drug and as a flavouring for 'pick-me-up' fizzy drinks. Look at the map below. See how drugs derived from and grown in the western hemisphere are mostly *stimulants* (such as nicotine and cocaine) while those that dominate in the eastern hemisphere are *depressants* (opiates such as heroin and morphine, see poppy, page 308). Note how well matched are the characteristics of these contrasting drugs with the traditional cultures of East and West. The developed Western world's thrusting, violent consumerism and urge for physical and commercial conquest contrasts with Far Eastern cultures' traditional desire for self-sufficiency (Japan until 1868, China until 1975 and North Korea still today) and spiritualism (such as the spread and appeal of Buddhism throughout Asia).

When Japan sprang out of the blocks in the late nineteenth century in a bid to bring Western industrialization to the Far East, its then colony Taiwan was sown with coca plantations – a counter-cultural drug – to feed the mother country. To what extent did Japan's shift to recreational stimulants (aided by Dutch industrialists) reflect or cause its cultural and economic transformation? To what extent are human cultures the outward expressions of powerful plants that just happen to grow in their gardens?

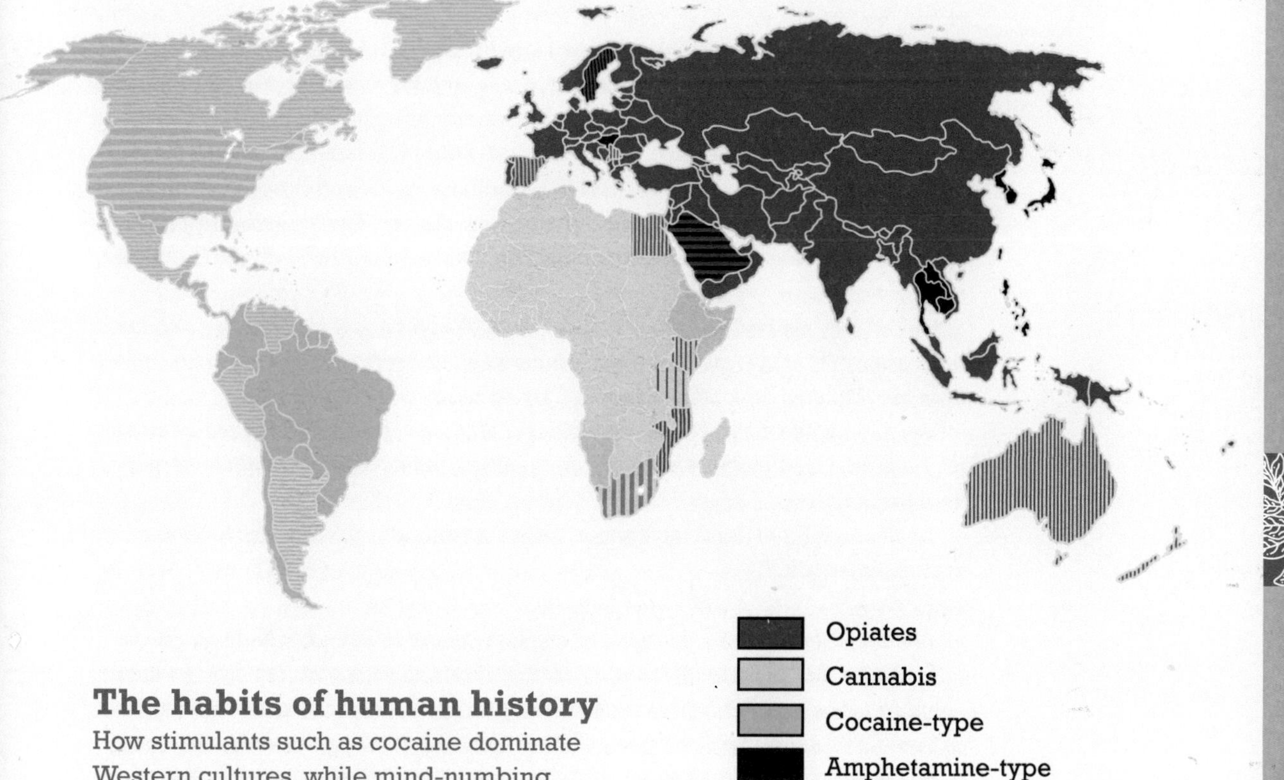

The habits of human history

How stimulants such as cocaine dominate Western cultures, while mind-numbing opiates are more traditional in the East.

Ergot

FAMILY: CLAVICIPITACEAE
SPECIES: CLAVICEPS PURPUREA
RANK: 69

A hallucinogenic parasitic fungus that has helped colour human culture from ancient Greece to The Beatles.

ALBERT HOFMANN was extremely fortunate not to have fallen off his bicycle one April evening in 1943. In a journal, the Swiss chemist from Basel described travelling home with a colleague. It was the most terrifying ride of his life:

> *'We went by bicycle, no automobile being available because of wartime restrictions on their use. On the way home ... everything in my field of vision wavered and was distorted as if seen in a curved mirror. I also had the sensation of being unable to move from the spot. Nevertheless, my assistant later told me that we had travelled very rapidly. Finally, we arrived at home safe and sound, and I was just barely capable of asking my companion to summon our family doctor and request milk from the neighbours ...'*

Hofmann, who died in 2008 at the grand age of 102, was describing his experience of ingesting a drug he had synthesized in a search for a new type of medicine. He called it lysergic acid diethylamide, more commonly known today as LSD. Hofmann later wrote how his bike ride was followed by at least six hours of extraordinary experiences in which even the tables and chairs in his room began to threaten him: '*Everything in the room spun around, and the familiar objects and pieces of furniture ... were grotesque, in continuous motion, animated, as if driven by an inner restlessness.*'

He felt like his body had been possessed: '*A demon had invaded me, had taken possession of my body, mind, and soul. I jumped up and screamed, trying to free myself from him, but then sank down again and lay helpless on the sofa.*'

He then had an 'out-of-body' experience: '*At times I believed myself to be outside my body, and then perceived clearly, as an outside observer, the complete tragedy of my situation.*'

Finally, feelings of terror gave way to relief and colourful, psychedelic hallucinations overwhelmed him:

> *'Now, little by little I could begin to enjoy the unprecedented colours and plays of shapes that persisted behind my closed eyes. Kaleidoscopic, fantastic images surged in on me, alternating, variegated, opening and then closing themselves in circles and spirals, exploding in coloured fountains, rearranging and hybridizing themselves in constant flux.'*

The ingredient in LSD responsible for these bizarre experiences comes from a particular species of parasitic fungus, known about since ancient times. It is called ergot.

Albert Hofmann, the Swiss chemist who went on a bicycle trip after synthesizing LSD from ergot.

Claviceps purpurea (the scientific name for the most common species) feasts on fields of rye, barley and wheat. When one of its spores lands on a floret of grass it mimics the behaviour of pollen by growing a tube that burrows its way down the flower's style. At the bottom it invades the flower's ovary and, virus-like, hijacks its seed-factory to produce its own hard-cased reproductive apparatus called a sclerotum. This grain-like growth eventually falls to the ground, waiting for just the right conditions to fruit into a series of microscopic mushrooms that generate new spores that can be blown in the wind to begin its life-process all over again.

Tucked inside the ergot's sclerotic sack is a series of chemicals manufactured by the fungus to deter grazing animals from interfering with its grass-subverting routine. These substances, called ergotamines, make animals (including humans) temporarily mad. They were also the base ingredients used by Hofmann to manufacture LSD.

The human and animal after-effects of eating ergot-infected rye, barley or wheat range from uninhibited convulsions to life-threatening gangrene. All domestic animals are susceptible, especially cattle. Since the discovery of LSD, a number of scholars and medics, including Hofmann, have begun to re-interpret key historical events with the hindsight of a better understanding of ergotamine drugs.

For example, ergot is thought to have supercharged the climax of ancient Greece's famous Eleusinian Mysteries, a late-summer festival celebrated in a temple dedicated to fertility goddess Demeter at the shrine of Eleusis, near Athens. During this annual ritual as many as 3,000 pilgrims drank a secret drink called *kykeon* concocted from a range of herbs, fermented wine and barley. Experts believe ergot from infected fields nearby was probably responsible for the drink's notorious LSD-like hallucinogenic effects. The participants' belief in a divine universe accessible by mortals via an imbibed sacrament is thought to have originated in Minoan Crete in about 1600 BC and lasted until the third century AD. The general idea has proved robust in subsequent human history and still lives on in many religions to this day.

The parasitic fungus also seems to have bewitched European people living farther to the north. Tollund Man, whose well-preserved body was discovered in a Danish peat bog in 1950, was a forty-year-old individual who died after being hanged from a tree in about 400 BC. Researchers who have examined the remains of his stomach have revealed that his last meal was a bowl of porridge heavily impregnated with ergot. The discovery has led some scholars to suppose that this man was perhaps part of a ritual sacrifice that involved the use of hallucinogenic drugs.

The effect of ergot on human history – especially in cults concerning witchcraft and Satanism – may have been a great deal more significant than surviving records suggest. In medieval times a fury called *ignis scaer* (holy fire) was known to cause wild behaviour among people in England, France and Germany. Several episodes have been recorded between the fourteenth and eighteenth century in Europe where thousands of people went about dancing and screaming wildly for no apparent reason. On one such occasion, in Aachen, Germany, on 23 June 1374, it is recorded that people came out on to the streets and twisted and contorted their bodies in a wild, uncontrollable manner. In July 1518 what has been described as 'a dancing plague' broke out in Strasbourg, France, where as many as 400 dancers took to the streets for up to a month, some of whom died from exhaustion or heart attack. Contemporary physicians ruled out supernatural causes, concluding it to be some kind of 'natural disease'.

The notorious hanging of fourteen women and five men in the early Puritan settlement of Massachusetts in May 1693 has also been attributed to the effects of ergot ingestion. Following the Salem Witch Trials these people were sentenced to death for behaving 'beyond the power of epileptic fits or natural disease', including screams, contortions and throwing objects. Descriptions of their behaviour, suggested to be supernatural forces by the local village doctor, have since been likened to ergot poisoning.

Outbreaks have become much rarer since medics first began to understand the effects of this fungus on the human mind and body from the mid 1850s. Farmers became increasingly aware of the danger of parasitic infection by ergot towards the end of the nineteenth century and started to manage their crops using fungicides based on copper sulphate solutions (see Bordeaux Mixture, page 53), crop rotation (ergot survives for only about one year in the soil) and grain cleaning techniques.

However, owing to Hofmann's discovery of how to synthesize LSD, ergot made a dramatic and unexpected comeback in the middle of the twentieth century. Since then its impact on the development of modern Western popular culture has been profound.

Right up until his death in April 2008, Hofmann believed LSD could have powerful remedial effects for people with mental health problems, act as a potential cure for alcoholism and even help inner-city people who have lost all contact with nature to reconnect with the universe. Indeed, LSD was a perfectly legal drug until the late 1960s. During this period extensive medical trials, in both the US and UK, were conducted by physicians, research students and even the American intelligence forces.

Fruiting fungus: *Claviceps purpurea* prepares to spread its spores.

In the mid 1950s the CIA conducted a series of clandestine experiments called MK-ULTRA in the search for a 'truth' drug that they hoped would let them take control of a person's mind. For a while LSD was hailed as a potential answer. Untrained officers, prostitutes and other volunteers were co-opted into a series of secret trials in which the drinks of unwitting people were spiked with LSD to see what effects the drug would have on their behaviour. Despite the deliberate destruction of all records to do with the project in 1973, a congressional report uncovered the agency's activities in 1976, reconstructing evidence from witnesses and testimonials. As a result, it became illegal for state authorities to involve US citizens in drug trials without taking the courtesy of asking their permission first.

Meanwhile, LSD trail-blazed new forms of cultural expression in the 1960s based on the liberating effects of psychedelic 'trips'. Literary works such as Ken Kesey's *One Flew Over the Cuckoo's Nest* (published in 1962), about the fate of patients in an American mental institution, were written under the influence of LSD. Popular music by bands from The Beatles ('Lucy in the Sky with Diamonds') to The Moody Blues became a symbol of the rejection of establishment politics concerning everything from nuclear disarmament to homosexual and racial suppression. LSD, made widely available by people in the psychiatric profession, was a trigger that seemed to unleash a new confidence in ordinary people, giving them the courage (or the gall) to raise two fingers to those in positions of authority.

The US government in particular was especially anxious to stamp out such subversive behaviour, championed during the 1960s by Harvard professor Dr Tim Leary (1920–96). But the justification for making LSD illegal was not as straightforward as might be supposed. LSD is not a typical drug. For a start there is little evidence of toxicity and it does not apparently lead to addiction. Provided it is taken in a controlled way, it simply makes people happy – and where's the crime in that? At least, that's what Leary argued.

In 1970 LSD was classified as a Schedule One drug by the US government, with its hero Timothy Leary branded by Richard Nixon as *'the most dangerous man in America'*. A year later the United Nations Convention on Psychotropic Substances was agreed, banning LSD and almost every other conceivable hallucinogenic substance for legal use. Recreational use went underground and an age of popular resistance to the political interference of the state in people's private lives was ignited by movements such as psychedelic rock and bands such as Pink Floyd and The Who. Acid parties became the focus of a sustained youth protest movement.

But since the 1980s ergot-derived LSD has been joined by other easier-to-produce hallucinogens such as ecstasy (produced from the dried root bark of sassafras trees, deciduous angiosperms that grow naturally in North America and eastern Asia). The effects on humans are similar, and the political issues much the same. As recently as February 2009, David Nutt, chairman of the UK's advisory body on the classification of illegal drugs, claimed that in terms of health risk *'there is not much difference between horse-riding and ecstasy'*. Both pursuits are associated with about ten deaths a year, according to Nutt, far fewer than occur with highly addictive heroin or cocaine, let alone legal drugs like nicotine and alcohol (see pages 289 and 295). Nutt's question as to why modern society permits the pursuit of *'harmful sporting activities but prohibits the use of relatively less harmful drugs'* was met with howls of protest by the UK government, which immediately dismissed any suggestions of re-classifying the drug.

The capacity of hallucinogenic drugs to enhance human creativity, challenge authority and even extend the parameters of mankind's cognitive potential suggests that species such as ergot may also have played a significant part in the prehistoric development of the human mind (see page 273). Unfortunately, unless clues lie buried somewhere in our junk DNA (see page 14), the chances are we will never know.

Poppy

FAMILY: PAPAVERACEA
SPECIES: PAPAVER SOMNIFERUM
RANK: 79

Nature's most addictive narcotic that has administered and relieved pain and pleasure in roughly equal measures.

JAMES MATHESON was a nineteenth-century drug dealer who got so rich from the profits of his trade in illicit narcotics that at the age of forty-six he retired, buying himself a large island off the west coast of Scotland. There he built a fairytale mansion for his Canadian bride, Mary-Jane. Lews Castle still stands on the Isle of Lewis today, although it has been empty for nearly twenty years and is in a considerable state of disrepair. The building's trustees are currently struggling to raise funds for some badly needed restoration.

The Hong Kong based company co-founded by Matheson and his business partner William Jardine is in much better shape, however. Established in 1832, this prosperous company still stands tall in the business of shipping and trading goods between China and the West. Today it is one of the largest private employers in the former British colony of Hong Kong.

Neither Matheson's fabulous riches, nor the long-term success of his trading enterprise could possibly have happened without poppies. In fact, the influence of these innocent, attractive-looking flowers is truly immense. Over the last 300 years poppies have not only determined the fate of political and economic relations between the world's western and eastern hemispheres, but they have also relieved countless humans worldwide from one of the greatest sources of suffering – excruciating pain.

The therapeutic effects of opium, dried extracts from the seedpods of the poppy species *Papaver somniferum*, have been known to people living in and around the Mediterranean for at least 6,000 years. Poppy heads have been found in a cave in Spain dating back to 4200 BC, which, according to experts, were probably used by Neolithic people for their intoxicating effects. The poppy's pain-relieving properties were well noted by the Greek medic Hippocrates (460–370 BC), the Roman physician Galen (AD 129–200) and the Persian polymath and scholar Avicenna (AD 980–1037). They all prescribed the use of the white latex contained inside poppy seedpods to relieve pain and suffering.

Codeine and morphine, two active compounds found in poppy latex, are brilliantly effective at blocking pain receptors in the brain. Even today they are among the most commonly administered medicines in the world, essential for everything from relieving pain from toothache to providing comfort and relief to those suffering terminal cancer.

While the enormous medical impact of poppies is beyond dispute, the political and economic ramifications of its recreational use as a highly addictive, pleasure-enhancing narcotic are a lot more ambiguous. The legacy of the use of opium and its

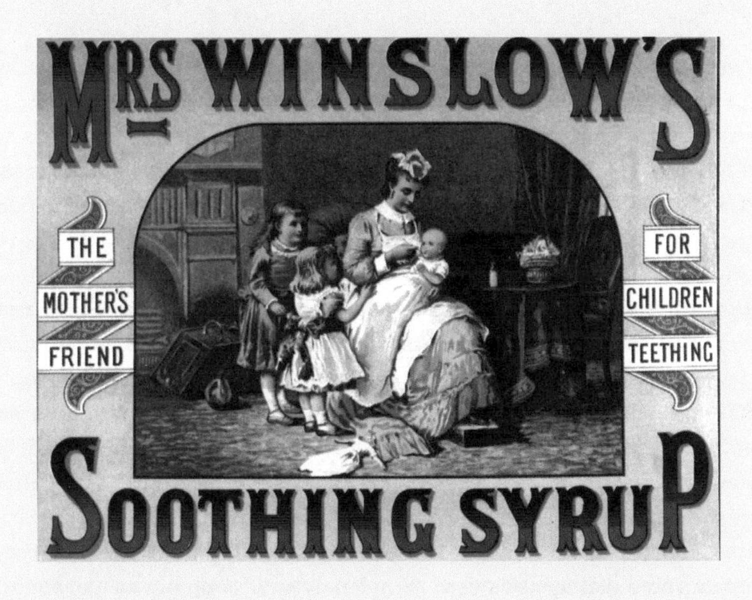

Killer remedy: Mrs Winslow's Soothing Syrup, marketed from 1849 as a medicinal aid for mothers coping with teething children, was laced with one grain of opium (65 mg) per fluid ounce.

semi-synthesized derivative heroin is a source of mistrust and resentment that still simmers menacingly beneath the surface of relationships between cultures from the East and West.

British schoolchildren seldom hear much about the nineteenth-century Opium Wars. Nor have they generally heard of Lin Zexu, one of China's most celebrated national heroes. In the same way, British business executives based in Hong Kong may be unaware that one of today's most successful UK trading companies, Jardine Matheson, was originally established by a group of Scottish drug smugglers. But when history is viewed in the context of non-human species such as *P. somniferum*, it is purified of national pride and prejudice.

Until the nineteenth century, the relationship between opium poppies and human beings was mostly positive on both sides. The drug was uniquely suited to helping settled people deal with their increasingly civilized, urban ways of life. Medics found that all manner of conditions from anxiety, boredom, fatigue and insomnia to squalling babies and diarrhoea, caused by unsanitary conditions, could be better managed and controlled with the use of opium. As a result Arab conquerors in the eighth century took poppy seeds with them as they travelled throughout Asia, spreading the therapeutic, painkilling properties of the poppy's dried extracts to India and the Far East. Islamic cultures preferred remedial opium as an alternative to alcohol, which was proscribed by the Prophet Mohammed (see page 271).

A Swiss-German alchemist called Paracelsus (1493–1541) was the first European to discover that the dried extracts from poppy latex (opium) could be dissolved in alcohol. He called his mixture laudanum (from the Latin *laudeo* meaning 'I praise') because of the satisfying effects it seemed to have on the mind. Paracelsus' potion was picked up by English doctor Thomas Sydenham (1624–89) who described its effects as *'a most delicious and extraordinary refreshment'* which he compared with *'that pleasure which modesty prevents the name of ...'* From then onwards this cocktail, known as an opium tincture, was being used increasingly by everyday people for everything from pleasure enhancement to pain relief and as a sleeping draught. Even into the late nineteenth century a wide range of prescription-free exotic opium-based potions were being promoted in UK pharmacists such as 'Mrs Winslow's Soothing Syrup', heralded as the perfect solution for soothing a baby's toothache.

Two technological developments upset the use of opium as a popular oral medicament. The first was the invention of the hypodermic syringe by Scottish physician Alexander Wood in 1849. Directly injected drugs bypass the body's digestive processes, which normally metabolize harmful chemicals into less damaging derivatives before they reach the bloodstream and the brain. The effect of a drug shot straight into the vein is therefore substantially more powerful than when administered orally, crossing the all-important blood-brain barrier within seconds and releasing a highly potent rush. The human body evolved over millions of years to deal effectively with ingested toxins (sickness and diarrhoea are two evolutionary defences). It has not yet had time to evolve a way that deals with a surprise injection.

The second technical change was the accidental synthesis of an acetylated form of morphine that turned out to be at least twice as potent as morphine itself. The substance was called 'heroin' after its 'heroic' effects and was promoted from 1898 by German pharmaceutical company Bayer as a non-addictive cure for morphine addiction. Twelve years later heroin was shown to be a far more potent and addictive form of the drug it was supposed to counter. By 1910, as a result of this unintended consequence (see pages 367–8), the heroin genie was thoroughly uncorked and an epidemic of addiction had taken grip across Europe and America. The situation was so bad that the world's first co-ordinated treaty prohibiting the use of certain narcotics (chiefly opium and cocaine) was signed by thirteen nations at The Hague in January 1912. The history of international co-operation by nation-states banning what substances individuals could or could not legally consume had begun.

Such technological 'advances' tipped what was once a mostly sympathetic relationship between man and poppy in Western society into a more one-sided affair. Drug-dependent citizens all over the world were increasingly turning into a criminalized sub-culture that required the growing of poppies and the trafficking of opiate products to feed their ongoing addiction.

Meanwhile an even more serious poppy problem had already broken out in the Far East. What China lacked in terms of syringes and heroin was more than compensated for in terms of the sheer volume of opium being smoked by its people in the nineteenth century. The reason for their addiction was the massive increase in supply stimulated by the sowing of poppy plantations in India as a way of re-balancing the unequal trading relationship between the British imperial authorities and China.

An underlying problem faced by all European explorers ever since they had found a maritime passage to the Far East around the Cape of Good Hope was that desirable products such as tea, china, silk and spices came from the Far East – but what could Europe offer in return? As the Chinese emperor Qianlong put it in a letter to King George III of England in 1793, China had little need for products from the Western world: *'I have no use for your country's goods ... Hence there is no need to bring in the wares of foreign barbarians to exchange for our own products.'*

British traders like James Matheson, who first ventured to the Far East in 1818, were acutely aware of the problem. In 1830 he wrote a paper arguing that providence had blessed the Chinese with the most desirable parts of the Earth and the fact they were trying to exclude foreigners as trading partners was a *'defiance of the laws of nature'*.

In the absence of other products to offer, silver became the primary currency of exchange in return for tea – by then an essential addictive stimulant for the people of Great Britain (see pages 283–4). China had a near monopoly. It wasn't until the British planted tea in India in the mid nineteenth century and opened markets to Japan (after Commodore Perry's visit in 1853) that China's stranglehold on the supply of tea was finally broken (see pages 284–5). How were European nations supposed to pay for such commodities if China wasn't willing to consume European products in return? By the late eighteenth century the bleed of silver bullion, the only currency acceptable to imperial China, seriously threatened the British economy.

The commercial genius of British traders like Matheson, who operated out of the Chinese trading port of Canton (Guangzhou), was to stimulate a Chinese mass-market for smoking highly addictive opium. Such tactics solved the Western trade imbalance at a stroke. British traders could sell opium, grown in poppy fields on nearby colonial plantations in India, in exchange for all the tea in China, if they so desired, without ever having to part with another silver coin!

Between 1767 and 1850 the quantity of opium exported from India to China is estimated to have increased seventy-fold. By 1799 the Chinese government had became so alarmed at the effects on its people of such a massive increase in supply (and fall in prices) that it banned the import of opium outright, making its possession a capital offence punishable by strangulation. But poppy-power, combined with the commercial determination of British merchants, ensured such attempts at prohibition were doomed to failure. The bribery of local officials (themselves addicted) helped further increase the quantities imported, regardless of any imperial ban.

To what extent the social consequences of rampant opium addiction in nineteenth-century China directly contributed to the breakdown of that empire's political control is still a matter of hot debate. But by 1850 China was embroiled in the bloodiest civil war the world has ever seen (as many as thirty million people died in the Taiping Rebellion). Imperial authority was further weakened by the intervention of British forces that twice invaded China in two Opium Wars (1839–42 and 1856–60).

The British declared war because of the action of a Chinese official called Lin Zexu (1785–1850), a figure who is increasingly seen as a national hero in modern China. In 1838 Zexu was charged by the Emperor to stand up to British smugglers and rid the country of the menace of drug dependency once and for all. Within a year Zexu had seized 1.2 million kilograms of opium from British smugglers and their local agents. So much illegal dope was taken that it purportedly took 500 workers twenty-two days to neutralize it (by mixing it with lime and salt) and throw it into the ocean. Zexu temporarily imprisoned a number of key British officials (including Matheson) and even wrote a haughty open letter admonishing the British government for allowing merchants to behave immorally and so blatantly to contravene the integrity of another sovereign empire:

'The barbarian merchants of your country, if they wish to do business for a long period, are required to obey our statutes respectfully and to cut off permanently the source of opium. May you, O King, check your wicked and sift out your vicious people before they come to China, in order to guarantee the peace of your nation ...'

Zexu's words fell on deaf ears. On 23 August 1839 a British advance party seized Hong Kong, which it used as a supply-station and trading bridgehead. Within a year, newly fashioned steam-powered gunboats arrived from Singapore and by 1842 Chinese forces had suffered a humiliating military defeat. Zexu was exiled in disgrace and China was forced to open up its ports and markets to the West in the first of a series of 'unequal treaties' that also granted Britain sovereign rights over Hong Kong for the next 100 years. By 1860, after a second Opium War ended with the burning of the Chinese emperor's palace in Peking, Western powers forced China to formally legalize the opium trade. By 1879 British traders were importing a colossal 103,000 chests of opium into China a year, equivalent to 5,300 metric tonnes.

People of influence in London were not blind to the British Empire's moral hypocrisy. In a speech to the House of Commons in 1840, future Prime Minister William Gladstone (1809–98) declared: *'A war more unjust in its origin, a war more calculated to cover this country with permanent disgrace, I do not know and have not read of ...'*

Even as early as 1750, the director-general of the East India Company, Warren Hastings, had warned against the dangers of opium usage, which he said could be pushed on to foreigners for the purposes of trade but should never to be encouraged as a product for use by the British people themselves: *'Opium is not a necessary of life, but a pernicious article of luxury, which ought not to be permitted but for the purpose of foreign commerce only, and which the wisdom of the government should certain restrain consumption.'*

Western hypocrisy was good news for poppies, which were planted in their millions throughout Bengal (in northern India). Cultivation then fanned out eastwards across south-west China, Burma and northern Thailand – designed to feed Asia's addicts. Poppies were aided further in their global propagation by the migration of tens of thousands of starving, opium-obsessed Chinese immigrants from their mother country during the Taiping Rebellion. Many were taken on by Western colonial powers as 'coolies', since overt slavery itself was, by then, mostly illegal. As a result, opium dens and drug-smuggling operations became entrenched in inner-city districts from London to New York, opening up as if in defiant revenge for what had already happened on mainland China. Once poppy plantations sprang up in Mexico and South America during the early twentieth century, the plant's global conquest was all but complete.

By the beginning of the twentieth century, Western governments were in the grip of an opium-inspired epidemic that still plagues their societies today. Wars, especially the Vietnam War (1965–75), increased the appeal, usage and level of addiction among serving troops who experimented with heroin as an escape from the traumas of jungle conflict. In a further twist of irony, drug supplies were increased by the deliberate cultivation of poppy plantations in Cambodia and Laos by pro-American, anti-Communist rebels as part of their bid to earn sufficient cash to pay for armaments with which to fight their Communist enemies.

Heroin addicts today (there are thought to be more than eleven million worldwide) have seldom been able to afford the cost of their habit, estimated to total, on average, £16,000 a year. Strict enforcement of prohibition of the drug, especially in America since President Nixon's 'War on Drugs' in the 1970s, means modern addicts and dealers resort to urban violence, thefts and gangland shootings to feed their addiction, features that have become commonplace in even the most salubrious Western cities from Berne to London, and New York to Rome.

There is even a line to be drawn that links poppies with the modern 'War on Terror' triggered by the attacks of 9/11. More than 92 per cent of today's worldwide supply of heroin is derived from the poppy fields of Afghanistan, a terrorist-harbouring country whose economy has been broken for more than 200 years by the repeated invasions of imperialist powers (British, then Soviet, now NATO). Other than growing poppies and trading opium for producing heroin, Afghanistan's economy has little to offer in exchange for the necessities for survival, be they arms or food. In 2007, according to a United Nations drugs report, global heroin production reached its all-time recorded high of 606 metric tonnes.

The globalization of the poppy species *P. somniferum* has created in its wake a list of winners and losers. While heroin addicts resort to crime to feed their habit, roaming the streets in addicted desperation or being locked up in prison to be kept from harming themselves and society, others are relieved from agony by the world's most effective poppy-derived painkiller, morphine. Meanwhile, the government and people of nineteenth-century China lost out to Western profiteers, made rich through corporations such as the East India Company. Then there was Sir James Matheson, the opium smuggler who spent part of his fortune – £190,000 – buying a Scottish island and more than £500,000 on building a crenellated mock-Tudor folly. He was certainly among the most spectacular individual winners in the great poppy lottery of life.

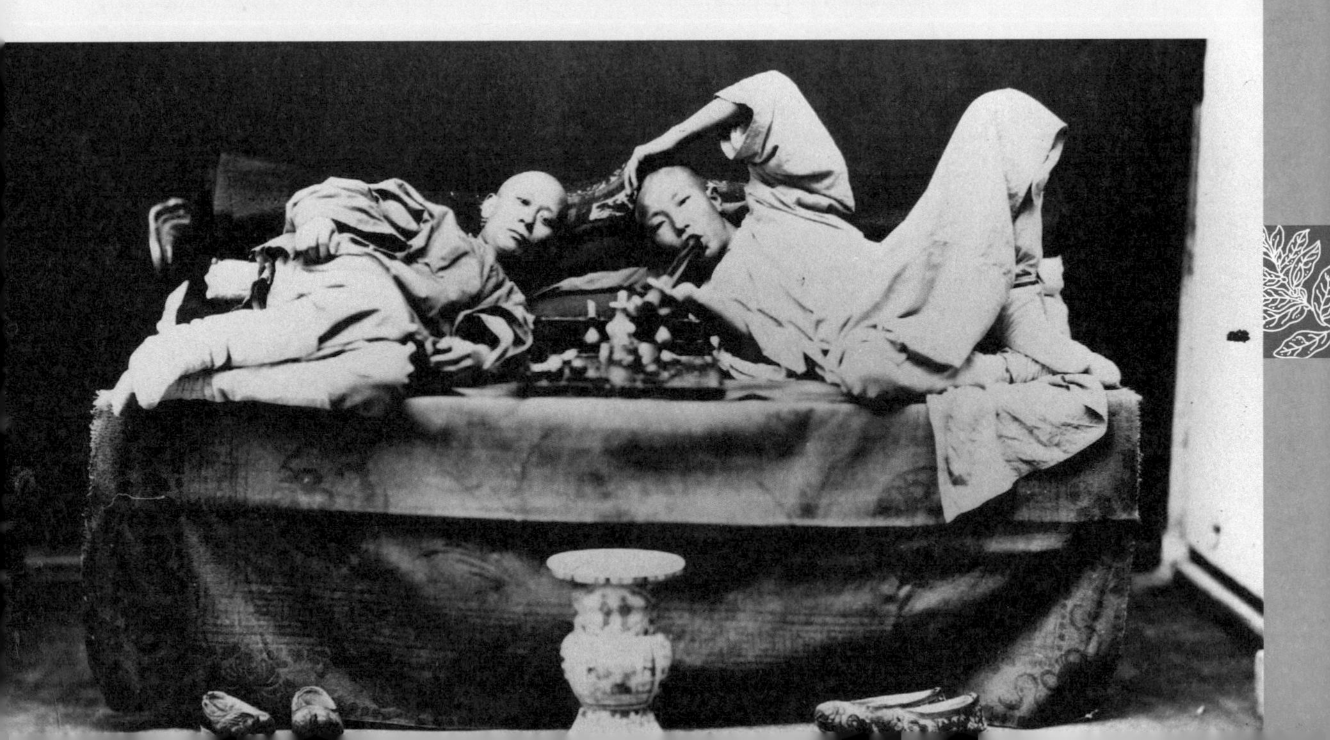

The high life: an opium den in nineteenth-century China.

Penicillium

FAMILY: TRYCHOCOMACEAE
SPECIES: PENICILLIUM CHRYSOGENUM
RANK: 10

A naturally occurring antibiotic that has transformed modern medicine and substantially increased human populations.

AN UNWASHED PETRI dish, an open window and three weeks of unusually cold weather were just a few of the random, chaotic circumstances that apparently led to Alexander Fleming's discovery of the antibiotic properties of a fungus called *Penicillium*. In the summer of 1928, the Scottish microbiologist had taken a three-week break from his research in the basement laboratory of St Mary's Hospital, London. On his return he casually began to clear up the left-overs of some previous experiments that he should have washed up before he went away. In one dish he noticed that a mould had contaminated a culture of the common infectious bacterium streptococcus. The mould was surrounded by a clear halo. *'That's funny ...'* said Fleming, gesturing towards his lab assistant, Merlin Pryce.

Whilst luck undoubtedly led to the mould contamination, it was only thanks to Fleming's eagle eyes and vast experience as a microbiologist that the enormous significance of that halo-effect was determined. Fleming realised that the mould must be secreting a compound that was killing off the bacteria. Fleming's discovery was the equivalent of biological dynamite because so many infections and diseases that afflict animal life (including humans) are caused by a group of bacteria classified as 'Gram-positive' – like those streptococci in the Petri dish. Fleming went on to write a paper about the antibacterial properties of the fungus, which he identified as *Penicillium notatum*. The antibiotic compound it produced he called Penicillin G.

Fleming was *not* responsible, however, for turning the discovery of *Penicillium*'s powerful antibiotic properties into something that had worldwide impact. That's why, when Nobel Prizes were awarded in 1945, he shared it with two other people – Australian scientist Howard Florey (1898–1968) and German-born biochemist Ernst Chain (1906–79).

Florey and Chain were part of a team of microbial researchers in Oxford, England. By chance Chain picked up the scientific paper written by Fleming in the 1930s that proposed using the *Penicillium* fungus as an antibiotic. Wounds sustained during World War II frequently became so badly infected that amputations were often the only way to save a victim's life. Even then, success was a hit-and-miss affair. In those days, infection was feared as much as cancer is today.

On Saturday 25 May 1940, Howard and Chain conducted an experiment on eight mice that were injected with a lethal dose of streptococci bacteria. Four were treated with penicillin and, as controls, the other four were left untreated. When Florey and Chain arrived at their lab the next morning they found, alongside the four dead untreated mice, the four treated mice restored to perfect health. Human trials began in earnest.

One of the main stumbling blocks was how to grow sufficient quantities of the fungus to administer a big enough dose to treat a human, rather than a mouse. One of the less well known (and less salubrious) production techniques initially used by the team was the deployment of hospital bed pans to grow mould. Liquid containing penicillin was drained from beneath the growing mould and filtered through parachute silk into milk bottles on nearby bookshelves. Such makeshift techniques could never produce sufficient quantities to make much impact on the masses of wounded soldiers needing treatment during the war.

It was another species – a most unlikely find – that came to the rescue thanks to some speedy transatlantic collaboration. An American cantaloupe, a type of melon originally taken to the New World by Christopher Columbus on his second voyage in 1494, was found to be the best medium for growing a *Penicillium* species called *chrysogenum* that produced up to 200 times more Penicillin G compound than Fleming's *P. notatum*. This surprising mould-growing habitat was discovered by Mary Hunt (also known as 'Mouldy Mary'). During 1943 a series of experiments that involved shooting x-rays through Mary's mould produced a mutant *Penicillium* variety that further increased the quantity of antibiotic compound to 1,000 times more than that in Fleming's original culture. Giant aerated 87,000-litre metal tanks (instead of milk bottles) were deployed to grow the mould so that by 1945 enough penicillin was being produced to treat as many as *seven million* patients a year.

For the genes of the fungus genus *Penicillium*, the partnership with modern human beings has been a propitious one. Antibiotic sales today are estimated at more than *£6 billion* a year. Infections that encounter penicillin suffer the bacterial equivalent of a holocaust, especially in domesticated animals regularly injected with antibiotics. A constant state of war now exists between different forms of artificially selected antibiotics and the super-fast evolutionary footwork of naturally occurring bacteria. With their short generation times, these microbes evolve at warp speed (see pages 84–5). So far humans and their fungal partners have generally had the upper hand – although thanks to the frequent overuse of antibiotics, resistant bacterial infections (such as MRSA) are reaching epidemic proportions.

Sir Alexander Fleming pictured examining a Petri dish six years after receiving the Nobel Prize for discovering the antibiotic properties of *Penicillium*.

The impact of penicillin on people at a personal level over the last sixty years has generally been one of enormous relief as millions of lives have been saved from otherwise fatal infections. On a global scale the development of penicillin may be more of a double-edged sword. The increase in the human population afforded by its widespread use has been estimated at more than 200 million individuals – not quite in the league of artificial fertilizers (see page 198) or Norman Borlaug's Norin 10 (see page 201) but very substantial nonetheless.

Florey, regarded as one of Australia's most important scientists, was personally extremely conscious about the global impact caused by the wonder-drug he helped bring to market. *'I am now accused,'* he said in 1967, *'of being partly responsible for the population explosion … One of the most devastating things that the world has got to face for the rest of this century.'* Florey became so sensitive about the issue that he dedicated much of his later research into developing improved methods of contraception.

Cinchona

FAMILY: RUBIACEAE
SPECIES: CINCHONA OFFICINALIS
RANK: 84

A tree that provides humans with their most effective defence against malaria.

PLANTS are victims of human discrimination and prejudice every bit as much as people. Suspicion is frequently propagated by religious tradition, as with Mohammed's intolerance of plants that ferment into alcohol, or the Christian Church's vilification of fungi and herbs such as cannabis, hemlock or *belladonna* as ingredients in 'witches' brews' (see page 271). Foods regarded as staples today, such as the highly nutritious potato, suffered hundreds of years of social stigma before they were accepted as good to eat in Europe (see page 207). Even when a plant offers an apparently miraculous cure from death by uncontrollable fever, prejudice can still intervene.

In 1658 Oliver Cromwell, a soldier who usurped power from the British monarchy during the English Civil War (1642–51), lay on his deathbed having contracted what many historians believe was malaria whilst campaigning in Ireland. A Venetian medic offered him a possible cure – a solution based on the powdered bark of a tree native to South America. Yet, despite his high fever, Cromwell dismissed the potion out of hand, fearing it was part of an elaborate popish plot. Instead he was bled, a procedure that inflamed his condition still further, and he died soon afterwards. We are therefore left to ponder how differently Britain's history might have turned out had Cromwell cast aside his Protestant prejudice.

Jesuit's bark: protection from malaria came from the Peruvian cinchona tree.

No one knows for how long the active ingredient quinine, found in the powered bark of the cinchona tree, has been used by people as a defence against the often fatal effects of the malarial parasite *Plasmodium falciparum* (see page 135). There is no explicit evidence of its use before Spanish invaders arrived in the Americas in the late fifteenth century, perhaps because, like smallpox, malaria travelled with them from Europe and Asia. But once this disease – spread by bloodsucking mosquitoes – broke out in South America, the indigenous people soon put their prodigious knowledge of herbal medicine to work.

By the 1620s Jesuit missionaries working in the forests of the Andes had learned of the healing effects of the powdered cinchona bark from South American natives. Its first noted success was in 1630 when the governor of Loxa, a thriving Peruvian city, drank an infusion made from cinchona bark and was cured of a life-threatening fever. Jesuit priests working in South America were the main conduit for the diffusion of the miraculous bark back to Europe where malaria sufferers in Italy and Spain made similarly spectacular recoveries. It is this association that led to decades of argument between Catholics and Protestants about the pros and cons of using a Papist remedy known as 'Jesuit's bark'.

It took a medic with exceptional cunning and entrepreneurial skills to break down this prejudice and finally to convince even Protestant nations that here indeed was an effective cure for a most pernicious disease. Robert Talbor, a Cambridge physician born in 1642, became famous for concocting a secret remedy for high fever. Aware of popular misgivings against the use of 'Jesuit's bark' in England, Talbor wrote a book in which he actively discouraged people from taking this remedy: '... *beware of all palliative cures and especially of that known by the name Jesuit's powder, as it is given by unskilful hands.*'

Talbor's fever-busting special remedy became so famous that King Charles II of England (ruled 1660–85) appointed him as Royal Physician in 1672 and knighted him in 1678. The following year Charles fell ill with fever but was cured by Talbor's remedy, as was the son of French King Louis XIV who paid Talbor 3,000 gold crowns out of gratitude for saving the French dauphin's life. When Talbor died in 1681, his secret remedy was finally revealed – a solution of opium, alcohol, lemon juice and ... guess what? Powdered bark from the cinchona tree.

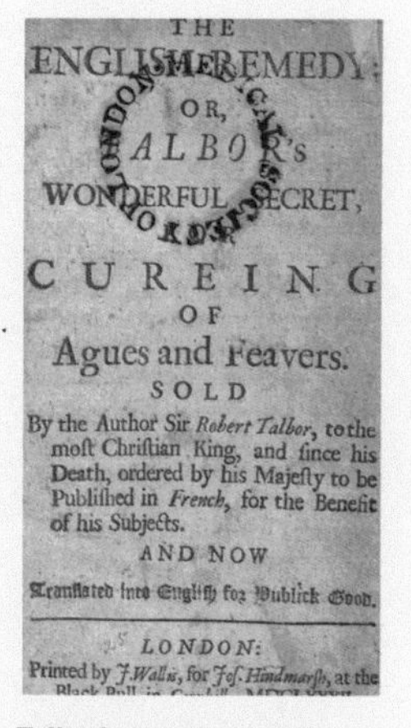

THE
ENGLI[illegible]EMEDY:
OR,
[illegible]ALBO[illegible]R's
WON[illegible]ERFUL [illegible]ECRET,
[illegible]
CUREING
OF
Agues and Feavers.
SOLD
By the Author Sir *Robert Talbor*, to the most Christian King, and since his Death, ordered by his Majesty to be Published in *French*, for the Benefit of his Subjects.
AND NOW
Translated into English for Publick Good.

LONDON:
Printed by *J. Wallis*, for *Jos. Hindmarsh*, at the

Talbor's secret remedy was published a year after his death 'for the public good'.

European efforts to cultivate the plant were long frustrated, however. When, in 1735, French naturalist and epic explorer Charles Marie de la Condamine travelled all the way down the Amazon river (see also rubber tree, page 256), he took the trouble to collect what he believed would be a vastly valuable cargo of cinchona seedlings, planted in boxes of earth, which he carefully nurtured during his eight-month journey through jungle, swamps and rapids. Just as he came within reach of the coast – in sight of the ship that would take him and his precious cargo back to Paris – his small boat was swamped by a wave and the cinchona plants were washed away.

Subsequent 'seed-napping' efforts by other Europeans failed to achieve successful cultivation of the cinchona tree outside South America until 1865, when an Englishman called Charles Ledger finally managed to break the Peruvian monopoly. He obtained seven kilograms of seeds from his native servant, Manuel Incra Mamani, for a price of about twenty dollars. Ledger's seeds were sold to Dutch colonialists who established plantations in Java on the other side of the world. British and German colonialists followed suit, extending the tree's reach into India and Ceylon (now Sri Lanka).

French scientists successfully isolated the active compounds in the cinchona tree – now known as quinine – in 1822. Along with a big increase in the worldwide availability of cinchona bark, thanks to the new plantations in Asia, the battle between humans and the *Plasmodium*-carrying mosquito began to tip in humanity's favour. The mass administration of quinine in the late nineteenth century made it possible for Europeans to colonize malaria-infested regions of equatorial Africa. The Scramble for Africa, sanctioned by the Congress of Berlin in 1884–5, therefore owed at least as much to the global spread of quinine as it did to the forces of European avarice, industrialization, steam power or mass-produced armaments.

Malaria, for which there is no prospect of a vaccine, is still a big killer today – especially in Africa. Attempts at producing an artificial alternative to natural quinine have been only partially successful. *Plasmodium* parasites have successfully evolved resistance to chemical alternatives such as chloroquine, daraprim and mefloquine leaving natural quinine as by far the most effective treatment. The cinchona tree has triumphed over human prejudice in the last 500 years, spreading itself with human help all over the world. In the process it has sustained and increased populations thanks to the remedial powers of its bark – a vital component in humanity's medical kit-bag, which we look set to rely on well into the future.

12 On Companionship

How some animals have prospered by being good company but have recently provoked a moral dilemma.

DIOGENES, an ancient Greek philosopher-cum-beggar, mostly preferred animals to people, declaring that *'dogs and philosophers do the greatest good and get the fewest rewards'.*

He lived with a dog in a barrel at the bottom of some steps leading up to a courthouse in Athens. Diogenes' philosophy, much like his way of life, mocked settled human societies with their craving for status, wealth and social prestige. Instead, he said, people should live more like dogs, self-sufficiently in the wild, removing themselves from the vices of social corruption.

The fact that a man could live in a barrel with a dog at all can be explained only by the many thousands of years of animal domestication that had already taken place before the rise of ancient Greece. Beginning at least 12,000 years ago, humans learned how to tame, capture and breed what were once wolves into domestic animals, originally for the purposes of helping them hunt and herd flocks of other animals, such as sheep (see page 218).

By the time of Diogenes (412–323 BC) the practice of cohabiting with dogs was well-established. Less well understood is the impact pets like Diogenes's dog may have had on the development of human philosophy and morality over the last 2,500 years. This mostly unwritten history is playing itself out today in the debate over animal rights, which fundamentally questions

It's a dog's life: Greek philosopher Diogenes made fun of human social values.

people's moral relationship with all non-human life – and by extension, the planet itself.

Taming animals

The dramatic domestication of grey wolves into domestic **dogs** (see page 323) is, in the animal world, as spectacular a transformation as was the cultivation of teosinte into maize in the realm of plants (see page 234). Humans have succeeded not just in taming aggressive carnivores into benign, trustworthy companions, they have also distorted a single species, *Canis lupus*, into a dazzling variety of forms. A trip to Crufts, the largest annual dog show in the world, now held in Birmingham, England, testifies how, over many generations of selective breeding, this species' gene pool has been twisted into every conceivable size, colour, shape and texture of canine. It ranges from the teacup-sized chihuahua to the metre-tall Irish wolfhound.

Some dogs have been bred to look cute or perform tricks (for dog shows) or to be good companions (for children, families or people living alone). Others have been genetically groomed for their usefulness, for example, to assist people who hunt and shoot or to search for drugs, as guide dogs for the blind or guard dogs to protect property against theft.

Cats (see page 327), canaries, **rabbits** (see page 330), **hamsters** (see page 334) and goldfish are other creatures that have taken up residence with people. In each case, their genes have responded to human fancies, either through physical modification – the bright colour of a canary or goldfish, for example – or by behavioural adjustments that have suppressed natural instincts for fleeing predators (such as in rabbits) or aggression (such as in cats) making them into compliant domestic companions. Mammal pets also display a level of intelligence and a capacity to learn (see On the Rise of Reason, page 157) so that humans have been able to adopt them, as if they were children or family members, teaching them a variety of simple calls and commands.

In return, these pets are usually given a name, fed regularly, allowed to dwell under the same roof as their cohabiting humans and are not generally farmed as food. The origin of this symbiosis is often ancient (as in the domestication of cats and dogs) although sometimes modern (the golden hamster). Nevertheless, the modern habit of keeping animals as household pets is a trend that began to catch on in Europe, especially England, only about 400 years ago. It has since spread worldwide with some remarkable consequences.

Pet power

Until *c.* 1600 people in Christian, Jewish or Islamic cultures generally believed that humans were ordained by God to have dominion over all other living creatures. Despite man's fall from grace in the Garden of Eden, Judeo-Christian scriptures were unequivocal in their teaching about his place in the great hierarchical chain of being. At the top was God, surrounded by the angels of heaven, and at the bottom lay inanimate earth and rock. In between there was life – plants, insects and animals – but with man ranked in a special place above the rest since only he had a spirit that could, on the Day of Judgement, be redeemed and join God and the angels in heaven.

Beginning about 400 years ago, this medieval ecclesiastical consensus, known as the 'Great Chain of Being', began to break down.

Eccentric author and playwright Margaret Cavendish, Duchess of Newcastle (1623–73) was one of the first Europeans to challenge the Judeo-Christian order of creation that regarded everything from cock-fighting to badger-baiting as morally acceptable.

It was man's ignorance of other creatures, said Cavendish, that lay at the heart of his cruelty and supposed superiority:

'For what man knows whether fish do not know more of the nature of water and ebbing and flowing and the saltiness of the sea? Or whether birds do not know more of the nature and degrees of air or the causes of tempests? Or whether worms do not know more of the nature of the earth and how plants are produced ... Man may have one way of knowledge ... and other creatures another way.'

And what gave humans the right, she asked in her *Dialogue Betwixt Birds* (1653), to shoot sparrows for taking cherries and then eat the fruit themselves?

Animal rights

Trace a line from the earliest advocates of animal welfare in the seventeenth century such as Cavendish, to today's modern animal welfare activists, and what they all seem to share is an inordinate fondness for pets.

People who love dogs, cats and rabbits were among the philosophers, politicians and naturalists who led the charge against traditional views regarding man's place in the world, challenging medieval assumptions about the essential superiority of human beings.

Jeremy Bentham (1748–1832), one of Britain's pre-eminent moral philosophers in the Age of Enlightenment, was both childless and wifeless. His love was channelled into his pet cat, Sir John Langbourne D. D. (which stands for Doctor of Divinity).

It is impossible to know exactly how far Bentham's feline feelings were reflected in his moral argument that animals were entitled to natural rights every bit as much as man. It is hard not to imagine that personal experience must have played some part in making up his mind that cruelty to sentient beings was morally reprehensible.

'The day may come when the rest of the animal creation may acquire those rights which never could have been withholden from them but by the hand of tyranny ... The question is not Can they reason? nor, Can they talk? but, Can they suffer? Why should the law refuse its protection to any sensitive being ... ?'

A similar moral position was proposed by Jean-Jacques Rousseau (1712–78), the Swiss philosopher who was so devoted to his pet dog Sultan that he refused to travel abroad without him. His famous theory on the best way to educate children stemmed from careful observations he made of the similarity of behaviour between young people and pets:

'Look at a cat entering a room for the first time. He inspects, he looks around, he sniffs, he does not relax for a moment, he trusts nothing, before he has examined everything and come to know everything. This is just what is done by a child who is beginning to walk. In addition to the vision which is common to both child and cat, the former has the hands that nature gave him to aid in observations and the latter is endowed by nature with a subtle sense of smell. Whether this disposition is well or ill cultivated is what makes children adroit or clumsy, dull or alert, giddy or prudent ...'

Laws designed to prevent human cruelty to animals first reached the statute books in England in 1822, thanks to the persistence of two British MPs, Richard Martin and William Wilberforce. Both were animal lovers. Among the many creatures living in Wilberforce's home in Kensington Gore (described as a zoo in everything but name), was a much-loved pet rabbit about which Wilberforce wrote: '*If "Love me Love my Dog" is an established axiom, much more "Love me, Love my Hare" holds true in this house ...*'

On 16 June 1824 a group of MPs including Martin and Wilberforce met in the stimulating environment of Old Slaughter's Coffee House (see pages 279–80) in St Martin's Lane, London, where they established a society '*instituted for the purpose of preventing cruelty to animals*'.

By this time popular zeal for pet ownership, already boosted by increasing European urbanization, had reached the highest echelons of society, thanks to the passion of British monarch Queen Victoria (ruled 1837–1901) for her household pets. Victoria adored having her portrait painted in the company of her faithful King Charles spaniel, Dash, or Eos, her greyhound, for whom she erected a monument at Windsor. Adamantly averse to vivisection, the Queen even addressed her nation during her Golden Jubilee in 1887, saying she had noticed with '*real pleasure*' that a sign of the spread of enlightenment among her subjects was '*the growth of more humane feelings towards the lower animals ...*'

It was thanks to Victoria's love of pets that Wilberforce and Martin's special society was granted a royal charter in 1840, becoming the Royal Society for the Prevention of Cruelty to Animals (RSPCA). The same organization is now the world's oldest and largest volunteer army dedicated to protecting animals against neglect or harm by humans.

Even scientific opinion about the relationship between man and other creatures was profoundly affected by the Victorian penchant for keeping pets. When Charles Darwin took his dog Sappho for a daily stroll around the Sandwalk at Down House in Kent, his mind never stopped considering the issue of the interconnectedness of all living things. '*For do not dogs have a conscience?*' he asked, rhetorically, in his book *The Descent of Man* (1871), '*and does not the deep love for their master approach religious devotion?*'

Arthur Schopenhauer (born in 1788), the German philosopher who is widely regarded as a leading proponent of the modern animal rights movement, was deeply devoted to his pet poodles Atma and Butz. They were his sole cohabitants in Frankfurt from 1833 until his death in 1860.

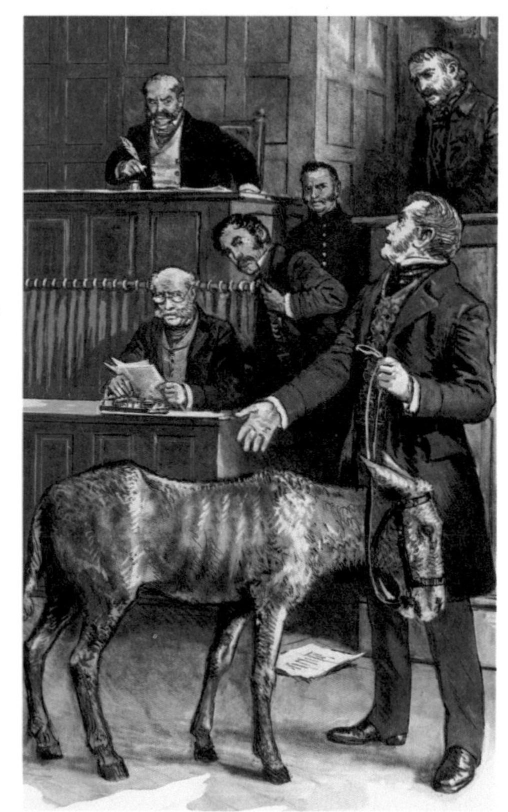

Richard Martin MP campaigning in court on behalf of a neglected donkey.

Such compassion can only have reinforced Schopenhauer's distaste for Jewish, Christian and Islamic traditions, which he blamed for the historically barbaric moral relationship between man and the natural world:

> '*Christian morality leaves animals out of account ... they are at once outlawed in philosophical morals; they are mere "things", mere means to any ends whatsoever. They can therefore be used for vivisection, hunting, coursing, bullfights, and horse racing, and can be whipped to death as they struggle along with heavy carts of stone. Shame on such a morality that is worthy of pariahs ... and that fails to recognize the eternal essence that exists in every living thing ...*'

Australian philosopher Peter Singer's book *Animal Liberation* (published in 1975), attempted to cut through the last links in the Great Chain of Being by proposing equal rights for animals and humans. His major campaign, the Great Ape Project, was established in 1993 and continues to lobby the United Nations to confer legal rights on all non-human Great Apes including chimpanzees, bonobos, gorillas and orang-utans.

Rights and wrongs

While the habit of sharing a household with a company of pets has undoubtedly transformed many aspects of the Western relationship towards some living creatures, it has done little to resolve the underlying moral contradiction as to why some animals are considered vermin, or suitable for the brutality of intensive farming, while others are cuddled and protected as pets.

Until *c.* 1900 very few creatures were used for scientific experimentation, yet today more than 2.5 million animals are kept caged in laboratories throughout the world for scientific research in an endeavour to benefit human (not animal) welfare. Over the last hundred years, moral dilemmas have become even more acute. While pet owners promote animal welfare, the needs of modern scientific and agricultural research often require animals to be considered unfeeling 'things'. Since the beginning of the twentieth century, experiments on animals in laboratories have been legitimized in the name of scientific necessity, factory farming permitted as an inevitable feature of contemporary human (not humane) living and 'anthropomorphism' invented as a label given by commentators (chiefly historians and scientists) to animal lovers who project human feelings on to non-human life.

Meanwhile, some Far Eastern cultures have few traditions when it comes to animal rights. For example, there are currently few, if any, laws against cruelty to animals in China, nor is there a taboo against eating dogs. It is no wonder, then, that pet ownership is also not well-established in this part of the world. China spends just less than $1 billion annually on pet food, compared to $43 billion in the United States, yet its population is more than four times the size.

But over the last decade, owning domesticated dogs, cats, rabbits, hamsters and guinea-pigs in China has begun to catch on as the country embraces Western lifestyles. If as a consequence this Eastern culture now begins to change its attitude towards animal welfare then it will further underscore the link between widespread pet-ownership and the changing moral relationship between human and non-human life.

To think that it is only humans who have altered the nature, shape and form of species such as wild dogs and cats into domestic pets through artificial selection captures at best only half of a much more complex saga. Pets have had their own major impact on human culture, too. As a result the legal, moral and emotional relationships between pet-keeping people and the animal world have changed profoundly, although any final global triumph for pets in their bid to secure full recognition for animal rights is still some way off.

Dog

FAMILY: CANIDAE
SPECIES: CANIS LUPUS FAMILIARIS
RANK: 42

A creature that has succeeded in becoming mankind's best friend thanks to its loyalty, learning, sensitivity and diversity.

GELERT was a thirteenth-century Irish wolfhound given by King John of England to medieval Welsh prince Llywelyn the Great. One evening Llywelyn returned home from hunting to find his baby son and heir missing. The child's cradle was overturned. Alongside lay Gelert who was licking off the remains of blood smeared around his mouth. In a fit of uncontrolled fury Llywelyn impaled the dog with his sword, only to hear, moments later, the cries of his baby child, unharmed, from behind the cradle. Next to the infant lay the body of a mighty, dead wolf. In his grief at killing the animal that had saved his son's life, it is said Llywelyn never smiled again. He later built a monument to his faithful dog, which can be seen today in the appropriately named Welsh village of Beddgelert ('the grave of Gelert').

A lack of corroborating evidence means the story of Gelert cannot be proved true or false. But its telling shows how throughout recorded history, and into the present day, domestic dogs are commonly regarded as loyal servants of humans. Popular culture is filled with stories of canine courage and heroism, from the name Fido to the fictional collie dog Lassie, distinguishing this species from all others, domestic or wild.

The first clues as to why dogs have become so incredibly successful in a world dominated by humans emerge about 15,000 years ago, with the earliest evidence for their domestication. Although the exact sequence of events that led to the cohabitation of dogs and humans is still hotly disputed, it is now widely accepted that today's domestic dogs all belong to one subspecies of the grey wolf (*Canis lupus*). Genetic diversity suggests east Asia as a location for their initial domestication and there is speculation that wolf puppies were initially taken in and looked after by humans so they could be groomed to pull sleds in what was then still an extremely icy world. Asian humans who migrated across the frozen Bering Straits to populate North America *c.* 13,000 years ago may have succeeded in the perilous crossing only because their

Gelert next to the wolf he had slain before his master, Prince Llywelyn, arrived home.

supplies were hauled over the ice by the ancestors of today's husky dogs. At about the same time, canine company was so precious to early agriculturalists living in the Near East that human graves have been found with the remains of their beloved dogs lying alongside the bodies of their owners.

Countless generations of breeding by humans have resulted in the evolution of a number of different races of dogs. Some have been groomed for herding sheep (collies), others for catching vermin (terriers), transporting goods (huskies) or fetching game (retrievers and pointers). By medieval times the reputation of a dog being a man's best friend was well established. Effigies of wealthy European aristocrats and soldiers are seldom seen without a dog at their feet, a symbol of chivalry, courage and fidelity.

Loyalty is commonly cited as one reason why domestic dogs have so successfully established themselves in a world dominated by humans where the human population has grown from approximately five million 12,000 years ago to nearly seven billion today. While many other species have suffered extinction or demise, dogs have benefited greatly from what seems like a naturally occurring inclination towards devotion, selflessness and self-sacrifice. However, modern scientific experts scoff at the notion that any animal has within itself the capacity to assist another species willingly without believing that it will derive benefit from such behaviour. Darwin said if such an occurrence in nature could be found, it would undermine the very foundations of his theory of natural selection, predicated as it is on the survival of species through a competitive struggle for existence.

Such loyal behaviour has undeniably proved a remarkably successful survival tactic for the domestic dog in an increasingly human-dominated world. Its population has now reached an estimated *400 million* individuals worldwide (compared to the numbers of wild wolves, which although still one of the most widely distributed animal species in the world, have dwindled to approximately 200,000).

A prodigious ability to learn is another attribute that has helped domestic dogs succeed in the human world. There is no established or standard method for measuring the intelligence of dogs (or humans, for that matter) but their capacity to comprehend, respond to and obey human demands is well established. A recent experiment conducted by animal psychologist Juliane Kaminski has shown how Rico, a border collie, can understand a vocabulary of more than 200 labels for different items, including being able to interpret commands such as 'fetch a sock' and give it to a particular person. A dog's capacity for learning about the human world and with it the commands of human beings has been, like that of horses (see page 246), instrumental to their success as a key domestic species.

Also like horses, dogs are extremely sensitive. With a listening range of between 40 hertz and 60,000 hertz, these creatures are far superior to humans in their hearing capacity at both ends of the audio spectrum. They use as many as eighteen muscles in each of their ears to focus sounds, enabling them to detect and pinpoint noises four times farther away than those heard by humans. While their vision is at least as good as that of humans (although they do not have the capacity to see as many colours, see page 337) it is their extraordinary sense of smell that makes all human/canine comparisons begin to look ridiculous. Relative to brain size, the olfactory cortex in a dog's brain is forty times larger than the equivalent processing centre for smells in humans, and while the number of smell receptors that feed information into a dog's brain totals 220 million, the equivalent in humans is approximately five million, making a dog's sense of smell vastly more sensitive.

This prodigious olfactory sense is derived from the habits of the domestic dog's wild grey wolf ancestors, creatures that define their hunting territories by leaving traces of urine as scent marks. Whereas humans identify each other through visual recognition (see pages 337–9), dogs, like wolves, secrete individual pheromones through glands in the base of their tails, between their toes, in the genitals and skin. Individual identification via smells explains why dogs, when they meet each other for the first time, typically begin their acquaintance by sniffing each other's bottoms. Throughout recorded history humans have compensated for their own olfactory inadequacy by employing dogs for everything from hunting rats and foxes to modern law enforcement (detecting drugs) and even in the search for the human survivors buried in rubble caused by disasters such as earthquakes and terrorist attacks.

No less essential to the enormous success of dogs is their extraordinary genetic malleability. Although several breeds had been successfully created over thousands of years to suit specific human demands, it is only in the last 150 years that most of today's 300-plus pedigree breeds have come into existence. From *c.* 1850 onwards some pet owners began to experiment by breeding dogs into particular shapes and forms for the sole purpose of attracting attention or participating in competitions. Lap-dogs became a social accessory for Victorian ladies while bulldogs became a symbol of British valour in war. Dog shows such as Crufts (founded in 1886) and organizations such as the Kennel Club (which took over Crufts in 1938) were founded in Britain but have since become worldwide catalysts for modern canine eugenics – selective breeding for the sake of achieving desirable traits in future generations. New breeds have been created through the generations by culling or sterilizing animals that do not conform to certain breeding rules (such as size, fluffiness, butchness and so on) while owners ensure the survival of those that do.

Now listen up! Breeding out wild instincts was just as crucial to canine domestication as ensuring dogs looked cute.

Unfortunately, inbreeding to exaggerate specific genetic traits has led to some of today's most prized breeds (such as British bulldogs and King Charles spaniels) suffering ill-health. The issue was given a public airing after the BBC broadcast a television programme in August 2008 that investigated the effects of modern dog-breeding methods. The broadcaster subsequently cancelled its coverage of

Patriotic propaganda: a World War I postcard reinforces British fondness for bulldogs.

Crufts in March 2009 and several top sponsors of the dog show withdrew their support. The Kennel Club has since begun a major review of the rules it sets for the world's 209 recognized canine breeds.

It was largely thanks to the huge success of Victorian dog breeders from *c.* 1850 onwards that Charles Darwin's cousin, Francis Galton (1822–1911), was inspired to propose that similar breeding techniques be used on humans. Galton's *Inquiries into Human Faculty and its Development* (published 1883) suggested human marriages should in future be based on identifying people with desirable traits (such as intelligence and honesty) and that incentives should be created for these people to marry and have children. *Kantsaywhere* was a novel Galton wrote about a utopian society where a religious order was responsible for breeding smarter, fitter humans. Only fragments of this work survived after Galton's niece burned much of it, so offended was she by its love scenes.

Galton coined the term eugenics for the science of selective human breeding. It rapidly became a popular political topic in the United States as a possible solution to racial divisions (as proposed by writers such as Madison Grant in 1916). Eugenics was later adopted with devastating fervour in Nazi Germany during World War II. Adolf Hitler attempted to revive a mythological master-race of humans (Aryans) through the culling of 'inferior' beings (mostly Jews) and the enforced sterilization of thousands of other 'undesirables'. Hitler's eugenic ideals were supported by a perverse belief in animal rights in which wolves were second to human Aryans in the natural pecking order of life. Other human races, such as the Jews, were subordinate to dogs. Historians are left to ponder to what extent Hitler's ideals were influenced by the love he had for his German shepherd, Blondi, which his secretary gave him in 1941.

The impact of domestic dogs on human culture therefore reaches far beyond their wide-ranging usefulness to mankind. From the rise in the movement for animal rights (see page 320) to the dark experiments of modern eugenics, these are creatures that have been, like horses, true protagonists in the turbulent history of human civilizations.

Cat

FAMILY: FELIDAE
SPECIES: FELIS SILVESTRIS CATUS
RANK: 46

An unlikely bargain struck between agricultural people and a small carnivore that gave rise to the world's most populous pet.

FEW ANIMALS in the history of civilization have been afforded as much hospitality as the domestic cat. In ancient Egypt they were thought to be divine. The cat-goddess Bast had her own picturesque shrine and cats were, like humans, frequently embalmed so they could return to their earthly bodies in the afterlife. In 1888 a farmer found an ancient burial site along the banks of the river Nile that contained *80,000* mummified cats. According to Greek historian Herodotus, who visited Egypt in 450 BC, cats were so venerated that when one died its owners were *'plunged into deep grief'* and would '*shave off their eyebrows*' to signify the loss.

In Islam, the Prophet Mohammed was a renowned lover of cats – which he regarded as clean animals, unlike dogs. Tradition has it that during one prayer time Mohammed found his pet cat *Muezza* asleep on the sleeve of his overcoat. The prophet cut off the sleeve so as not to disturb the cat's slumber. Mohammed's legendary love for his feline friend meant that a high regard for cats has continued to run throughout Islamic cultures.

One cat is said to have waved its paw at a passing Japanese nobleman. As the man turned to approach it, a bolt of lightning struck the ground where he had been standing. The beckoning cat that saved this unknown aristocrat's life has since become Japan's most widely recognized sign of good fortune. Even today *Maneki Neko* ('beckoning cat') dolls welcome people into shops and homes throughout the country.

In Europe, cats became common household creatures following the Roman conquest of Egypt in 31 BC. They were transported by early-modern European maritime explorers to the Americas and Australia where they previously had no native presence. The consequence of such cultural alliances and global transportation is that domestic cats have recently eclipsed even dogs in terms of their populations, now estimated at *600 million* individuals worldwide.

The extraordinary success of cats presents a conundrum because these are not creatures that have been trained or domesticated by humans to provide any obvious kind of service. They do not transport things – like dogs, horses and camels do – nor do they provide humans with a source of food, drink or clothes – like cows, sheep and pigs. Even as a badge of social status, cats have mostly failed to impress other people in the same way as a pedigree dog or thoroughbred race-horse. What is it, then, that accounts for the domestic cat's present and historic success? How come hundreds of millions of people all over the world and throughout the history of civilizations have welcomed carnivorous cats into their homes?

The earliest archaeological evidence for domestic cats goes back about 9,500 years to graves in Cyprus where cat remains have been found buried alongside those of humans. Recent genetic analysis of 979 domestic cats suggests they are all a subspecies of wildcat (*F. silvestris*), which itself evolved from a common ancestor about 230,000 years ago. Today's domestic cats are thought to have derived from as few as five *F. silvestris* females that originated in the Near East at about the time humans began experimenting with a lifestyle based on farming *c.* 12,000 years ago.

Unlike other domesticated farm animals, cats do not seem to have been 'selected' by humans for any specific purpose. In fact, cats are just as likely to have chosen to live with humans in a process known as 'self-domestication'. Experts believe that once people began farming in the Near East, it was a survival advantage for cats to become less fearful of humans because stores of grain and other produce were a constant magnet for the animal's favourite prey – rats and mice. A good life could be had – a life of fun, playing, teasing and killing rodents – in the granaries and barnyards of human farmers.

Egyptian cat goddess Bast, cast in bronze, dating back to *c.* 600 BC.

One reason why humans initially welcomed these exotic-looking creatures into their living quarters may also be due to a curious ancient genetic accident. Some time between the origin of *F. silvestris* and their much more recent domestication, cats lost the use of taste buds that detect sweetness (was it a mutation or viral infection, perhaps?). As a result, domestic cats have evolved into 'obligate' carnivores, creatures that feed exclusively on flesh. Their teeth and digestive systems have since adapted to this specialist diet so that, unlike dogs or humans, they cannot eat or digest grain, vegetables or fruit.

That cats are unable to eat fruit or seeds must have given early human farmers sufficient comfort that these creatures would not harm precious stores of agricultural produce. Instead, they would focus their advanced, stealthy hunting skills on whatever invading vermin came into range. This human–cat 'self-domesticating' symbiosis seems to have begun in Egypt – the bread basket of the ancient world. Cats tolerated humans, becoming tame and friendly towards them, while humans tolerated cats as the chief guardians of their food.

The idea that Neolithic humans deliberately domesticated cats through selective breeding is also as impractical as it is improbable. Female cats allow a dominant male to mate with them, but then frequently encourage several others to follow suit in rapid succession. Selective breeding, when each kitten in a litter may have a different biological father, is no straightforward affair!

Without the focus of selective breeding it is remarkable that humans have become so close to cats for, unlike domestic dogs (see page 323), they are mostly solitary, self-sufficient creatures. While cats sometimes live in colonies (today as many as 300,000 cats roam freely in Rome), they almost always hunt alone and fiercely defend their own territory. Such behaviours are usually regarded as the antithesis of those characteristics required to live with humans.

But cats compensate by behaving in ways that are curiously reminiscent of human habits, perhaps explaining why they have got along so well with people over the last 12,000 years. Cats, like people, hunt other animals for pleasure (we call it sport), triumphantly bringing trophies of dead birds or mice back into their homes to show off to their human foster parents. Also, like many people, cats have a penchant for psychoactive drugs – their favourite being catnip (*Nepata cataria*), a common herb that they chew, sending them into a euphoric state for up to two hours. Nor do cats, like humans, respond well to being enslaved. They prefer a degree of personal freedom and to live semi-independent lives, coming and going as they please – sometimes indoors, sometimes outside, where, like entrepreneurs – or nation-states – they establish and defend their own territorial patch.

The physiology of cats also mirrors human beings in other striking ways. Inherited and immune deficiency diseases in cats provide realistic models for scientists trying to develop solutions to human ailments. Feline Immunodeficiency Virus (FIV) is analogous to HIV/AIDS (see page 18) and so a search for an HIV vaccine is under way, based on experiments to find a cure for FIV. Detailed modelling of the mechanics of the human anatomy, especially of the eye, has been based on research on cats. Eight separate Nobel Prizes in physiology have been won by scientists who, through their research on cats, have decoded the inner workings of the human nervous system.

The question as to whether cats bring good or bad luck has fascinated humans throughout recorded history. While Egyptian, Islamic and Japanese cultures have generally venerated cats, in medieval Europe their image took a hammering from their association with gypsies and witchcraft. Cats were frequently cast as accomplices of the devil because of their solitary habits as night-time hunters, which often seemed coolly detached from human dependence.

By Victorian times, as the Age of Reason prevailed over the age of witchcraft, cats recovered from such satanic associations. Breeding programmes were established to improve their looks, distancing most varieties from any association with a witch's black cat. From the mid nineteenth century, cats became much-adored companions for everyone from pet-loving children to lonely adults such as Victorian author and poet Thomas Hardy, who famously commemorated the death of his beloved pet cat in his poem *Last Words to a Dumb Friend* (1904):

Never another pet for me!
Let your place all vacant be;
Better blankness day by day
Than companion torn away.

Rabbit

FAMILY: LEPORIDAE
SPECIES: ORYCTOLAGUS CUNICULUS
RANK: 28

A domesticated species, globalized by humans, that has recently been decimated as its population soared out of control.

NO DOMESTIC MAMMAL has been subjected to such a roller-coaster ride of fortune and misfortune as has the domestic rabbit. Despite their cute looks and lack of aggression, these creatures have experienced everything from soar-away success to catastrophic failure as a result of their 2,000-year association with humanity.

Rabbits belong to a family of rodent-like mammals called the Leporidae, which also includes hares and pikas. Although more than twenty species of wild rabbit are known, domestic rabbits all belong to a single species of European rabbit – *Oryctolagus cuniculus* – which is native to the Iberian peninsula.

For most of their history, domesticated rabbits have been kept as food, not pets. The Romans – notorious for flinging every species from grapes to cats all over their empire – brought rabbits from Spain to Italy about 2,200 years ago, eyeing these fast-reproducing herbivores as a good source of easy-to-farm meat. Multi-tiered amphitheatres, triple-decker aqueducts and long straight roads are what people generally perceive as the most enduring impacts of this once-mighty European civilization. But, in many parts of the modern world, the consequences of their initial transportation of rabbits across Europe far outstrip these remaining marvels of ancient engineering.

It is clear that rabbits from Spain were being cultivated in Italy by 36 BC thanks to a book on agriculture written by a Roman writer called Varro:

> *'There is a recent general practice of fattening these [rabbits] by taking them from the warren and shutting them up in hutches and fattening them in an enclosed space … one species … is native to Spain – like our hare in some respects, but with short legs – which is called a cony. The conies are so named from the fact that they have a way of making in the fields tunnels [cuniculos] in which to hide …'*

The burrow-making habits of European rabbits explain in large part why this species has been so successful compared with hares and other types of rabbit (like cotton-bobtails) that live only above ground. Females dig tunnels and underground chambers where they give birth to their young. The young bunnies benefit from the sheltered environment in the first few weeks of their lives. The brilliant burrowing antics of European rabbits also turned them into expert escapologists. Not only are they able to jump high walls using their strong hind legs but they can tunnel deep beneath fences and other artificial enclosures. Therefore, wherever rabbits have been transported by humans for the sake of farming, they have also managed to escape into the wild to seed feral populations.

A second success factor is their prodigious rate of reproduction. Unlike in other mammals, ovulation in female European rabbits is triggered by copulation, which means that instant fertilization is almost guaranteed. With a gestation period of just thirty-one days, and litter sizes of up to twelve bunnies at a time, rabbits represent one of nature's most prolific reproductive systems (see also chickens, page 226).

From Italy, rabbits spread throughout medieval Europe, reaching Britain shortly after the Norman Conquest. Outdoor warrens were established by farmers all over the country, eager to cash in on the sale of rabbit meat and fur. Harvesting was accomplished by sealing burrow entrances with nets and putting a predatory animal such as a ferret into the maze of tunnels to flush the rabbits out (lobsters and crabs were used closer to the coast). Although wild populations in Europe grew prodigiously as more rabbits escaped from warrens, natural predators such as foxes, stoats and lynxes kept their numbers reasonably in check. By the nineteenth century, shooting rabbits became a popular British pastime, too, as population control became a competitive blood sport.

During the nineteenth century, however, the transportation of rabbits to other habitats proved environmentally catastrophic. Mirroring the ecocide inflicted on parts of Europe and Asia by the introduction of species such as the Australian eucalyptus (see page 264), European rabbits inflicted unquantifiable damage to ecosystems, especially where there were no predatory creatures to control their numbers. Islands from Lundy to the Falklands were turned into rabbit-infested ruins because European sailors had treated these places as breeding grounds – since rabbits cannot swim they could not escape. Meanwhile, the most notorious introduction took place in October 1859 when homesick Australian

Rabbits have had a love/hate relationship with humans.

settler Thomas Austin asked his English nephew to send over 24 breeding rabbits as game on his 12,000 acre estate at Winchelsea, near Geelong in the southern Australian state of Victoria. Soon after the creatures arrived Austin released them into the wild, declaring to his friends that: '*... the introduction of a few rabbits could do little harm and might provide a touch of home, in addition to a spot of hunting ...*'

Without the threat of natural predators, and with their extraordinary reproductive prowess, rabbits progressed across Australia at a rate of about 110 kilometres per year. By 1865 Austin had killed 20,000 on his own estate alone. By 1883 the government of New South Wales had spent £361,492 on killing off 7.8 million rabbits. By 1950 the rabbit population of Australia had reached *600 million*. It was the fastest spread of any mammal ever recorded anywhere in the world.

The effect on the country's habitat was devastating. Soil erosion (from burrows), deforestation (from eating young plants and seedlings), and a decline in native creatures (such as the Bilby, a rabbit-like marsupial) were direct consequences. Crop destruction was of greater concern to humans, however. A Royal Commission was established in 1901 to try to find a solution, following which a fence was erected in 1907 – the longest in the world – stretching for more than 1,700 kilometres across western Australia in a bid to keep the invading rabbits at bay.

Of course, it never really worked. Fences, traps, shooting, poisons – none of them provided any significant impediment to the spectacular burrowing and reproductive success of escaped domesticated European rabbits living in a world bereft of natural predators. Ironically, by this time, domestic rabbits back in Europe – the source of all the trouble – had become one of Europe's most treasured household pets. The Victorian fancy for household animals led to the establishment of as many as 150 different cosmetic breeds based on the colours of their coats and the length of their ears. Since 1975, pet rabbits have even been bred for competitive sports – Sweden and Britain lead the world in rabbit 'show-jumping' events.

Until 1950 the planet Earth had looked kindly on the genes of *O. cuniculus*. In the southern hemisphere the species had successfully overwhelmed ecosystems while in its indigenous northern homeland it was given additional protection as a popular household pet. As the Western pet-keeping craze developed, even the traditional habit of eating rabbit meat, or wearing rabbit fur, became a twentieth-century taboo in countries such as Britain.

Neither nature nor man guarantees a species prosperity for ever. Frank Fenner (born 1914), an Australian virologist, offered humans their first real chance of striking back at the rabbit's soar-away success. Fenner discovered that the *Myxoma* virus infected *O. cuniculus* specifically, causing the disease myxomatosis, which had an extraordinary 99 per cent mortality rate. Fenner proved the virus was perfectly safe to humans by injecting himself with it before it was released into the Australian wild in the summer of 1950.

The effect was devastating. The disease quickly spread through the feral population via flea and mosquito bites and within two years more than *500 million* Australian rabbits had been killed. As the rabbit population declined, so the mortality rate decreased (a viral evolutionary tactic to ensure numbers do not reduce so much that the virus no longer has a host population in which to thrive). Of course, some rabbits were naturally immune (thanks to the insurance system of sexual reproduction, see page 47) so that there was no real risk of total extinction. The disease spread to Europe in 1952, thanks to its illegal release in France. Similar devastation occurred, this time destroying so many rabbits that other wild animals that depend on them for food (such as the Iberian lynx and Spanish imperial eagle) are today threatened with extinction.

Since the 1950s there has been a growing ethical debate about the continued use of myxomatosis. Rabbit death from the *Myxoma* virus is an especially unpleasant, drawn-out, affair. It begins with acute conjunctivitis that leads to blindness. Then the creature develops a high fever, during which pneumonia and inflammation of the lungs typically take hold, leading to a long, painful death after about thirteen days. Meanwhile, owing to the advance of another virus, rabbit haemorrhagic disease (RHD), which first appeared in feral populations in 1984, the number of European rabbits in their native Spain, Portugal and North Africa has fallen by 95 per cent. They are now featured in the 'Near Threatened' category of the IUCN Red List of Threatened Species (2008).

The rights and wrongs of deliberately trying to wipe out mammals such as rabbits have become more complicated recently with the development of effective vaccines against both myxomatosis and RHD. *Shope fibroma*, an inactive virus, provides rabbits with immunity but it has to be administered orally or via injection, making its use impractical for large-scale feral populations. However, virologists in Spain have recently isolated a live strain of a *Myxoma* virus variant (called MV 6918) that confers immunity to these diseases in rabbits. If released, its protective properties would spread by flea and mosquito bites in just the same way as the disease itself – potentially ridding the wild population of this human-induced plague. The possibility of immunizing the rabbit population today as effectively as it was decimated in the 1950s presents a major dilemma regarding the human management of ecosystems. At present, the vaccination, as well as its live virus alternative, is illegal in Australia.

The history of domesticated rabbits shows how chaotic the relationship between man and other species can be, especially when rational human minds try to harness a species for their own ends. Loved as pets, loathed as pests and depended on as a viable source of meat and fur, rabbits provoke attitudes in man that are riddled with contradictions. However, rabbits can claim to have proved capable of surviving even the most devastating turns of fortune.

Hamster

FAMILY: CRICETIDAE
SPECIES: MESOCRICETUS AURATUS
RANK: 90

Unsung wartime heroes that have since become one of the world's most popular pets.

WHEN SAUL ADLER challenged his Jewish zoologist friend Israel Aharoni to go on a hunting trip to the Syrian desert in 1930, little could he have imagined how extraordinary the consequences of the expedition would turn out to be. Adler, a Russian-born researcher into tropical diseases, was desperate to find a cure for the parasitic disease leishmaniasis, spread by bloodsucking sand flies. His research depended on the discovery of a new species of desert hamster.

Records concerning this sometimes fatal illness stretch back at least 2,500 years. Stone tablets found by archaeologist Austin Layard in the royal palace of Assyrian King Ashurbanipal (died 627 BC) describe the symptoms almost perfectly, which include blistering sores on the skin. Muslim medic Avicenna (AD 980–1037) also sought a cure in tenth-century Baghdad. Blotchy people depicted on pots made in pre-Incan South America show the global reach of this protozoan parasite. Leishmaniasis is contracted by hundreds of thousands of people each year in areas as far flung as Colombia and Afghanistan. After invading the body in a sand fly's saliva, the *Leishmania* parasite spreads through the bloodstream infecting a number of different tissues, eventually causing severe inflammation of the liver and spleen. Skin sores usually develop weeks or sometimes months after initial infection. If left untreated, the disease leads to a particularly unpleasant death.

Saul Adler cradles a precious golden Syrian hamster.

Drugs that can kill the parasite have been developed over the last eighty years mostly thanks to that Syrian expedition undertaken by Aharoni. Adler had been using Chinese hamsters in his Jerusalem laboratory to carry out research into a drug cure for leishmaniasis. Through repeated infections these hamsters were immune and no longer fell victim to the disease. What Adler needed was a similarly small, fast-breeding hamster species that was just as content to live alone in confinement so he could continue with his research. But it must be a species that was not immune.

The unsung heroes of the medical breakthrough that finally helped find a cure for the ancient disease leishmaniasis were a single, wild, female golden hamster and her litter of eleven babies, living in a burrow in the desert near Aleppo, discovered by chance by Aharoni. Here was a new species, never before exposed to the disease, which researchers could use to continue with their experimentation. On the way back to Jerusalem the mother died, leaving Aharoni the task of feeding and caring for the babies himself. By the time he got back to Adler's lab, four had survived. Recent DNA analysis has confirmed that it is from these four individuals that all the golden hamsters kept in captivity today, and sold in pet shops throughout the world, are descended.

In Sicily during World War II, large numbers of allied troops contracted leishmaniasis from sand fly bites and were at serious risk of death. A cure was desperately sought. In 1943 British disease expert Leonard Goodwin (1915–2008) contacted researchers in Israel for a consignment of Syrian golden hamsters on which to test some new drugs. Thanks to their lack of immunity to the *Leishmania* parasite, Goodwin was able to calculate the correct dosage for administering the drugs to humans without risking their death from side-effects of the treatment itself. Within a year the drug pentostam was approved and used to treat the troops.

After the war, Goodwin donated the research animals in his laboratory off Euston Road, London, to some pet fanciers. The fast-breeding, easy-to-keep, solitary, handsome golden-coated creatures quickly became favourite first-time pets for children. Selective breeding from this original colony has introduced a number of other breeds, with different-coloured coats, adding to the range and choice of hamsters available in pet stores today.

The golden Syrian hamster, which lived in the wild until the 1930s, is now one of the most popular pets in the world.

The present-day impact of these creatures, countless millions of which are now caged up in homes all over the world, is less easy to quantify than the heroic contributions of their wartime ancestors. Often these tame rodents are an urban child's first close-contact experience with an animal. Such experiences can have profound effects – Goodwin himself was inspired to pursue a career in biology after his grandfather, who was a gamekeeper, took him out of London on nature holidays to Rutland in the East Midlands.

In the wild, golden hamsters are rated as a 'vulnerable' species because of habitat loss. But, thanks to their chance discovery by a man on a mission near Aleppo in 1930, the attractive, tame, golden-coated Syrian variety *Mesocricetus auratus* has become one of the most successful domestic species in the modern world.

13

On Beauty

How good looks, strong smells and powerful flavours in some species have proved irresistible to humans.

WHEN IT COMES to identifying species that have most changed a world dominated by humans, it is impossible to ignore those that excite our senses. By far the most important are living things that appeal to the delights of sight, smell and taste.

The fragrance of a woman's perfume, the beautiful stripes of freshly cut grass and the zesty taste of freshly squeezed orange juice are all sensations – to choose just a few at random – that many people find appealing and that have boosted the survival prospects of different living things.

But which sensual qualities have most helped species thrive in the human world? If the old adage that *'beauty is in the eye of the beholder'* has the ring of truth, then it would suggest that success is little more than random chance.

Visual pleasure

Beauty, at least for humans, does seem to be a predominantly visual phenomenon – it really is in the *eye* of the beholder. Sound, touch – even fragrance and flavour – are never quite as important when it comes to making an initial judgement on how pleasant (or ugly) something seems to be. Good-looking species have therefore had a significant survival advantage in the human world.

But humanity's obsession with visual beauty is odd. The primary sense used by most mammals for everything from food detection to mate selection is via their sense of *smell.* The olfactory system, which is so astonishingly powerful in creatures ranging from mice to dogs (see pages 324–5), is evidence of the pivotal role that odour and fragrance play in helping mammals perceive their world.

Indeed, the default mammal system for avoiding inbreeding is based on pheromones, chemicals usually found in sweat or urine that secrete smells which affect the behaviour of the opposite sex. When dogs mark their territory they leave an individual chemical signature in their urine that indicates their genetic makeup, rendering their scent attractive to potential mates whose genes are sufficiently *dissimilar* to their own. Their offspring will reap the benefits of such a match through stronger immunity to disease, which comes from having greater genetic diversity (see pages 47–8).

Smell was a vital part of the mammalian survival kit from about 200 million years ago. An enhanced olfactory capability may have been a critical selective advantage for finding food and identifying mates when it was safest to do so under cover of darkness.

Since humans are mammals, why don't human females also go to the trouble of sniffing a potential lover's urine before deciding whether or not he represents a good match? While the idea may sound rather absurd to us, the concept of smothering a female face with make-up would probably seem equally bizarre to a dog.

Multi-coloured mandrill: vivid visual features in African primates may have helped compensate for a weaker sense of smell.

Smelling bad

Recent genetic research suggests that at least two events seem to have occurred in the evolutionary history of humans that have made *looks* supreme amongst the attributes that count to us.

Clues are emerging from all that junk DNA found in the process of decoding the human genome (see page 14). The first event happened approximately twenty-five to thirty million years ago during the evolution of primates, long before humans appeared. For reasons not yet well understood, at about this time 'Old World' monkeys (early members of the group Catarrhini) experienced a significant reduction in their capacity to detect smells. Some genes responsible for the working of the olfactory system were incapacitated. At about the same time a genetic mutation occurred in these same primates that enabled trichromacy – the ability to see a full range of colours.

The effect of a reduced olfactory system but an enhanced capacity for colour vision came to distinguish this line. Instead of them sniffing out the best food, their better colour vision is thought to have helped them distinguish red from green more clearly when foraging among trees for berries and edible leaves. When looking for a mate, colourful faces and pretty bottoms (characteristic of these monkeys) were adaptations that helped compensate for their diminished capacity for detecting pheromones. Visual communication, for example by hand gestures and facial expressions, was another innovation common to the Catarrhini, such as gibbons, orang-utans and gorillas, that helped make up for their less effective sense of smell.

Recent studies of the human genome suggest that a second event occurred sometime after the hominid line split from chimpanzees (between seven million and four million years ago). This further reduced the capacity of humans to smell pheromones. So severe was this genetic infection/mutation that it left humans with a substantial olfactory handicap. As a result, the size of a human brain's olfactory epithelium – the part that interprets smells – is on average about two square centimetres. The equivalent region in a rabbit's brain is typically nine square centimetres and contains 100 times more smell receptors. Geneticists have recently discovered that as many as 60 per cent of the human genes responsible for smell are dysfunctional.

The consequences of having almost no capacity to detect chemical scents such as pheromones have been profound, not just for human evolution but also for other species that touch the human world.

One result was that sound and vision, not smell, became the predominant mechanisms for human communication. Also, since interpreting visual

Galton's gallery of criminal faces looks a lot less menacing when layered one on top of another, since a composite image looks more 'normal'.

cues and facial expressions requires a great deal more mental processing power than picking up scents, humanity's poor sense of smell may partly explain some of the natural selection pressures that led to the development of new mental capabilities. It is an intriguing speculation that visual processing, pattern recognition, associative learning and ultimately the evolution of language, the written word, art and culture in humans may be distantly linked to our comparatively lousy sense of smell.

Species with features that appealed to humans' advanced visual systems were therefore especially well equipped for survival in the era of human societies. Good-looking flowers including tulips, **roses** (see page 341), irises and peonies have thrived alongside brightly coloured fruits such as lemons, **oranges** (see page 361) and – the apes' great favourite – **bananas** (see page 364).

Why average is best

But how, in a visually dominant world, do humans decide if another living thing is good to eat or, for that matter, if a member of the opposite sex is fit to mate with? What is the equivalent of a dog's urinal genetic marker for species, such as humans, that cannot rely on a sufficiently sensitive nose?

Part of the answer, it seems, has something to do with the visual phenomenon of symmetry. Francis Galton, the cousin of Charles Darwin, whose mind was so taken with the idea of how to breed evil out of humanity (see page 326), was one of the first people in modern times to propose that beauty is not, as the old proverb suggests, a purely subjective matter.

Galton wanted to catalogue facial characteristics that could be used to indicate if a person was likely to be a criminal. In his attempt to find the ultimate 'villainous countenance' he used a series of photographs of convicted prisoners, overlaid one in front of the other, to produce a composite portrait. The results astonished him because what emerged was an image …

> *'… much better looking than those of the components … because the average portrait of many persons is free from the irregularities that variously blemish the looks of each of them …'*

Similar experiments have been conducted in more recent times by creating an average face based on digitally layering a series of real faces, one on top of the other. When researchers asked people to rate the most attractive face, they found that the composite image was more frequently chosen than the individual faces of real people.

Personal human beauty, therefore, seems to have something to do with the closeness to which a person's visual image approximates the average in a given population. Blemish-free looks and a lack of distinguishing asymmetrical features are used by humans to form an initial value judgement of biological fitness. Deviations from this norm (warts, birthmarks, skin defects and so on) are markers suggestive of genetic abnormality, stressful birth or susceptibility to disease.

Looks good to me …

People who buy fruit and vegetables at the supermarket do not generally sniff each product before they make a purchase. Rather, they pick it up, and judge its fitness for eating based on how closely it conforms to a symmetrical, ideal specimen. Crooked carrots, bendy bananas and knobbly apples are rejected by fastidious consumers on the basis of their looks. Such are the criteria by which humans generally judge if an item is fit for consumption.

Species that exhibit features of genetic perfection with precise symmetry and bright, attractive colours are therefore frequently favoured in the world of artificial selection.

Sunflowers demonstrate this elegantly by following ratios employed in the mathematically pure Fibonacci sequence (namely 0, 1, 1, 2, 3, 5, 8, 13, 21, 34, 55 and so on, with each successive number being the sum of the previous two), with their florets packaged in symmetrical spirals of thirty-four and fifty-five around the outside. A similar display occurs on the fruitlets of a pineapple and the spiral whorl of a pine cone.

The **lotus** flower (see page 354) is famed for its beauty thanks to the highly symmetrical way its petals unfurl, a quality that has captivated Eastern cultures for more than 2,000 years.

Now imagine a butterfly without identical patterns on each wing, a peacock with a lop-sided fan of tail feathers or a human with only one eye. These images would appear as mutations, the outward expressions of faulty genes, which have meaning only because they deviate from the average design.

Symmetry is one of nature's most ancient methods for demonstrating how close any individual living entity is to genetic perfection. Its biological origins go all the way back to the body plans established in the earliest instances of vertebrate and invertebrate life by those Hox genes hundreds of million of years ago that determined the symmetry of every living creature from a house-fly to a human being (see page 61).

Judging things by how beautiful they look has been used throughout the history of civilization as a way of measuring perfection. Sometimes it has taken the form of sculpting nature to conform to human ideals of visual beauty. In this light, the perfectly symmetrical gardens of palaces from Versailles to Hampton Court could be interpreted as expressions of a biological fitness for royal or aristocratic government.

Meanwhile, golfers (like my father) tend to find beauty in manicured fairways and greens with automatic watering, providing a constant benchmark for well-pitched perfection.

Architectural gems from the ancient Egyptian Pyramids of Giza to the Islamic Taj Mahal in Agra, India, are built according to natural symmetrical principles based on what the ancient Greeks called the 'Golden Ratio' – a formula first elucidated by Greek geometrician Euclid in 300 BC. The same ratio, which generates a rectangular shape, is ubiquitous throughout human culture today in everything from the shape of wall paintings and advertising billboards to television screens, fireplaces and credit cards.

Carolus Linnaeus devised his comprehensive system of taxonomy by organizing the natural world based on how it looks, separating living things into kingdoms, families and species. He followed a tradition established by the ancient Greek philosopher Aristotle, which was to base relationships between living things on their relative appearance. Only since the acceptance of Charles Darwin's concept of the relationship between living things being based on *inheritance* and *common descent* have these categories begun to change. As a result, today's classification of species (taxonomy) has become a complete mishmash of systems (see page 112).

... don't judge by looks

Although looks are the predominant system used by humans for organizing, ranging and selecting desirable objects – organic or otherwise – they are fairly inefficient when it comes to choosing a mate. Unlike the pheromones used by most mammals in mate selection, outward appearance does not itself indicate if one individual's genetic code is sufficiently diverse from another's to produce disease-resistant offspring. In fact, symmetry that aspires towards an average has an innate tendency to disguise an individual's genetic differences, increasing the risk of inbreeding.

Perhaps this is where the biological origins of human culture truly lie? Language, music and the performance arts may have evolved as way of compensating for a recognition system that, because it is visual, is poor at revealing genetic diversity. Hence it has become natural for a female to search beyond mere looks when considering a male mate. Conversation about interests, habits and family background is usually indulged in prior to long-term partnerships, often leading to lasting relationships in which it is commonly said 'opposites attract'. Such cultural due diligence may improve the chances that offspring will be fitter for harbouring genes that derive from across varied parts of the human genetic spectrum.

Despite humanity's obsession with looks, fragrance and flavour are still significant because these are senses that *reinforce* choices based on initial appearance. Perfumes and incense have been used since ancient times to disguise foul smells or create an inspirational atmosphere. Plants such as roses (see page 341), **lavender** (see page 348), jasmine and sandalwood are species that have prospered through human selection as a result of their powerful scents.

Taste also reinforces or deters initial visual selection. Species that provide a pleasing stimulus to the five main detectable flavours – sweet, salty, bitter, sour and savoury – have generally prospered under the influence of human societies and have been spread across the globe. **Apples** (see page 344), oranges (see page 361), lemons and olives (see page 209) have succeeded alongside strong-flavoured spices such as **black pepper** (see page 351), **vanilla** (see page 346), peppermint and ginger. Other species have profited from the human fad for cooking with flavours that open up and soften when sizzled on a stove – mushrooms, **chilli peppers** (see page 356), onions and garlic are just a few.

Some sensitive conclusions

The old saying that *'beauty is in the eye of the beholder'* therefore seems about half right. Good looks are what count initially for species that thrive in the human-dominated world, although taste and smell are powerful secondary forces.

But – despite tradition suggesting otherwise – beauty does not seem to be simply a matter of subjective whim. While individual preferences vary (because sexual reproduction confers genetic variety), people also express a natural preference towards other living things that demonstrate biological fitness through colour, shape and form. A symmetrical, youthful, blemish-free face mirrors the perfect shape, form and vibrancy of a deep red rose. Both contain that essence which we visual people tend to call beauty.

Rose

FAMILY: ROSACEAE
GENUS: ROSA
RANK: 62

An ancient symbol of love and romance that has recently become unsustainably successful.

LAKE NAIVASHA is a 130-square-kilometre stretch of African water that has been a paradise for wildlife throughout the history of humanity. Located in the heart of Kenya, one hundred kilometres north of Nairobi, the lake straddles the Great Rift Valley, a region richly endowed with volcanic soils where some of the earliest hominid fossils were discovered in the early 1970s (see page 177). An abundance of fish, lions, antelope, leopards and hippos thrived in this habitat alongside more than 350 recorded species of wild birds. Until the 1970s, about 7,000 people lived around the lake's shores, many of them belonging to the Maasai, settlers from the expansion of West African Bantu herds-people shortly after their cattle developed immunity to a disease carried by the tsetse fly (see page 140).

But in the last thirty years the region has been transformed as if someone had struck gold. More than 300,000 people now live or work on newly established farms around the lake's shores. Aside from the frequent use of harmful pesticides (such as DDT, see page 99) that damage the area's ecosystem, the lake's water level has fallen three metres below minimum safety levels. If current trends continue, says Dr David Harper, a biologist at the University of Leicester, the lake will become a *'turgid, smelly pond with impoverished communities eeking out a living along the bare shores'.*

As the lake becomes smaller and shallower it will become warmer, says Harper, fuelling the growth of microscopic algae. *'It is only a matter of time before the lake turns toxic.'* Already the plummeting supply of water is having an effect on wildlife. The number of hippos living in the lake decreased by 25 per cent between 2004 and 2006.

The reason for this impending ecological collapse defies common sense. Kenya, one of the driest places in the world, is exporting vast quantities of its precious water to some of the planet's wettest countries in Europe. The main reason for this is, in a word, the rose – a flower that for thousands of years has been admired for its beauty and fragrance.

Water accounts for about 90 per cent of the mass of a cut flower. Roses are cultivated in their millions on the commercial flower farms that have sprung up around Lake Naivasha over the last thirty years. These flowers are exported by airfreight to Europe in ready-tied, well-watered bunches where they are sold by florists and supermarkets. The Kenyan flower industry, commercially viable thanks to plentiful supplies of cheap labour, is responsible for generating precious income amounting to some $350 million a year, 74 per cent of which comes from growing and selling roses. One farm, owned by an Indian enterprise, produces *600 million* roses a year. Every stem accounts for the export of that vital commodity, water, out of arid Africa 365 days a year and into rain-rich Europe – and all because of the human obsession with these fragrant, mythical flowers.

Roses are loved by humans because they provide both a visual feast and a fragrant bouquet. One of the basic ingredients for perfumes sold throughout the world is oil made from damask roses, which originated in Persia. They were brought over to Europe by crusaders in the thirteenth century, along with Arabic oil-distilling techniques, and the practice of growing roses was eventually established in Bulgaria and the Netherlands thanks to their well-watered soils.

Today Bulgaria's Rose Valley is the world's largest rose-growing region, producing oils for the perfume industry. Its 300-year-old rose-growing tradition is centred around the town of Kazanlak. Thanks to its mountainous terrain and wet climate, there are no ecological concerns about growing roses here. Traditions live long. Their agricultural methods have so far resisted the ecologically damaging practices of modern intensive farming and on the first weekend in June the locals, dressed in traditional costumes, flock to the annual flower festival. The highlight of the event is a beauty pageant in which one girl, chosen from among many, is crowned 'Queen of the Roses'.

Roses were a powerful symbol of beauty throughout the ancient civilizations of Persia, Greece and Rome. In Greek mythology, Aphrodite, the goddess of love, presented a rose to her son Eros

who then gave it to Harpocrates, the god of silence, to try to persuade him not to gossip about Aphrodite's immoral behaviour. The rose as a traditional symbol of beauty, love, secrecy and discretion has continued with the annual Valentine's Day celebration. On average 10,000 metric tonnes of roses are sold in the UK each year as gifts to loved ones on Valentine's and Mother's Day.

Five petals that unfurl into myriad shapes and colours, some highly scented, some not, underlie the other great secret of the rose's enduring success. These are flowers, like oak trees (see page 145), that so easily hybridize that the broader term 'genus' is really more appropriate than the narrower 'species'. Botanical enthusiasts are able to cultivate flowers that suit almost every nuance of human fancy in colour, shape, form or fragrance. Once a desirable variety has been grown, horticulturalists can replicate its genetic formula *ad infinitum* by grafting (cloning) a parent plant on to the root stock of another plant (see apple, page 345). When varieties of roses that bloom throughout the year came across from China to Europe in the 1800s, they triggered a craze for horticultural experimentation with rose genes that has never lapsed since. Such genetic pliability has provided botanists with a broad palette of possibilities with which to adapt to changes in human taste, style and fashion.

The cost of sustaining our fondness for these flowers doesn't just threaten the ecology of countries like Kenya. As the lake's waters dry up, large cut-flower producers are busily buying up land in other nearby cheap labour markets, such as in Ethiopia, incentivized by a ten-year tax break from the government of a country that is also chronically short of water. Will the appeal of the rose prove so enduring that African farmers will continue to cultivate them and European consumers continue to purchase them, despite the ecological consequences? The problem of short-term gain over long-term pain seems to be as perennial as the rose itself.

Apple

FAMILY: ROSACEAE
GENUS: MALUS DOMESTICA
RANK: 53

How the genes of a highly nutritious fruit with greater than average diversity were reduced by humanity's desire for visual perfection.

JOHN CHAPMAN is one of America's oldest and most famous iconic heroes. The eighteenth-century frontiersman (1774–1845) earned the nicknamed 'Johnny Appleseed' because of his obsession with planting apple trees. He would pre-empt the arrival of new settlements by establishing fenced-in apple plantations before moving on to a new area, leaving his orchards under the stewardship of the locals. Every few years he would return to pick up a payment made in kind or as rent. Therefore, every time settlers pushed west, Johnny, the nomadic holy-man-cum-horticulturalist, had ensured they were welcomed by a dependable supply of rich, juicy fruit.

Apples were everything to North American prospectors following their successful war of independence from Britain (1775–83). Sweetness, a flavour humans have enjoyed throughout history, largely came from fruit. Sugarcane was rare, and honey supplies were limited. It was mostly thanks to 'Johnny', the thoughtful apple-grower, that essential fruit supplies were available to the settlers of Ohio and Indiana.

It wasn't just his curious nomadic lifestyle (he never wore shoes) or his clear foresight that make Chapman's story intriguing. More remarkable was his subversive style of horticulture, which embraced, rather than competed with, the evolutionary systems of the natural world: he *always* planted his orchards from seed.

Apples are remarkable for their huge genetic diversity. Buried inside the apple's core are up to a dozen seeds. Grow trees from all twelve seeds and each one will have remarkably different characteristics. This is the effect of sexual reproduction working at its explicit best. Regardless of where apple seeds are carried by their target transporters – mammals – such diversity ensures that at least one or two are able to respond favourably to the prevailing environmental conditions. This statistical bet has helped apple trees adapt to habitats far removed from their native origin in central Asia.

But by the time Chapman set about establishing his orchards, a rival human system for propagating fruit trees was already well established. No one knows who first discovered that it was possible to graft trees together by carefully tying a branch or cutting from one growing tree on to the roots or stem of another. The technique was definitely used in ancient Greece and Rome where apples, transported from Asia via the silk route, were widely propagated as a favourite fruit. In 323 BC Greek writer Theophrastus, a pupil of Aristotle and a keen botanist, wrote a comprehensive practical treatise that included advice on the best techniques for grafting trees.

American hero 'Johnny Appleseed' commemorated on a US postage stamp in 1966.

Trees propagated in this way avoid the lottery of sexual reproduction's genetic mix 'n' match. Choose a specimen that produces the shape, form and flavour of fruit you like, then graft one of its branches on to the roots of another tree of the same (or similar species) and, hey presto! Identical fruit as produced by the first tree will grow back every time. This propagation technique works for most fruits and flowers that grow on vines or trees, from peaches and almonds to apples and pears (see also roses, page 341, and grapes, page 299).

Despite this, Chapman insisted on growing his trees from seeds that originated in Europe. The upside of Chapman's intransigence was that those seeds that survived were those best suited to the conditions of their new American home. The downside was that it was impossible to predict the quality, quantity, size, shape or form of fruit from one tree to the next. Some were small and sour, others were large and sweet. Some were susceptible to disease, others mostly immune. There was absolutely no standard or consistency – just as nature intended.

For more than a century these quirks didn't matter because American settlers never ate Chapman's apples. Instead, they were crushed, fermented and turned into a nourishing, refreshing brew called cider – a beverage consumed even in preference to water (as was beer in early-modern England, see page 279) because it was a clean, healthy, safe drink for all the family.

But when at the start of the twentieth century apples began to be consumed as food, the modern fate of the apple tree changed dramatically. A marketing campaign was launched to get Americans to *'eat an apple a day – it keeps the doctor away'* and the onset of Prohibition, which made the consumption of alcohol illegal, stunted the appeal of cider. Since human consumers judge the fitness of food on its visual symmetry, colour and proportion (see pages 338–9) random sizes, shapes and variable taste were no longer acceptable. Grafted trees that produce predictable fruit became all the rage. They have remained so ever since.

As a result, the number of varieties of apples grown across America (and now the whole world) has declined dramatically – in defiance of Chapman's best-practice plantation programme. Out of the 7,500 known apple cultivars, almost all of the fifty-five million tonnes of apples grown worldwide each year are derived from a core stock of just six varieties, specially selected for their looks, shape and sweetness.

The risks of farming so few mass-produced varieties (see also water mould, page 52) resulted in a bacterial near-miss in the late 1980s when a disease called fire-blight (caused by the bacterium *Erwinia amylovora*) ripped through the fruit orchards of North America. As a result, many growers faced collapse in a similar way to French vintners just over a century before (see pages 298–9). Only thanks to a revolutionary approach to pest control was resistance to the disease established. Researchers at Cornell University successfully pioneered the use of a 'gene gun' to transfer an antibacterial gene from a species of giant silk moth (see page 260) to an apple tree. Cuttings from this tree have since allowed farmers to propagate fire-blight clones in a bid to immunize their orchards.

The need for such techniques would have American hero 'Johnny Appleseed' spinning in his grave. Apples earn their place as the third most popularly grown fruit (after bananas, see page 364, and oranges, see page 361) but artificial genetic intervention has become a new ingredient vital to ensure the survival of inbred crops specifically designed to satisfy the innate human desire for beautiful, healthy-looking food.

Vanilla

FAMILY: ORCHIDACEAE
SPECIES: VANILLA PLANIFOLIA
RANK: 97

How the dried beans of a Central American orchid became one of the most popular flavourings in the modern world.

ANNE SHIRLEY, an eleven-year-old, red-haired orphan, was utterly determined not to miss her Sunday school picnic tea-party. But she had a problem. Her guardian Marilla was under the distinct impression that Anne had stolen a precious amethyst brooch. As a punishment Anne had been banished to her bedroom until she owned up to the crime. So desperate was precocious little Anne not to miss the party that she made up a false confession: *'Punish me any way you like,'* pleaded Anne to Marilla, *'but please, please let me go to the picnic! Think of the ice cream ... For anything you know, I may never have the chance to taste ice cream again ...'*

Later that day Marilla found her brooch, which she had carelessly mislaid, the confession was exposed as false and Anne skipped off gleefully to the picnic after all. *'Words fail me to describe that ice cream,'* Anne told Marilla, after she returned home to Green Gables, *'I assure you it was sublime.'* (From *Anne of Green Gables*, 1908, by Lucy Maud Montgomery.)

Exotic taste: the vanilla orchid, native to Mexico, appealed to the palates of people from Europe.

The popularity of this cool, refreshing and typically irresistible dessert can be put down to a powerful flavouring extracted from the cured beans of an exotic species of orchid called vanilla. Native to Mexico, the plant was known only to the Totonac and later Aztec people of the region until Spanish conquistadors led by Hernán Cortés arrived in the early sixteenth century. Mixed with hot chocolate, vanilla sweetened the flavour of this royal drink, fit for emperors and the nobility. Taxes in the form of vanilla beans were paid as tributes to the growing Aztec Empire in the fifteenth century, but only those who lived in the surrounding hills knew the complex art of cultivating the plant and extracting its precious essence.

European attempts at growing the plant in other parts of the world ended in abject failure for as long as 300 years. Even when the plant flowered, for some reason it never seemed to produce its delicious-smelling pods anywhere other than its native Mexican habitat. Until the mid nineteenth century supplies were therefore restricted to a small area where the orchid grew naturally and the European fancy for adding vanilla to cocoa (an addition generally preferred to the Native American alternative, chilli, see page 276) was strictly limited to an aristocratic elite.

The vanilla orchid's mysterious lifestyle was finally exposed thanks to two discoveries that followed each other in quick succession. The first was in 1836 when Belgian botanist Charles Morren (1807–58) spotted a particular species of black bee buzzing around a vanilla orchid that decorated the patio of the house where he was staying in Veracruz, Mexico. After close observation he saw that in the process of finding nectar, these pollinators, called Melipone bees, lifted up a transparent flap before depositing their load of pollen (for other examples of bee and orchid co-evolution, see carpenter bee, page 130). This flap, called a rostellum, is nature's way of ensuring that only the pollen from other orchids visited by these bees gets rubbed on to the style, thus avoiding the hazards of self-pollination and inbreeding. Vanilla orchids produced their aromatic beans only in Mexico because this was the one place inhabited by Melipone bees. Flowers cultivated elsewhere remained unfertilized. They simply withered and died without ever bearing fruit.

Perhaps of greater surprise to vanilla ice cream lovers is that they enjoy their favourite flavour only thanks to the ingenuity of a second pioneer – an African slave called Edmond Albius (1829–80). At the tender age of twelve, this orphan boy developed a quick, ingenious way of pollinating vanilla flowers by hand, without the need for Melipone bees. Using nothing more than a thin bamboo cane and a quick flick of the thumb, Albius found he could manually pollinate up to 1,000 orchid flowers an hour. Vanilla orchid production suddenly became big business. So successful was Albius's pollination technique that it is still used today.

The Indian Ocean islands of Réunion (where Albius lived) and Madagascar, as well as South America and the West Indies, became new habitats in which vanilla orchids were subsequently cultivated and harvested. By 1898 more than 200 tonnes of cured beans were being exported from Madagascar alone, still the world's largest vanilla producer.

Given the enormous demand for vanilla flavouring today, not just in chocolate bars and ice cream, but also for use in perfumes and aromatherapy oils, it might be expected that the cultivation of vanilla orchids would have followed the same pattern of intensive agriculture as other high-demand crops such as sugarcane (see page 202), roses (see page 341) and coffee (see page 278).

But no. The lasting impact of the genes of this awkward-to-cultivate orchid manifested itself in a spur to human scientific endeavour rather than in the reality of a reshaped landscape. The flavour of vanilla proved so popular in Europe and America that only thirty-three years after Albius discovered how to propagate the plant, German scientists worked out how to recreate the chemical structure of its flavour-producing compound, vanillin, artificially. Of the 12,000 tonnes of flavouring used in vanilla-flavoured products today, only about 10 per cent comes from genuine, naturally cured vanilla beans. The rest is produced artificially using a base resin derived from the left-overs of woodpulp processing or from a chemical called guaiacol, extracted from the tar of beechwood.

Sometimes the economics of cultivation are such that it is cheaper for humans to turn their hands to organic chemistry rather than agriculture (see also synthetic rubber, page 259) to satisfy popular demand. Nevertheless, had humans not experienced the taste and smell of natural vanilla, they would never have tried, let alone been able to copy, nature's chemistry. While the widespread use of artificial flavourings may be worse for the dissemination of the genes of *Vanilla planifolia*, it is probably a lot better for the health of a global ecosystem, dominated as it is by so many people with a love of sublime ice cream.

Lavender

FAMILY: LAMIACEAE
SPECIES: LAVANDULA ANGUSTIFOLIA
RANK: 93

An alternative-therapy-cum-placebo that has wafted its way into the history books thanks to its extremely powerful smell.

THAT HUMANS have a lousy sense of smell was not lost on Charles Darwin, who said in his book *The Descent of Man* (1871):

> *'The sense of smell is of the highest importance to the greater number of mammals – to the ruminants in warning them of danger, to the carnivore in finding their prey ... but the sense of smell is of extremely slight service, if any, to men ... He inherits the power in an enfeebled and so far rudimentary condition, from some early progenitor ...'*

Therefore only plant species with the very strongest odours were ever likely to grab people's attention through their olfactory pipe – the nose. Pretty flowers that grow around the Mediterranean are two-a-penny. But what really distinguishes a herb such as purple lavender from most other plants is its unforgettable fragrance.

People living in Europe and North Africa from the days of ancient Egypt to the present have all shared a common belief that something as aromatic as lavender must have a therapeutic, medicinal or other powerful effect. The catalogue of uses this single herb has been put to, and the many effects ascribed to it, are extraordinarily diverse. Traces of lavender fragrance were found by archaeologists when they began to rummage through the treasures of Tutankhamen's tomb after its discovery by Howard Carter in 1922. Lavender oil was also used in Egyptian and Phoenician purification ceremonies and was added as an ointment during the all-important process of mummification. Cleopatra, the stunningly beautiful last pharaoh of Egypt, is said to have seduced Roman generals Julius Caesar and Mark Antony with the aid of perfumes laced with lavender.

The name lavender derives from the regular use of this strong, fresh-scented herb by the people of ancient Rome. They added it as a fragrance when washing their linen or taking a bath – *lavare* is a Latin word meaning 'to wash'. According to the Gospel of St Luke, Mary Magdalene, a disciple of Jesus Christ, anointed his feet with oil that contained spikenard, a herb thought by some scholars to have been a type of lavender.

As with so many other products of the natural world (from grapes, see page 296, to rabbits, see page 330) the Romans transported powerful-smelling lavender all over their empire, where it took root throughout Spain, Britain and France. By medieval times it was being used as a miracle cure for a wide range of ailments and diseases from head-lice to warding off the devil. Author, linguist, scientist and philosopher Hildegard of Bingen (1098–1179), who founded a monastery on the river Rhine, was under no illusions as to the mystical power of the herb: *'If a person, with many lice frequently smell lavender, the lice will die ... It cures very many evil things and because of it malign spirits are terrified ...'*

The use of lavender-oil-soaked blotting paper as an apparent antidote for head-lice continued as a popular remedy until as recently as the 1870s.

Fragrant lavender became associated with royalty from the time of Queen Elizabeth I of England (ruled 1558–1603), who believed its medicinal properties helped cure her frequent headaches. Henrietta Maria, wife of King Charles I, introduced lavender soaps and potpourri from France just as the French perfume industry was establishing itself under the patronage of Catherine de Medici

Lines of lavender, favoured for its fragrance in Europe for more than 2,000 years, grown in southern France.

(1519–89), mother of a triad of French Kings: Francis II, Charles IX and Henry III. Renato Blanco, an Italian perfume expert and personal employee of Catherine, is credited as being the first professional perfume maker in modern times. His laboratory was connected to Catherine's palace via a secret passageway to prevent others from knowing the exact compositions of her sweet-smelling fragrances. Soon after this, the Pilgrim Fathers took lavender with them to the Americas where it thrived despite the harsh winter weather.

By the time of Queen Victoria (ruled 1837–1901) lavender had become a royal obsession. It was used in everything from an eau-de-toilette to a disinfectant solution for washing down the floors and furniture of her many royal palaces. She even appointed Lady Sarah Sprules as '*Purvey of Lavender Essence to the Queen*'.

But the powerful scent of lavender wasn't just the focus of people's desire to smell good. It was also thought that something so fragrant could fight off plague, disease and infection – a popular belief that originated in ancient Rome but which was revived during the plagues that struck Europe from the fifteenth century onwards. In London, during the Great Plague (1665–6), people tied bunches of lavender around their wrists hoping the smell would ward off infection. The use of lavender oil in solution as an antiseptic was widespread even into World War I before the advent of drugs such as penicillin (see page 314).

YARDLEY'S
Old English
LAVENDER
THE FINEST LAVENDER PERFUME MADE.
For generations it has been famous for the charm of its beautiful, fresh, old-world fragrance.
YARDLEY
LONDON

Perfectly charming perfume as promoted during the 1920s.

In modern times lavender not only remains a major constituent of perfume but has also become the premier essential oil in the global market for alternative medicine. Aromatherapy – a practice that emerged in the first half of the twentieth century – claims to be able to improve mood, health and spiritual wellbeing.

It was developed by René-Maurice Gattefossé (1881–1950), a French chemist who specialized in manufacturing perfumes. While working in his laboratory, Gattefossé accidentally set fire to his hand and, in an effort to ease the pain, instinctively thrust it into the nearest vat of liquid he could reach. The container was filled with lavender oil. The remarkable speed with which Gattefossé's hand healed inspired him to explore more fully the restorative effects of naturally occurring aromatic compounds. His book *Aromathérapie* (1937) became the cornerstone of the modern alternative medicine movement that is now a multi-billion-pound business.

Modern Western science has little, if anything, to say about the medicinal efficacy of lavender apart from its obvious properties as a highly pungent fragrance. Yet it is beyond dispute that for thousands of years this aromatic herb has regularly convinced people of all social classes that its strong odour has beneficial medicinal or aphrodisiac effects. Placebo or ancient remedy, lavender's ability to power its way through the mostly dulled capacities of the human olfactory system undeniably accounts for its worldwide success.

Black Pepper

FAMILY: PIPERACEAE
SPECIES: PIPERACEAE NIGRUM
RANK: 91

A vine with an ugly, shrivelled berry that people were convinced was essential for their personal wellbeing.

IT IS CURIOUS to think that black pepper was once one of the most sought-after natural products in the world – more precious even than gold. As long as 2,000 years ago the people of ancient Rome would pay a fortune to purchase the spice, which they knew came from a land somewhere far beyond their vast empire in the mysterious, exotic East. According to Pliny, author of a thirty-seven-volume natural world encyclopaedia written in AD 71, such was the demand for pepper that *'there is no year in which India does not drain the Roman Empire of fifty million sesterces [silver coins]'*.

Move on 1,500 years and the European zeal for the spice was no less intense. By 1496 black pepper was in such demand that in one shipment Venetian merchants were said to have imported as many as four million pounds of spices from the Far East, most of it pepper. In the following year, Manuel, King of Portugal, patronized feisty mariner Vasco da Gama to go on a perilous quest *'in search of spices'* so that His Majesty could avoid having to pay exorbitant prices to the cosy cartel of Italian and Islamic merchants who maintained a tight stranglehold on overland supplies. His gamble paid off. In 1498 da Gama successfully pioneered the first direct overseas trading contacts between India, China and Europe by finding a way to sail around the southern tip of Africa. Meanwhile, Christopher Columbus had embarked on a similar quest westwards, on behalf of Spanish monarchs Ferdinand and Isabella, to try to find a rival route.

Piperaceae nigrum, a vine that grows up the sides of trees, is native to the monsoon region of south-west India (Kerala). Its fruit grows in clusters of red berries that, once harvested, are left to shrivel in the sun, turning them black. When crushed, these light, easy-to-transport peppercorns can be added to any sort of food, giving it a hot, exotic, spicy flavour.

One popular theory as to why pepper was so precious in ancient, medieval and early-modern Europe was that, in the pre-refrigeration age, people needed a way of disguising the taste of off-meat. But modern commentators have discarded this popular idea as myth. Cured meat doesn't need a fridge to keep it from going off and the age-old skills of salting, pickling, desiccating and conserving are sufficiently robust to debunk the idea that pepper was used as a medieval mask for mould.

Rather, a triumph of mind over matter best accounts for the historic popularity of black pepper. Ancient Greek philosophy was heavily influenced by the Aristotelian concept that the secret to maintaining good human health lay in the proper intermixing of the four essential elements that comprised the natural world – earth, air, fire and water. Scholars such as Hippocrates (460–370 BC) and Galen

(AD 129–200) said that when applied to the human body these elements manifested themselves as four humours: black bile (melancholy), yellow bile (choleric), phlegm (phlegmatic) and blood (sanguine). Until the birth of modern Western science in the late seventeenth and early eighteenth century, medical practice was primarily concerned with finding the correct balance between these humours to aid wellbeing.

Pepper, it was widely believed, was vital to this process. It was one of the few natural ingredients with a hot, dry effect that could temper a diet of sanguine (red) meat. Today's rump steak with peppercorn sauce is a modern relic of a medieval passion for spicy food, long since forgotten with the emergence of a taste for sugar (see page 204) that became widely available from the seventeenth century onwards.

The need to balance the effects of different foods on the stomach was thought vital for healthy, balanced living. That's why spices like pepper were in such demand by those who could afford them, from monks to merchants, and kings to courtiers. *Circa Instans*, a medieval recipe book written *c.* 1160, lists 270 substances thought to help balance the four essential humours. The spiciness of black pepper was especially effective, it said, for removing wet phlegm from the chest, as a remedy for asthma. Pepper would also heal bloody sores when ground into a powder.

Such humours didn't regulate only physical health. Sexual appetite was stimulated by the use of spices, according to an ancient Greek work that was translated from Arabic into Latin by eleventh-century Benedictine monk Constantine the African. According to the text entitled *De Coitu* ('On Sexual Intercourse') a man's impotence was due to an excess of liquid humours (phlegmatic and melancholic) that could be corrected by the use of hot spices such as pepper, ginger and cinnamon.

The fact that spices such as pepper came from a mostly unheard of corner of the world only added to their mystique. Scarcity and myth – possibly perpetrated by enthusiastic Islamic merchants – were additional psychological factors that helped make this hot flavour seem desirable to medieval palates. Isidore of Seville (AD 560–636), the Archbishop of Milan, was the author of another authoritative natural encyclopaedia called the *Etymologiae*. In it he explained how Indian pepper trees were 'guarded' by poisonous snakes. He said that in order to harvest the valuable berries, natives would light fires between the trees to ward off the scary serpents. His colourful explanation also explained why peppercorns were always black and shrivelled.

European determination to gain direct access to supplies of pepper was boosted by the thirteenth-century globetrotter Marco Polo's popular account of the bountiful supplies of pepper produced in the spice island of Java. He described it as a rich island: '*... producing pepper ... and all the precious spices that can be found in the world. This is the source of the most spice that comes on to the world's markets. The quantity of treasure ... is beyond all computation.*'

Columbus and da Gama were ready to risk everything to find an unfettered route to these fabled lands. When da Gama finally reached Calicut on the west coast of India in May 1498, he began negotiations for rights to trade spices, even though he had little of value to trade in return (see also page 310). He intended his overtures to be taken seriously. On his return three years later he devastated the Indian town with his cannon, imprisoned the locals and burned them alive before returning to

Portugal with more than one million kilograms of pepper, equivalent to what Venice obtained during an average year. By 1530 the Portuguese had established a capital at Goa with a string of fifty forts along the west coast of India and a fleet of more than one hundred ships. The era of unequal 'trading' between the nations of Europe and the East had begun.

As a result of these adventures, modern historians are minded to credit pepper with some of the most extraordinary achievements in human history. According to one: *'The starting point for the European expansion out of the Mediterranean and the Atlantic continental shelf had nothing to do with, say, religion or the rise of capitalism – but it had a great deal to do with pepper ...'* Another cites spices, of which pepper was by far the most popular, as having *'dramatically affected history because they launched Europe on the path to eventual overseas conquest whose success and failure affects every aspect of contemporary world politics'.*

In the last 500 years black pepper, a catalyst for European conquest, has retreated into relative obscurity. During the seventeenth century, the rise of new medical knowledge and the arrival of other, more addictive substances in Europe (such as laudanum, see page 309; sugar, see page 204; tea, see page 283; and coffee, see page 279) reduced the demand and allure of pepper from a medicinal necessity to a more occasional culinary flourish. But by then its impact had been felt worldwide. From the discovery of the Americas to the colonization of East Asia, world history, as a result of pepper, tacked into a new head of wind.

Pepper harvesting in southern India as depicted in Marco Polo's widely read thirteenth-century travel journal.

Lotus

FAMILY: NELUMBONACEAE
SPECIES: NELUMBO NUCIFERA
RANK: 99

A flower that has inspired Asian people's search for inner peace and is now regarded as a model material by Western science.

FOR 3,000 YEARS people from the Far East have held up the sacred lotus flower as the ultimate model of natural perfection. Hinduism, the most ancient existing human religion, and Buddhism, its more recent offshoot, draw great inspiration and power from the shape, form and function of this flower – which is not to be confused with the Egyptian blue lotus that belongs to a different species of water lily.

Indian goddess Laksmi, wife of Vishnu, rises up on a lotus flower holding her infant Ganesha (illustrated *c.* 1860).

On one level, the lotus flower, like most other flowers, represents visual, symmetrical perfection – its unfolding petals spreading out evenly to present the plant as a genetically fit and attractive prospect for birds, insects or other passers-by. But it is the habitat of the lotus, as a perfectly clean flower that arises as if by magic from out of the murky, muddy water, that has most captivated the human imagination. In the Bhagavadgita, the holiest of Hindu scriptures, the lotus flower is used to demonstrate what humans should strive for in order to achieve inner perfection:

'One who performs his duty without attachment, surrendering the results unto the Supreme Lord, is unaffected by sinful action, as the lotus is untouched by water ...'

Shortly after his enlightenment, Siddhartha Gautama, known as the Buddha (*c.* 563–483 BC), showed a lotus flower to the audience that had come to hear him preach. All he did was lift the flower towards the assembled crowd – that was it. The Flower Sermon, as the occasion has been remembered, contained no spoken words. Part way through a man in the assembly smiled at the Buddha, showing that he understood the potency of the message that was being portrayed. Mahakasyapa went on to become one of the Buddha's most revered disciples and the Flower Sermon, as a way of reaching inner perfection that transcends words, is now considered the birth of the branch of Buddhism called Zen.

The noble path towards human perfection in Eastern culture concerns stilling the mind, freeing it from desire, angst, worry or guilt. Just looking at the lotus flower, suggested the Buddha, without speaking a word, was key to finding this inner calm. The approach was in stark contrast to the ancient Greek philosophical tradition, founded by Socrates (470–399 BC) and his pupil Plato (428–347 BC), which sought perfection in human society by stimulating argument, rhetoric and debate. It is in this culture that Western society has its roots.

Tibetan monks perfect a mandala made out of multi-coloured sand, with a lotus flower at its centre.

Tibetan monks continue the Buddha's tradition today through the symmetry of a mandala – a concentric image – which they use as an aid to meditation and trance. These geometric images are created out of multi-coloured sand, funnelled into lines and shapes using tubes and scrapers. At the centre lies the perfect image of a lotus flower. Meanwhile, ancient Hindu traditions for maintaining human spiritual health are focused around the seven *chakras*, lotus-shaped vortices that are thought to channel and circulate *prana*, the universal life-giving force, in and around the body.

Nature's mirror, celebrated in Eastern cultures for thousands of years in the form of a lotus flower, has recently begun to be appreciated by Western scientists in their own pursuit of perfection based on the material world. Biomimetics is a relatively new branch of science that aims to copy the way nature solves problems and apply it to the human world in a sustainable way.

In the early 1970s German botanist Wilhelm Barthlott uncovered the mystery of how lotus flowers are able to stay so pristine despite growing in some of the muddiest, water-logged habitats on Earth. After careful observation of the plant's leaves through an electron microscope he discovered they were covered in tiny protrusions called papillae. This design ensures that when a droplet of water falls on to a leaf, its surface tension stays intact. This causes the droplet to roll across the surface of the leaf, gathering dirt or other unwanted particles along the way, before falling off the plant altogether. Even sticky honey and glue are unable to attach to the leaves of the hydrophobic lotus plant. This simple, natural, self-cleaning system protects the plant against attack from bacteria, fungal spores and other harmful pathogens.

The science of biomimetics uses the study of natural systems to transform the world of artificial materials science. Dirt-repellent, self-cleaning coatings, embedded with rough microscopic surfaces similar to those found on lotus leaves, are being trialled on products including surface paints, roof tiles and waterproof clothes and shoes.

Today the sacred lotus can be appreciated in all its glory by both Eastern and Western cultures. Whether seeking inner peace or material perfection, many millions of people continue to find the lotus a source of natural inspiration.

Chilli Pepper

FAMILY: SOLANACEAE
SPECIES: CAPSICUM ANNUUM
RANK: 92

A brightly coloured, easy-to-grow, adaptable fruit that became popular throughout the world thanks to its ability to spice up the flavour of food.

WHEN A WARNING light suddenly illuminates in an airline pilot's cockpit does it mean there is a genuine malfunction with the plane's systems or is it simply a faulty sensor? A similar question might occur to a person who accidentally bites into a raw chilli. Is the searing pain that explodes in the mouth causing actual bodily harm or is this fruit somehow triggering a powerful false alarm?

Chilli pepper, native to South America, but transported by Portuguese colonists to India where it has been a hot hit ever since.

Chilli peppers are thought by archaeologists to have been among the first crops to have been domesticated by humans living in the Americas. Capsicum grains, used in cooking, have recently been found in south-west Ecuador dating back to 4000 BC. Their discovery, on the opposite side of the Andes mountain range from their natural habitat in Peru, suggests that chillies were a highly valued and widely traded supplement for early farming people who enjoyed eating these fruits to spice up their lives.

Quite why humans enjoy the flavour of chilli peppers is something of a mystery. These plants evolved to appeal to the palates of birds, not mammals. The active chemical produced in the fruit of the chilli plant, capsaicin, evolved as a defence against fungi and grazing herbivorous mammals. Rather than stimulating one or more of the five standard tastes (see page 340), this compound binds to the same receptors in a mammal's mouth as those that respond to hot temperatures or tissue damage, triggering a sensation of heat and pain that persuades a mammal's brain that its mouth is being ripped apart by a surge of something violently hot to the touch. As anyone with a fancy for vindaloo curry knows, the result is an increase in heart rate and sweat.

This feeling is a complete illusion, however – a natural false alarm. Devour a hot chilli and the temperature inside your mouth will be no different from how it was before you took the first bite. Birds are insensitive to the chemical, allowing them to eat the seeds of raw chillies without discomfort. Yet mammals recoil with pain.

There is a simple evolutionary explanation. When chilli seeds are eaten by herbivores they get damaged by their grinding molars. But when the seeds are eaten by birds they pass through the digestive tract unharmed, allowing them to be spread successfully in their droppings. Therefore, over millions of years, natural selection has seen to it that the bright yellow and red colours of mature peppers attract birds, while the hot, choking chemical capsaicin wards off grazing mammals. So why did early American agricultural humans begin to cultivate these obnoxious fruits as a food of *choice*?

One reason may involve the side-effects of capsaicin. In humans, this substance triggers the release of endorphins – chemicals that reduce stress, relieve depression and raise the spirits. Also, humans are unlike other mammals in that they have learned how to cook. Sizzling peppers alongside other foods mixes the hot taste of chilli, both diluting and enhancing its effects. A third factor may be psychological. It is a well-known phenomenon that humans enjoy the sensation of pushing their bodies beyond standard operating limits, as when thrill-seeking by riding a roller-coaster (see also cannabis, page 286). Experiencing pain without actually being physically harmed is thought by some experts to have similar appeal.

It wasn't until the arrival of European explorers with the voyages of Christopher Columbus in the late fifteenth century that chillies had the opportunity to experience habitats beyond their Native American origins. Spanish and Portuguese people became their new-found champions. They called them chilli 'peppers' after the similarly spicy pepper-like effect they had on the palate. All the medicinal benefits afforded to black pepper (see page 351) as a way of balancing the body's humours were now also bestowed on chillies, as a Spanish physician writing in the 1570s was eager to explain: *'It doth comfort much, it doth dissolve the winds, it is good for the breast and for them that be cold of complexion it doth heal and comfort, strengthening the principal members ...'*

Spanish people began to use chilli peppers in their cooking instead of the traditionally expensive and difficult to obtain black pepper. Meanwhile, Portuguese explorers settled off the west coast of India soon after the pioneering voyages of Vasco da Gama (see pages 352–3). By the 1530s they had established India's first chilli pepper plantations. Seeds were probably brought from Brazil via Lisbon to India. Once cultivated in the traditional spicy-black-pepper-growing region of western India, chillies became an instant hit – a perfect antidote to the blandness of rice.

Indian spice farmers dried and crushed these new culinary delights, turning chillies into easy-to-transport powder. By the late sixteenth century capsicums grown on the west coast of India were being traded along the medieval spice routes into Persia and around the Black Sea where they were quickly incorporated into Middle Eastern and Turkish cuisine. From there it was just a short hop over to Central Europe, completing the spice's global journey. Following the Turkish invasion of Hungary in 1529, ground chilli powder in the form of paprika became a new European favourite, giving additional colour, taste and flavour to everything from pepperoni sausages to Hungarian goulash soup.

Even today's European and American consumers who dislike the spicy effects of traditional chilli peppers can enjoy the common bell-shaped pepper, *Capsicum annuum*, which has been bred to contain little or no pain-producing capsaicin.

With as many as thirty varieties catering for all manner of tastes, from benign to fiery hot, the global adoption of chilli peppers into so many different cuisines throughout the world will ensure these fruits continue to be widely propagated for as long as humanity remains a dominant global evolutionary force. It's a somewhat ironic result for a fruit that succeeded in the wild by developing defences against being eaten by mammals with molars.

Grass

FAMILY: POACEAE
SPECIES: POA PRATENSIS • LOLIUM PERENNE • AGROSTIS STOLONIFERA • FESTUCA RUBRA
RANK: 24

A velvet carpet of green that humans have cultivated beneath their feet to please the eye and soften the fall.

ONE WAY of judging the impact of living things is to imagine what the world might be like without them. That's pretty much what British science-fiction writer John Christopher did in his 1950s classic *The Death of Grass*, which told the story of a deadly virus that progressively swept across the world destroying all living things in the family Poaceae. The fate of humankind, unable to come up with an effective antidote, was predictably apocalyptic.

Modern humanity depends on grasses for its survival. These vigorous plants contain the third largest number of species among the flowering plants (angiosperms) after orchids and asterids (the daisy family). Several have already featured in our top one hundred (see wheat, page 199; rice, page 230; sugarcane, page 202; maize, page 234; and bamboo, page 150). There are, however, other types of grass that have been even more widely cultivated, not because they are good to eat or useful for construction, but because they appeal so greatly to our senses of sight and touch.

Gardens, lawns, city parks, football pitches and golf courses are just a few of the countless horticultural features dominated by cultivated species of cut green grass. According to recent research conducted by NASA scientist Cristina Milesi, satellite images reveal that there are *three times more* acres of grass lawns in the United States than irrigated corn (maize). The total surface area covered by cut grass (including golf courses and sports fields) is estimated to cover an area in excess of 128,000 square kilometres.

Gardening is as old as recorded history. Ancient Chinese records tell how it was in a garden that Leizu reportedly made her silkworm discovery (see page 260) and it was in a garden that God spoke to Adam and Eve, warning them not to eat of the forbidden fruit, according to Jewish scribes who wrote the Torah in about 500 BC.

But the roots of cultivating a garden dominated by a velvety grass lawn go back to an aristocratic fashion that caught on in eighteenth-century England. Lancelot 'Capability' Brown (1716–83) was a landscape-gardening guru who designed more than 170 gardens, many of which surrounded imperial Britain's finest country houses. Brown's passion was to use vast swathes of closely cut, undulating grass to provide a soft, pleasing velvet-like carpet that drew an onlooker's eye inwards towards the central artificial architectural masterpiece. Newly sown green grasses became a symbol of aristocratic status, wealth and prestige because keeping such extensive areas of closely cut lawns required a veritable army of garden staff, equipped with an arsenal of specialist cutting tools including lifters, edgers and scythes.

As European social ideals were transplanted to the newly independent United States of America, the finely manicured lawn became that new nation's most potent symbol of man's triumph over nature.

President George Washington (ruled 1789–97) cherished the lawn at his Mount Vernon estate, which symbolized his country's ongoing conquest of a barren, hostile landscape into a well-cultivated continent, civilized by man. In his diary, dated 1785, he wrote of his pride for his grass: *'I sowed what was levelled and smoothed of it with English grass seeds; and as soon as the top was so dry as not to stick to the roller, I rolled and cross-rolled it.'*

Washington's desire to see his lawn criss-crossed and rolled with stripes underlies the power of grass to please people's love of visual geometrical perspective and symmetry in nature. What better display of the genetic fitness to govern than a meticulously groomed, well-rolled lawn? It was the perfect adornment for a ruler intent on stamping his authority over nature and society. The extensive well-cut grasses that still surround the American presidential residence, the White House, were themselves inspired by Washington's Mount Vernon estate.

American garden designer Andrew Jackson Downing (1815–52) took Washington's love of finely cut, freshly rolled grass to a mass audience. In his many writings about the theory and practice of landscape gardening, Downing made specific recommendations about the types of grass best suited to make the perfect lawn which, he said, should be cut with an English scythe edged like a razor: *'No expenditure in ornamental gardening'*, he wrote, *'is productive of so much beauty as that incurred in producing a well-kept lawn. It is a universal passport to admiration.'*

Popular admiration for perfectly cut grass exploded soon after Jackson's death with the invention of the mechanical lawn-mower. American inventor Elwood McGuire, from Indiana, pioneered the first human-powered machines in 1870. Now anyone could show off their own ability to civilize nature with a weed-free, finely cut, perfectly striped lawn surrounding their ideal home on all four sides.

Golf, cricket, lawn-tennis, croquet, football and rugby were all sports that took off towards the end of the nineteenth century thanks chiefly to the powerful combination of mechanical cutting equipment and new varieties of soft, spongy grass. Such surfaces not only looked good but also proved perfect cushioning for everything from the bounce of a golf ball to a fall from a badly timed tackle. In fact, so ideal was cut grass as a surface for European and American sports that by the early twentieth century, agronomy – the business of cultivating, breeding, growing and protecting grasses – became established as its own branch of science, leading to the founding of the American Society of Agronomy in Chicago in 1907. Twenty-one years later, an otherwise unlikely alliance between the US Department of Agriculture, the US Golf

An eighteenth-century cutting machine complete with roller to ensure a stripy, symmetrical finish.

Association and the Gardening Club of America pioneered the growing of as many as 500 separate plots of different varieties of grass in an effort to find a combination that consistently produced the best terrestrial coverings for human deployment and enjoyment.

Such research confirmed that the attempt at breeding a perfect species of grass was a false pursuit. Much better was to sow areas with a mixture of varieties, some of which are tolerant to drought, others tough enough to withstand being smashed by balls or churned up by feet. As a result, a mixture of species from four families of grasses has most commonly been used since the early twentieth century to create the purpose-sown lawns, parks, playing fields, golf courses and grasslands of the modern world. Bluegrasses (such as *Poa pratensis*) are well suited to shady, colder temperatures while rye grasses (such as *Lolium perenne*) work best in dry conditions. Others such as *Agrostis stolonifera* and *Festuca rubra* provide a fine, soft texture, ideal for bowling lawns and putting greens.

Between 1947 and 1964 an average of more than 1.25 million new homes were built in the United States every year, each one established within its own garden plot. Social convention dictated that they were surrounded by cut grass. From 1963, the air-cushioned Flymo, invented by Swedish engineer Karl Dahlman, allowed even females to cut grass easily because it was 'a lot less bovver with a hover'. Meanwhile, a huge industry sprung up around the ideal American suburban home. Automatic sprinklers, grass-seed, fertilizers, pesticides and machinery for tending to the upkeep of grass all aided the growth of a new multi-billion-dollar industry. In Britain, the post-war building boom was no less spectacular – with an estimated total of sixteen million lawns being rolled into regular service by 1989.

Brown, dormant grass looks like the product of a failed proprietor so humans tend to insist on extensive watering to keep their lawns looking healthy and green. NASA's Milesi has calculated that as much as 900 litres of fresh drinking-quality water are needed per person per day to keep America's lawns looking fresh. This is a significant factor, she says, behind the rapidly falling water table along North America's eastern Atlantic seaboard.

Without irrigation or artificial fertilizers most of American's cut grass would perish. The only places lawn-quality grass could grow in the US without human assistance are a few areas in the north-east and the Great Plains. Just picture how different a world without cut grass would look. Would mankind's relationship with nature be different today if deprived of its most conspicuous stamp of human authority? Would the mass-adoption of competitive sports from golf to football have been possible without this supple, robust, naturally rejuvenating surface? Could people bear to live in urban environments if totally deprived of their parkland patches of velvet green?

John Christopher's fictional world fell apart because a virus wiped out all types of grass. But just remove the Earth's soft carpet of green, which modern humans have so extensively cultivated under their feet and around their homes, and much of the world we know today would look and feel very different indeed.

Orange

FAMILY: RUTACEAE
SPECIES: CITRUS SINENSIS
RANK: 95

A globally dominant, brightly coloured berry that humans count on to help them compensate for an ancient genetic weakness.

EBOT MITCHELL, a housewife from Cornwall, England, wrote a remarkable book in the early eighteenth century which recently surfaced in the attic of a Gloucestershire home. It detailed more than one hundred household recipes, one of which was found to be the first known description of a cure for a disease that has plagued human beings from at least the time of ancient Greece.

Sweet oranges originated in China, but became popular in Europe by the mid sixteenth century from where they were transported worldwide.

Scurvy is a particularly nasty illness, traditionally contracted by sailors on long voyages. Symptoms include excessive bleeding, loss of teeth and hair, bone deformation, hallucinations and blindness, eventually leading to death. Mitchell's book described the earliest-known effective remedy (nearly fifty years before James Lind, the Scottish naval officer usually credited with discovering a cure). Her answer, like Lind's, was perfectly simple: always eat plenty of citrus fruits, like oranges. This is because they are rich in vitamin C.

Vitamins occur naturally in a wide range of fruit and vegetables, but vitamin C is highly concentrated in the small family of fruits called citrus. By far the most popular species of citrus fruits today are oranges. Sweet varieties called *Citrus sinensis* are grown in huge orchards all over the world from Florida to Spain, California to China and Australia to the South African Cape. Other citrus crops of significance, but far less intensively grown, include other varieties of orange (in particular mandarins) but also lemons, limes and grapefruits.

It is something of an evolutionary quirk that humans need to eat a diet of fresh fruit stuffed with vitamin C at all. Most animals have enzymes that automatically synthesize vitamin C to keep their bodies healthy – in the same way humans automatically manufacture their own vitamin D in their skin using little more than sunlight. But for some ancient evolutionary reason primates (including humans) have lost this capability when it comes to the manufacture of vitamin C. Instead, it is vital for their good health to find naturally occurring, external sources to supplement their diets.

Not that this was a problem for most of human history. As hunter-gatherers, eating wild berries ensured vitamin C deprivation never occurred in the Stone Age. Only with the emergence of settled human societies did scurvy begin to become a problem, and then only if people failed to eat sufficient quantities of fruit and veg. The disease became most acute from the sixteenth century onwards when European maritime explorers went on epic voyages all over the world (see page 352). Not realising their diets had to be supplemented with sources of vitamin C, many sailors died appalling deaths as a result of a mysterious disease called scurvy.

Oranges, Ebot Mitchell's solution to the scurvy problem, have been enjoyed by humans for almost all of recorded history itself. Their first recorded appearance was in China – hence their botanical name *Citrus sinensis*. These sweet oranges, commonly grown throughout the world today, are a hybrid cross between two species, the grapefruit-sized *Citrus maxima* and the mandarin *Citrus reticulate.*

It is still unclear exactly which people may have transported oranges across Asia to Europe and when. The Jewish Torah makes no obvious references to citrus fruits, although figs, olives and dates appear throughout. But it is clear that by the time of the later Roman Empire, species of oranges were being grown around the Mediterranean. They were regarded mostly as exotic, luxurious and usually ornamental fruits because these varieties had a much more bitter taste than the sweet oranges popularly grown today. The same sour varieties were also spread by Arabs following the Islamic conquests of the eighth century. They were used by herbalists, such as Avicenna (AD 980–1037), to make a type of medicinal syrup called *alkadare.* The fruit therefore spread throughout Egypt and North Africa to Spain, Sardinia and Sicily.

It was only when sixteenth-century Portuguese maritime explorers brought back seeds from the sweet orange hybrids grown in China (*C. sinensis*) that the fruit began to feature more fully in the minds and stomachs of people living in the Middle East, Europe and North Africa. By 1646 sweet orange trees were growing in Italy, introduced from Portugal (but originally imported from China). Giovanni Ferrari (1584–1655), a highly regarded Italian botanist, could hardly contain his excitement at their sight and taste:

> *'Just recently there has been sent to Rome ... from Lisbon a beautiful tree with golden fruit. The fruit is decidedly round in shape with a skin that is most glowingly and delightfully yellow. The pulp has a sweet and most pleasingly spicy taste ... and is so golden in colour one would think gold had been melted away into its juice ... this fruit is a sweet and fragrant morsel for anyone's palate.'*

This hybrid species with its large, juicy, good-looking berries became so visually appealing that even its name came to be used to describe an everyday colour. The appearance, symmetry, smell and taste of this sweet orange provided it with excellent growth prospects in an increasingly interconnected human-centric world.

Columbus and other European colonists took seeds of the sweet orange with them as part of their overseas survival kits. They were planted on the island of Hispaniola (Haiti) where Columbus founded his first colony in 1493. Spanish settlers then took the fruit to Florida sometime between 1513 and 1565, when their first colonies were founded around St Augustine. Once introduced, these fruits relished the hot, subtropical American climate. Wild orange and other citrus fruit groves also sprung up around Florida's many lakes.

Meanwhile, the delightful taste of sweet oranges grown around the Mediterranean started to make its way into preserves for transportation and consumption further north. Marmalade became a favourite at the British and French breakfast tables from the eighteenth century. A dedicated factory opened up on the east coast of Scotland in 1799 that specialized in turning the whole fruit of sweet Seville oranges into delicious Dundee marmalade.

A FULL BIG GLASS

This much Florida Orange Juice every day

'An essential source of vitamin C' was how US advertisers in the 1950s persuaded consumers to adopt a juicy new habit.

At about the same time greenhouses called orangeries began to grace aristocratic estates as *C. sinensis* became fashionable to cultivate in cold climates because of its attractive golden fruits.

The twist that turned sweet oranges from an exotic, aristocratic fancy into the mass-consumer product we know today came about thanks to a nineteenth-century trilogy of education, transportation and refrigeration. Knowledge of the health benefits of eating citrus fruit that started with the cure for scurvy, as written about by Mitchell and Lind, was quickly followed by the railway revolution that gathered steam in the mid nineteenth century. By 1869 freshly picked oranges grown in sunny climates such as California (planted as a result of the gold rush 1848–55) could be transported quickly by train all the way to New York. Within twenty years the trains were upgraded with refrigeration, preserving the fruits' freshness. By 1892 the first steamers loaded up with oranges set off for London. Queen Victoria, when she tasted one, is said to have pronounced them 'palatable'.

Advanced mechanical processing techniques emerged soon after World War I that automatically converted fresh oranges into easy-to-transport, concentrated juice that could be reconstituted close to its place of consumption. Such innovations have meant that as human populations have surged over the last sixty years, people's genetic weaknesses have been compensated for by the ubiquitous consumption of vitamin C – mostly provided by sweet oranges.

More than sixty million tonnes of oranges are produced each year, two-thirds of which are turned into juice, with Brazil, the United States, Mexico, India and China the world's top orange-producing nations. These good-looking, sweet-tasting, vitamin-enhancing fruits have therefore been transformed from once-exotic luxuries first cultivated in China into the modern world's premier healthy drink.

Banana

FAMILY: MUSACEAE
SPECIES: MUSA ACUMINATA
RANK: 98

The world's most popular tropical fruit, little known until the rise of modern commerce, which now faces mass extinction.

QUESTION: what town used to host an annual fruit festival that featured the consumption of a one-tonne banana pudding and a beauty pageant that crowned the world's one and only International Banana Princess? A place in Papua New Guinea, perhaps, where bananas grow naturally in the wild and where humans first cultivated the fruit as long ago as 1000 BC? Or maybe a country in Central America or an island in the Caribbean, home to many of the twentieth century's so-called 'banana republics', states with economies driven by the growth and export of long, bendy, yellow-fingered fruits?

The world's tallest herb, which hates the frost, was perfect for companies that thrived on transportation.

No. This unique jamboree was the pride and joy of a small railway town called Fulton in the US state of Kentucky. Today a population of just over 2,000 people resides in this mid-eastern town, but for about fifty years – between 1930 and 1980 – Fulton was known as the 'banana capital of the world' and became renowned for its eccentric international festival.

Fulton's rise to fame goes to the heart of the story of how soft yellow bananas have become one of the world's most widely consumed fruits. Americans today eat more bananas than oranges and apples combined – an average of more than seventy-five per person, per year. How is this when bananas are almost impossible to grow on US soil (except in Hawaii), while oranges and apples, just as good a source of healthy vitamins, can be grown all over the country from California to Florida?

Bananas are so popular to eat today because about 120 years ago they were identified by a number of American shipping magnates as an ideal food for transportation from one part of the world to another for the pursuit of profit. Their success depended on farming the fruits cheaply enough and selling them to the mass market in sufficient quantities – challenges that bananas have risen to extremely well.

Until 1880 few people outside Asia and Africa ever ate bananas. Initially cultivated in Indonesia, bananas had reached Africa by the birth of Islam in *c.* AD 600. Muslim warriors then spread these nutritious fruits to the Near East, Egypt and Spain as well as down the coast of West Africa, where they were discovered by Portuguese merchants in 1482. From here European explorers planted them on the Canary Islands and took them to the Americas, seeding them on the island of Hispaniola (now Haiti and the Dominican Republic) in the early sixteenth century. Bananas were found to grow especially well in the tropical climates of countries such as Panama and Mexico in Central America as well as throughout the islands of the Caribbean.

Bananas belong to the tallest family of herbs in the world, sometimes reaching a height of ten metres. Their lack of wood means they are completely frost intolerant, strictly limiting their range to tropical climes. The plants need between fourteen and twenty-three months of frost-free, sufficiently wet conditions to bear their fruit, which grows in bunches like long yellow up-pointing fingers. The name banana is itself derived from the Arabic word *banan*, which means 'finger'.

Despite their globalization, by the year 1900 bananas didn't seem set to change the world. Europeans and Americans were unsure what to make of the phallic-looking fruit, which no self-respecting lady could possibly peel and eat in public without first cutting it into pieces using a knife and fork. By the turn of the twentieth century they were still regarded by most people as rare and exotic. An article about bananas in *Scientific American* magazine in 1899 even included a list of instructions on how to eat them: *'The fruit is peeled by splitting the skin longitudinally and giving it a rotary motion with the hands ...'*

Ten years later the fate of the banana had changed beyond all recognition. American shipping entrepreneurs pioneered a new business model by encouraging the growth of bananas on plantations in South and Central America so they could be transported in refrigerated steamships to ports such as New Orleans. From there they were packed on to ice-trains, 70 per cent of which went through the railway town of Fulton, Kentucky, where the main line divided into five tributaries, each of which headed out to different regions of the northern United States. The people of Fulton were kept busy servicing the steam trains and refreshing the wagons with ice. Keeping the fruits cool prevented them from ripening prematurely. So important was the banana trade to the people of Fulton that they marked each passing year with their annual banana festival – the first one staged in November 1963 – a mixed celebration of North, Central and South American culture.

Corporations such as the United Fruit Company, founded by Lorenzo Baker, a fishing captain from Boston, and the Standard Fruit and Steamship Company, headquartered in New Orleans, Louisiana, found that bananas were perfect for marketing as an essential food. Of particular importance was the way that their ripening process could be postponed by refrigeration during transportation and then reactivated in heated rooms close to their point of sale. It was discovered that ethylene, a gas produced naturally by the fruit, could be used to accelerate the process, turning green bananas into a more visually attractive yellow.

American banana merchandizing companies successfully convinced everyday folk that life was incomplete without eating bananas. Everything about them was presented in ways that made bananas look and feel good: their thick skin was promoted as a natural protection against germs; the fruit was advocated as a source of vital vitamins; their convenience was emphasized in the form of a between-meals snack; and their soft texture made them ideal for mashing and feeding to small children. Best of all, bananas were 100 per cent safe to eat since domesticated varieties contained nothing as potentially hazardous as an easy-to-choke-on seed.

But man's dependence on a single mass-produced variety of seedless banana called Gros Michel (or Big Mike) very nearly ended capitalism's love affair with this golden-fingered fruit. A fungal wilt known as Panama disease (*Fusarium oxysporum*) first appeared in Asia in the early twentieth century. By the 1950s it had spread on the wind to the Americas, decimating plantations that sustained the world's banana exports. Growers' businesses were wiped out (see also grape, page 298, potato, page 200, and apple, page 345). Even drowning crops in Bordeaux Mixture (see page 53) provided only limited protection.

By the 1960s, Gros Michel bananas had become extinct. It was only thanks to the discovery of a new, disease-resistant variety in Vietnam that the prospects for the genes of mass-produced bananas were revived. Cavendish bananas (*M. acuminata*) let growers replace their rotten stocks and by the mid 1970s the business of profiting from the growth and carriage of this tropical fruit to consumers in the non-tropical world was once again restored.

By now new technology, driven by the efficiencies of markets, had rendered steam trains redundant and electric refrigeration had replaced the need to restock wagons with ice. Fulton, the Kentucky town once famous for its banana festival, was no longer at the cross-rails of banana commerce because goods trains seldom had the need to stop there. Its festival limped on until the early 1990s after which it finally bit the dust.

Almost all bananas sold in shops and supermarkets throughout the world today are of the Cavendish variety. Connoisseurs say they taste less good than the once ubiquitous Gros Michel and traders agree that Cavendish bruise more easily in transport. But these are minor inconveniences, far outweighed by the enormous benefits of bananas with natural disease resistance.

But then about ten years ago, almost as predictably as the return of the Terminator in a Schwarzenegger movie, a new variant of the banana-decimating fungus began devouring its way through Cavendish crops in Asia. Experts predict its inevitable arrival in the banana-producing countries of Central and South America within twenty years. As yet there is no cure. A chronic lack of genetic diversity means that the chances of finding another disease-resistant substitute are remote. Genetic engineering, as with the apple disease fire-blight (see page 345), may be mankind's only escape route if it wishes to continue its love affair with the banana. Meanwhile, about seventy million tonnes of Cavendish bananas are consumed worldwide each year. For how long these cheap, affordable, plentiful fruits will retain their premier place rather depends on which side gains the upper hand in the ongoing tug-of-war between the natural and human worlds ...

Eighty-five thousand bananas being taken from the Caribbean to Vancouver, Canada, by steamship and train.

BANANA SPECIAL
85.000 STEMS - WEST INDIES TO VANCOUVER
VIA C.N.STEAMSHIP LADY RODNEY
AND CANADIAN NATIONAL RAILWAYS SPECIAL TRAIN

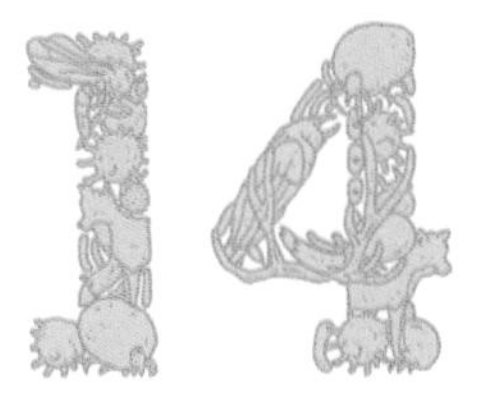

On Rivalry

How certain species have thrived with the rise of human civilizations, often despite attempts to keep them at bay.

THOMAS MIDGLEY is famed for being one of the world's most unlucky inventors. Born in 1889, this American from Beaver Falls, Pennsylvania, made quite a name for himself in the world of chemistry and mechanical engineering. In the 1920s Midgley came up with a way to make petrol engines run smoothly by adding a chemical to gasoline called tetra-ethyl lead (TEL). The enormous damage this poisonous lead compound inflicted on both human health and the environment was an unforeseen consequence that has taken more than fifty years to put right.

Later in his career Midgley discovered a new method of electric refrigeration using an artificial chemical called freon. By the early 1980s it was discovered that freon, a chlorofluorocarbon (CFC), reacts with oxygen in the upper atmosphere destroying the Earth's ozone layer, which shields all life from the harmful effects of the Sun's ultraviolet radiation. The world's first global environmental treaty, the Montreal Protocol, came into force in January 1989, banning the use of such chemicals in a desperate attempt to repair the hole in the ozone layer unwittingly caused by the manufacture of Midgley's CFCs.

Unintended consequences

Human history overflows with unintended consequences. Many have involved the inadvertent introduction of species to alien habitats that then wreak havoc on native ecosystems.

The deadliest example in human history is the spread of modern humanity itself. Although native to Africa, thanks to *Homo sapiens*' control of fire and use of tools (see pages 184–5), our species has managed to migrate to every continent on the globe except Antarctica. One unintended consequence of

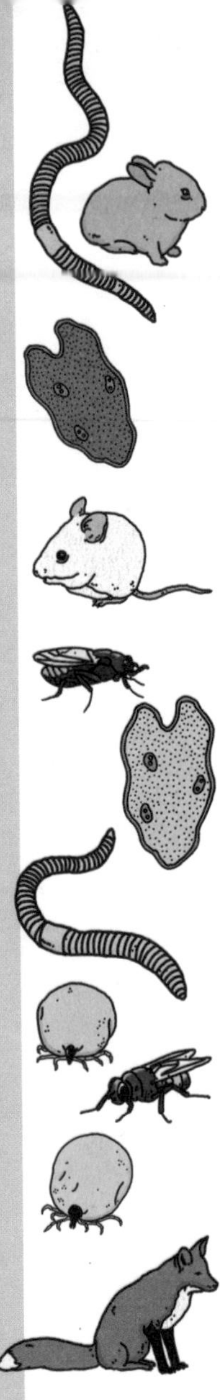

the radiation of farming people from Europe to the Americas was the spread of the smallpox virus, which decimated Native American populations in the sixteenth century (see page 26).

In the early 1950s, the brown tree snake (*Boiga irregularis*), native to Australia, Indonesia and Papua New Guinea, hitched a lift in the undercarriage of a US military aircraft, which then landed on the Pacific island of Guam. Lacking any natural predators there, the brown tree snake population has since exploded, numbering as many as 13,000 snakes per acre, and wiping out twelve of the island's eighteen native bird species. The snake has recently been found in the United States and Spain, having taken other illicit rides across the world. It is now listed as one of the world's most threatening invasive species.

Similar unforeseen consequences include the explosion of a type of seaweed in the Mediterranean called *Caulerpa taxifolia*, released into the sea following the clearing out of fish tanks at an aquarium in Monaco in 1984. This prolific weed has since colonized vast tracts of the sea floor, smothering habitats used as nurseries by marine species and threatening the health of the whole Mediterranean ecosystem.

Other notorious incidents include the release of Iberian rabbits in Australia (see page 332), the planting of eucalyptus species in Asia and Europe (see pages 265–6) and the progress of European earthworms across the Americas (see page 99). In each case, the human transportation of organisms from one environment to another has led to the unwitting triumph of certain alien species over others. Nowadays, thanks to human intervention, a lack of predation rather than biological fitness is often what matters most for a species to achieve biological success.

Feeding off man

Other species have thrived as a direct result of the growth of the human population. They range from the smallest forms of genetic replicators, existing inside the human body, to birds and mammals that profit from human waste.

Probiotic bacteria such as **Lactobacillus** (see page 386), which live inside the mammal digestive system, have thrived as a result of the explosion of human population numbers. Trillions of these bacteria feed on sugars inside our guts where they lower cholesterol levels, boosting immunity and reducing susceptibility to bowel cancer. Better human health increases the number of anaerobic habitats in which these bacteria thrive (see extremophiles, pages 30–31).

Others microbes are not so beneficial. *Borrelia burgdorferi*, a spirochaete bacterium spread by deer **ticks** (see page 384), causes Lyme disease, an illness with symptoms ranging from extreme fatigue to arthritis. It is currently spreading uncontrollably through human populations in the northern hemisphere. Other infections, although treatable, thrive in the wild owing to a lack of effective vaccines. They include *Plasmodium falciparum*, the malaria-causing protozoan (see page 135); *Trypanosoma brucei*, the cause of sleeping sickness (see page 140) and *Toxoplasma gondii*, a protozoan spread in the faeces of domestic cats which can cause pneumonia but resides, mostly in a benign condition, in as many as four billion people worldwide.

Ectoparasites are small creatures that live on the surface of animals and humans. Head-lice, body lice, fleas, flies, mites, bed bugs and ticks have enjoyed considerable success alongside the booming human population.

Some larger wild species have also done well. Populations of rats (see page 172) and mice have exploded over the last few thousand years by feeding on the left-overs of human agriculture and by hitching rides all over the world on every manner of conveyance from canoes and carts to trawlers and trains. They benefit from a shared fondness for the same types of food as humans. The modern habit of discarding edible waste suits them well. The profusion of rodents is also good news for birds such as **crows** (see page 378) and magpies that feed on carrion.

Some larger, more solitary creatures have prospered thanks to their excellence at playing hide and seek, skills honed over generations of natural selection in the wild. **Deer** (see page 372) and **red foxes** (see page 375) have proved their ability to survive despite being relentlessly hunted by man for sport. These creatures dominate habitats as far-flung as North America, Europe, Asia, North Africa and Australia. Like rats and squirrels, urban foxes have adapted to feeding on the copious scraps of the human world.

growing of cotton in Russia and eucalyptus in India (see pages 255 and 266–7). Greater affluence has allowed many people to switch to eating more meat, leading to an ever-increasing population of intensively farmed animals such as cattle, sheep, chickens and pigs – all requiring yet more crops to feed them and all adding to the world's spiralling levels of carbon dioxide (see page 225).

Tipping point

For at least the last 12,000 years humans have, like Midgley, excelled at focusing the centralized, rational, problem-solving skills of their larger-than-average brains on ways of trying to improve the survival prospects and material comforts of their species.

By almost any measure humanity has, to date, been spectacularly successful. The population is approaching seven billion today – with an awesome 211,000 more humans added worldwide each day. Longevity has increased, too, in many parts of the world so that the average European today lives until nearly eighty years of age, compared to just seventeen at a low point in *c.* AD 1350.

An almost superhuman effort is now required to ensure that the resources of the natural world can be tailored to support the human population.

The birth of organic chemistry in the mid nineteenth century gave humans the ability to mix their own cocktails for stimulating life in the form of artificial fertilizers, mass produced from 1916 onwards. As much as 40 per cent of the world's population is now dependent on crops that survive only thanks to this unnatural fix. Norin 10 wheat (see page 199) and IR8 rice (see page 233) are so-called 'miracle crops' genetically modified to produce ten times the normal yield. But these are crops that need constant watering, fertilizing and ongoing protection against infection to survive.

While climate-damaging fossil fuels increase atmospheric carbon dioxide, fresh water tables diminish owing to the unlimited use of automatic watering systems on everything from the manicured golf courses of America (see page 360) and the cultivation of roses in Kenya (see page 341) to the

Man versus nature

Human actions like these have happened because mankind and nature are locked in a battle for control of the operating system of life. What began with the first domestication of species *c.* 12,000 years ago has accelerated dramatically over the last 1,000 years with the globalization of crops (begun by Islamic traders and completed by European mariners) and the innovation of Western approaches to medicine and science.

A violent wrestling match has emerged between these two contrary systems. Artificial selection tailors a thin sliver of natural species to suit man, while natural selection champions the broadest possible biological diversity as an insurance policy against ecological wipe-out. This conflict has become the hallmark of our age, a period defined by geologists as the 'anthropocene'.

Natural selection, founded on sexual reproduction and inheritance, harbours no sentimentality for the survival of any one species over another. Diversity is its ultimate measure of success. Its simplicity lies in its self-organizing, self-correcting symmetry – infinitely adaptable, remarkably robust.

Artificial selection is fashioned out of the rational, centralized, problem-solving human mind, replete with its inevitable myriad unintended consequences. Modern civilizations abhor biodiversity. Species that enhance man's welfare are vigorously encouraged while those that threaten him are violently destroyed. Living things that look symmetrical, taste good, provide shelter, warmth, transport or other benefits are bred as monocultures that work for man. Cloning, grafting and selective breeding provide the reproductive plumbing

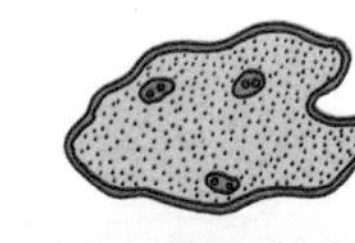

necessary to focus gene pools to ensure that it *isn't* the fittest that survive, just those most useful to humans.

Scientists, medics, geneticists and agriculturalists are humanity's foot-soldiers. The fate of modern mankind depends entirely on his ability to protect the precious, ubiquitous monocrops and carefully bred animals that can no longer live in the wild world.

Pitched against them are fast-reproducing microbes operating under the self-organizing laws of natural selection, rapidly adapting themselves by means of sexual reproduction in ways that circumvent man's genetically modified crops and disease-resistant cultivars.

Natural selection never rests. While vaccine-elusive viruses such as HIV/AIDS (see page 18), avian influenza (see page 16) and **herpes** (see page 388) threaten the human population, other infections keep up a constant attack on crops. Potatoes are under threat from the resurgent form of *Phytophthora infestans* (see page 51), for example, while the fungus *Fusarium oxysporum* will almost inevitably decimate the popular Cavendish variety of banana (see page 366).

Balancing act

It is easy to forget that most of the human conduct we take for granted in the modern world is part of this constant battle against natural selection. Think how normal it seems to apply weedkiller to garden grass, buy a bunch of cut roses from Kenya, drink coffee from sun-drenched South American plantations or for a doctor to prescribe a course of antibiotics to relieve a sore throat. These activities are all examples of humanity's war on nature.

It is a sobering thought that the law of unintended consequences, so graphically exemplified by Thomas Midgley, may also be applicable to traditional human heroes. Alexander Fleming (see page 314), Norman Borlaug (see page 199) and Fritz Haber (see page 198) respectively pioneered antibiotics, high-yielding wheat varieties and artificial fertilizer. All were innovations that helped prevent mass starvation and death by disease, but all had the unintended side-effect of creating an unprecedented boom in the human population which, as a result, has more than *tripled* in less than eighty years.

What's to be done? Could humanity somehow apply its rational mind to scaling back the number of people on the planet? If so, perhaps there is a way the two rival systems of artificial and natural selection could co-exist in a kind of evolutionary stable state?

Clear and sufficient boundaries would have to be drawn between the wild and cultivated worlds. Maybe each could benefit from the other in a symbiosis, a mutual collaboration as seems to have happened at the start of complex cellular life (see On Symbiosis, page 43), against the common enemies of climate change, or 'selfish' viral genes that devastate diversity.

Unfortunately, today debate about managing the human population is mostly taboo. Who dares stand up and say that the increase in human numbers from fewer than a billion to nearly seven billion in just 200 years is causing such a conflict between the modus operandi of nature and of man that the foundations of life itself are cracking? What nation, politician or religious leader will lead the charge in finding ways to *manage down* the level of the human population when everywhere people want to live longer and have the unlimited individual 'right' to produce as many children as they wish?

Current United Nations projections suggest numbers will continue to rise for at least the next fifty years, reaching between nine billion and twelve billion individuals before levelling out. How intense will man's war with nature have to be by then to support such a crowded Earth?

There is hope. The People's Republic of China has a policy of actively managing population numbers with its one-child-per-couple law, applied from 1979. It has reduced the rate of population growth from 15 per cent every five years in 1970 to 5 per cent every five years in 2005. Western governments with large welfare states are also finding that people are opting to have fewer children, provided they are guaranteed a pension in later life, thereby reducing the need for relying on children to provide an income and look after them in old age. The link between welfare state provision and population stability is already well established in Europe, where the population even declined marginally between 1995 and 2000 from a total of 727,361 to 726,586.

Colin Tudge in his excellent book *The Variety of Life* (2002) proposes an expansion of the welfare state in less affluent parts of the world, where population growth is most acute, as an incentive for people to choose not to have more than, say, two children. If properly managed, he suggests, such techniques could stabilize populations at current levels within about 1,000 years, and even see them fall to a more sustainable two billion *'a few centuries after that'*.

Could such a system be politically and economically engineered on a global scale? Perhaps it could become part of a similarly challenging climate change initiative in which national targets are set not just for limiting carbon emissions but also for reducing population levels. Living standards do not have to decrease in a world of declining populations in which existing resources can be shared out between fewer individuals. Is this a viable route out of the vicious cycle of poverty, famine, inequality and ecological catastrophe we find ourselves in today?

Nature, with its self-organizing systems of sexual reproduction, biodiversity and natural selection, has been successfully evolving for 3.7 billion years. During that time it has survived even the harshest environmental catastrophes such as the annihilation of up to 96 per cent of marine life on Earth (see pages 118–19). Humans, on the other hand, began the process of crafting evolution to suit their species soon after the end of the last ice age and then became seriously proficient at it during only the last 200 years – that's less than *one hundredth of a second to midnight* on the scale of Earth history when measured on a 24-hour clock.

Whatever approach we take, humility is key. In the world of artificial selection, man's survival will always, at some level, be dependent on nature. Conversely, in the world of natural selection, nature has absolutely no need for man.

1
2
3
4
5
6
7
Children per couple

Human fertility

How reproduction rates vary dramatically in different parts of the world, ranging from 0.9 in the Chinese administrative province of Macau, to 7.34 in Mali, West Africa.

Source: CIA World Factbook

Deer

FAMILY: CERVIDAE
SPECIES: CERVUS ELAPHUS
RANK: 55

A fleet-footed, half-venerated, half-hunted herbivore that has succeeded through the constant pressure of natural selection.

IT TAKES a very special type of large mammal to prosper in the wild in a world populated by nearly seven billion humans. The few species that have succeeded (see red fox, page 375) have done so only thanks to some especially powerful attributes acquired through natural selection. Deer are stealthy, herbivorous animals that currently boast healthy populations in every continent except uninhabitable Antarctica and Sub-Saharan Africa. The secrets of their present success were sown in the wild, millions of years ago.

Modern species of deer evolved from a common ancestor about twelve million years ago. The world then was full of wild predators including lions, lynxes, bears and wolves, just to name a few. The best survival strategy for these even-toed ungulates (ultimately related to camels, pigs, giraffes, sheep, goats and cattle) lay in speed and nimbleness. In such an environment, the slowest to flee were those most likely to be caught by hungry predators, leaving the fastest, fittest individuals to live on and reproduce. With the passage of time, natural selection therefore saw to it that deer genes were honed into making this creature one of nature's most powerful long-distance runners, capable of speeds exceeding fifty miles per hour and of scaling heights greater than two metres.

The other trade secret belongs to the unique set of disposable antlers that are the hallmark of male European red deer (*Cervus elaphus*), usually called stags. A highly polygamous society underpinned by violent rutting contests ensures that only the strongest, fittest males earn the right to mate with a harem of up to twenty different females. This pattern of wild behaviour has also focused the deer's gene pool in favour of size and strength. Every year, males grow a fresh set of antlers to display their current level of biological fitness. During the mating season males parade past each other comparing their body and antler sizes in a bid to determine

Stuck in a rut: two male red deer fight to determine which is the fittest to mate.

which will have the right to mate. In the event they cannot decide, they lower their antlers and resort to a violent duel, a method that invariably sorts out which of them is fittest. Over thousands of generations such rituals have resulted in a super-race of strong, swift creatures.

It was a doe that led the Franks across the river Vienne to victory.

A stag's speed and strength have been enshrined in myths and legends that have come to characterize the relationship between these wild animals and humans over the last few thousand years. An association with the mystical, god-like power of deer permeates many cultures. In the mythology of ancient Greece, the stag was sacred to Artemis, goddess of the hunt, while Heracles was tasked with capturing the Ceryneian Hind, with its golden antlers and hooves of brass, in the third of his twelve epic labours. It was said this sacred creature could outrun an arrow. In northern Europe a remarkable, intricately decorated silver Celtic cauldron, dated to the first century BC, depicts a stag looking across at a human deity seated with his legs crossed and antlers growing out of his head. Scholars still debate the origin and use of this receptacle, called the Gundestrup cauldron, found in a Danish peat bog in 1891, but they are in broad agreement that the antlered figure represents Cernunnos, Celtic Lord of the Hunt and god of the underworld.

The French nation owes a substantial debt to the magical powers of a female deer sent by God in answer to the prayers of Clovis I (AD 466–511), founder of the Franks, who was unable to find a way to lead his army across the river Vienne. A huge doe is said to have entered the river, showing Clovis and his men the best place to ford. As a result the Frankish army was able to attack the Visigoths, and their leader Alaric II, in 507 at the battle of Vouillé, leading to Clovis's capture of south-western France and Toulouse.

By medieval times stags' antlers had become prized by the landed gentry as trophies, symbolic of man's power over nature. They still grace the halls, dining and billiard rooms in many of the stately homes and palaces of Europe. These were not just a Western fancy. Honda Tadakatsu (1548–1610), a samurai general, never once wounded in fifty-five battles, was famous for his lucky helmet, which was adorned with deer antlers. Honda's headgear possibly also reflected the traditional Far Eastern belief in the power of the stag's horns as an aphrodisiac. The soft-velvet skin from a growing stag's horn, as well as the powder made from the crushed horns themselves, are still used in Chinese medicine today as a cure for impotence.

Musk deer (*Moschus moschiferus*), native to Tibet, China, Korea and Siberia, also have powerful sexual connotations. These deer have been hunted at least since the sixth century for the small gland positioned delicately between their genitals and tummy button. The world's most valuable perfumes are made from these glands, which are dried in the sun before being ground into powder. The resulting 'musk grain' is then dissolved in a tincture of alcohol to make perfume. By the beginning of the nineteenth century, musk grains from Tibetan deer were worth more than twice their weight in gold. Although synthetic substitutes are mostly used today, these deer are severely endangered because of over-hunting, despite a worldwide ban. A musk grain shipment of 700 kilograms was seized in Japan in 1987, representing the harvesting of more than 100,000 deer.

Cousins of these deer, belonging to the family Cervidae, have no such problems. Not only have these creatures benefited from their natural endowments of speed and size, but their worldwide mythological associations have gripped the human imagination – to their considerable advantage. The sika deer (*Cervus nippon*) is a Far Eastern species, closely related to the native red deer, which has been transported all over the world to populate the ornamental deer parks of kings and aristocrats. Thanks to their unrivalled jumping skills, feral populations have since become widely established. Consequently this species is now common almost everywhere, from Britain to New Zealand and from Morocco to the United States.

Native populations of red deer, elk, caribou, moose and reindeer – all of which belong to the Cervidae – thrive today thanks to the demise of most of their wild predators at the hands of man. Wolves, jackals, bears and big cats – aggressive carnivores that present a threat to people – have either been eliminated, domesticated or sidelined so that wild, well-adapted deer have few enemies left to check their growing numbers.

Most deer are unsuitable for transport, with the notable exception of the semi-domesticated reindeer herds of Arctic people such as the Sami and Nenets – otherwise how could Santa Claus conduct his annual global tour? Meanwhile, some countries still hunt deer for their meat. More than 150,000 are hunted each year in Sweden. In other places herds have been introduced as a source of income for landowners. Most venison consumed in the United States today comes from deer hunted on enclosed estates in New Zealand.

However, the vast majority live in the wild where the selective forces of nature jostle uneasily alongside the hazards of man's modern world. Today's deer 'predators' are mostly fast and metallic with dire consequences for all involved. More than 1.5 million car–deer collisions are reported in the United States each year, resulting in more than a billion dollars' worth of damage and the loss of 150 (human) lives. Meanwhile an estimated 13,000 moose have been killed in collisions with Norwegian trains since the year 2000. In other countries, such as Argentina, deer are listed as a highly invasive species owing to their soaring populations and voracious appetite for young seedlings.

Deer have a long, contradictory history with humans, with their status shifting variously from demi-god and ornamental marvel to prize game and vermin. Since neither life-form shows any natural limit in population size, the human–deer relationship is one potential flash-point where the ongoing conflict between the evolutionary and ecological powers of man versus nature may find future expression.

Red Fox

FAMILY: CANIDAE
SPECIES: VULPES VULPES
RANK: 65

A highly adaptable predator that has permeated human cultures all over the world and continues to prosper despite man.

WHAT'S THE BEST way to eat hedgehog? Most predators would think twice before assuming they were on to a tasty morsel when faced with a creature curled up into a tight defensive ball with razor-sharp spines pointing menacingly outwards. But not the fox, whose strategies for survival in the wild few other creatures on Earth can match. Calmly, it cocks its leg and squirts a jet of concentrated urine directly on to the hedgehog's back. As the animal recoils in shock its vulnerable underbelly is momentarily exposed, allowing the fox to go in for the kill.

Foxes are exceptional creatures. Over the last 2,500 years they have held their ground in the face of intense human encroachment by adapting, fleeing, double-crossing and manipulating their way to reproductive success. So great an impact have these creatures had on human culture that foxes have become worldwide icons in folklore, myths and legends from Europe to America, China, Korea and Japan.

Foxes are everywhere depicted as cunning. *Huli jing* are Chinese fox spirits that possess the bodies of beautiful young women in order to tempt and seduce married men. The Japanese fox spirit *Kitsune,* first recorded in AD 794, also assumes human forms. *Kitsunetsuki* refers literally to the state of being possessed by this fox spirit, which typically enters a young woman through her fingernails or breasts. The spirits bring about a state of madness that apparently causes its victims to run naked, shouting through the streets. Other symptoms include foaming at the mouth or eating

'*Kitsune*: The Fox's Wedding' from a series of prints by Tachibana Minko (*c.* 1770).

copious quantities of tofu. Until the early twentieth century, possession by fox spirits was a common diagnosis for mental illness in Japan. *Kumiho* is a Korean fox spirit that transforms itself into the identical likeness of a bride at a wedding. No one can spot the change until the wedding night, when she finally takes off her clothes.

The fox's association with deception and guile also permeates Western culture, from Ben Jonson's *Volpone* ('The Fox'), the story of a man who tricks his friends and relatives into thinking he is dying so that they will bring him expensive gifts, to Joel Chandler Harris's *Tales of the Wily Br'er ('Brother') Fox*. Calling someone 'foxy' or a 'vixen' today is to suggest slyness or sexual mischief. This animal's distinctive reputation in human culture stems from the fact that foxes have, without deliberate encouragement from humans, prospered over the last 2,000 years while other similar species have either become dependent (domesticated dogs) or are in danger of extinction (from the European lynx to the grey wolf).

By far the most prolific fox species (there are twelve true fox species in all) is the red fox (*Vulpes vulpes*). These creatures have easily the largest worldwide distribution of any member of the dog family (Canidae) with an estimated range of seventy million square kilometres, nearly half the total landmass of the Earth. It thrives across all northern and eastern hemisphere habitats, with the exception of Greenland and Ireland. Populations in Europe stand at a healthy one million, with 250,000 in the UK where they (nearly) all live in the wild.

The success of the red fox in the modern age is a triumph for natural selection. Without needing to be tamed as a pet, or farmed as food for human consumption, this creature has been spectacularly effective at adapting to a world dominated by people without condescending to live by their rules. Perhaps such a spirit of defiance helps explains why, for nearly 1,000 years, humans have tried their level best to hunt foxes to oblivion using packs of specially bred, well-trained domesticated foxhounds and servile horses (see page 246).

When William the Conqueror invaded Britain he brought with him the European habit of hunting wild animals with dogs. By the sixteenth century foxes were the most widely hunted targets (rather than deer or wolves). Against all the odds (at least thirty hounds in a typical pack take up the chase) foxes have outwitted, out-manoeuvred and outrun their adversaries and have survived in good numbers into the present day. Double-backing to confuse the hounds, diving into water to obfuscate the trail of their scent and running over long distances at speeds of more than thirty miles per hour are characteristics of creatures determined not to lose their independence in the wild.

When humans began their ostentatious programme of building towns and cities on 'greenfield' sites following World War I, foxes did not simply yield to the inevitable and flee into what was left of the woodland. They adapted by hiding in the day and feeding on the waste discarded by humans in urban areas at night. There are as many as 33,000 urban foxes in the UK (around 15 per cent of the overall population) that have migrated from their traditional woodland territories and now patrol leafy suburbs, raiding ponds for fish, bins for titbits and bird tables for seed. So flexible are the diets of these creatures that they can eat almost anything that lives, or has lived, from earthworms, which they feed to their pups, to snakes, rodents, rabbits, eggs, berries ... and, of course, chickens.

Like cats and humans, foxes are popularly thought to enjoy killing for pleasure. Witness a fox raiding a chicken coop and no bird is spared, even though its killing spree usually results in far more dead meat than a single fox can possibly consume in one sitting. However, what humans may project on to the fox as a penchant for killing for fun is more likely a wild instinct that causes it to kill whatever meat is available in case of scarce future supplies. One by one it takes the chickens in its mouth, burying them in different locations (a technique called scatter-caching) to minimize the chances of another creature discovering the location of a single, large hoard. Chicken owners who discover their dead animals apparently discarded by the fox interrupt this process, hence the common perception that foxes, like people, kill for the thrill.

Perhaps foxes have been able to cope with adapting to a human-dominated world without compromising their wild ways because they are, mostly, solitary creatures. Each individual is used to fending for itself, unlike dogs or wolves that hunt in packs for food. When myxomatosis struck the UK in 1953 (see page 333), decimating the rabbit population, foxes simply switched eating habits and hunted voles instead. In a fast-changing environment versatility, resourcefulness and independence of mind are characteristics favoured by natural selection, each generation becoming just a little more foxy than the last.

British fox-hunting has only managed to slaughter up to 25 per cent of the wild fox population even when the sport was legal (it was outlawed in Scotland in 2002 and England and Wales in 2005). So successful was the fox at maintaining its numbers – despite its persecution – that the British gentry exported these creatures to the Americas and Australia to give them a sufficiently well-matched adversary in the wild so they could continue their fox-hunting traditions overseas (see also rabbit, pages 331–2). US Presidents George Washington and Thomas Jefferson both owned packs of foxhounds. Since then the American fox has been displaced by the European red fox. Meanwhile, owing to the lack of natural predators in Australia, special fox eradication programmes have recently been established to try to bring back as many as thirty-one indigenous animal species from the brink of extinction.

While foxes have adapted to the human infestation of their natural habitats, their predators all over the world have suffered. The European lynx has been decimated owing to its penchant for rabbit (killed by myxomatosis), and the wild wolf has been both marginalized through a shortage of open spaces in which to hunt in packs and deliberately modified by humans into a subspecies that hardly resembles its ancestors either in looks or instincts (see dog, page 323). Golden eagles are now the single major natural threat to foxes in Britain, but these giant birds survive only in the Highlands of Scotland.

There is a new predator, though, which even foxes are unable to find a way of avoiding. Cars are thought to kill as many as 100,000 foxes on the roads of Britain each year, mostly at night. Natural selection has not yet taught foxes the highway code. Even though the arrival of these new perils is not endangering the entire species, it is an effective unforeseen consequence of modern motoring to control fox populations in ways that used to be accomplished by hunting with hounds. There are few better illustrations of the often cataclysmic collision between the systems of natural and artificial evolution than the sight of a bloody red fox splayed across a country road with its mechanical killer speeding off into the dark, oblivious of its deed.

Crow

FAMILY: CORVIDAE
SPECIES: CORVUS BRACHYRHYNCHOS
RANK: 64

A global survivor that has recently taken up residence in towns and cities but which could, in the future, inherit the Earth ...

COULD OTHER animals one day become more intelligent than humans? The logic, according to the system of natural selection, isn't hard to grasp. Wild creatures best able to survive a world populated by seven billion humans will be those endowed with the highest levels of wit, brains and cunning that enable them to turn today's rapidly changing environments to their advantage. While it took humans about two million years to develop into intelligent beings with their own art, language and culture (see page 184), the challenges of living with humans today may mean other creatures develop similar ways of life a lot more quickly. Foxes represent such species on the land (see page 375) and birds such as crows lead the way among animals that fly.

Humans and crows evolved alongside each other in antiquity. Shamans, holy men from the hunter-gather world, painted images of crows in the caves where they contacted the spirit world. Some of these images, dating back 30,000 years, still remain in Lascaux, southern France. Ravens – larger, solitary members of the crow family – appear in legends contained in texts such as the *Epic of Gilgamesh* and the Jewish Torah, where their navigational wits were used by humans to divine for land. When Noah released a raven that failed to return to his Ark, he knew in which direction land lay. In the same way, eleventh-century maritime Vikings are said to have discovered Iceland. By the later Roman Empire, the association with crows and intelligence was a well-established fact of life. One of Greek writer Aesop's most famous fables tells how a crow found he was unable to reach the water in a pitcher. He solved the problem by dropping pebbles into the pitcher one by one until the water level rose sufficiently so he could quench his thirst. The moral, of course, is that intelligence, rather than brute force, is sometimes the surest means of survival.

Dramatic changes in the natural habitat since the end of the Roman Empire haven't worried these creatures. On the contrary, like foxes, crows have adapted and prospered as a result of human population growth. Until medieval times their penchant for eating dead flesh was highly venerated in cultures throughout the world from Europe and Native America to the Far East and aboriginal Australia, where totems were frequently established in their honour. Excarnation is the practice of leaving human remains in the wild to be eaten by birds and other scavengers. Crows and vultures, which picked at dead human flesh, were regarded by Neolithic pastoral people in Europe and the Near East as intermediaries with the spirit world, part of nature's endless cycle of birth, death and new life. Until the 1950s it was a tradition in Tibet to place the dissected remains of a loved one on to a ceremonial altar for crows and ravens to carry off into the next life.

By medieval times, however, crows had become associated in the popular mind with the spread of disease, witchcraft and macabre death. Human bodies were frequently left to decay across the terrain of Europe and around the Mediterranean as Islamic and Christian civilizations pitched headlong into battle for control of valuable overland trading routes, and crows gorged on these victims of warfare. Plague, Black Death and famine afflicted Europe from the fourteenth to nineteenth centuries. They reinforced the association of these creatures with bad luck. Crows are an integral part of the opening scene involving witches depicted in William Shakespeare's great Scottish tragedy *Macbeth*.

About forty members of the crow family have been identified, ranging from the larger ravens to the smaller jackdaws and rooks. They are thought to have evolved from a common ancestor in Asia approximately five million years ago. As Charles Darwin deduced when he visited the Galapagos Islands, identical creatures when distributed over different, isolated habitats eventually develop small changes in anatomy and instinct that ultimately lead to the creation of related, but genetically distinct, species. When the Bering Strait was a swathe of coniferous forest *c.* two million years ago, crows are thought to have migrated across to the Americas. The common crow, *Corvus brachyrhynchos*, has since become one of that continent's most successful avian species.

Survival for this native American crow has come from adapting its traditional meat-eating diet to the consumption of grain, in response to human farmers turning vast tracts of wild terrain into cultivated fields and pastures. Over the last 2,000 years crows, like foxes, have learned to eat almost anything. One recent survey found as many as 650 different types of food in the bellies of 2,118 crows. Their diet ranged from eggs, snakes and worms to noodles, chips and even fluorescent orange cheesy puffs.

Haphazard attempts at destruction by humans have ended in abject failure, from the relatively benign worldwide craze for scarecrows to the more pernicious crow hunters of Oklahoma who, between 1934 and 1943, blasted 127 rook roosts with dynamite, killing an estimated 3.7 million birds. Despite such slaughter there was no noticeable reduction in either crow populations or agricultural damage.

As a result of urbanization in the late nineteenth and twentieth centuries, crows have adapted to living in every location from chimney stacks to the tops of advertising billboards. Walk down a typical city street today in most parts of the world and – apart from a few ornamental trees and perhaps a patch of grass – the only signs of nature will be crows. These are birds that have learned to live off mankind's most prolific by-product, urban refuse.

Reproductive success, flexible diet and adaptability of habitat all help a wild animal prosper in the modern world. But there is more to the crow's success. As naturalists spend time observing how crows live, the traditional perception of birds as dumb, instinctive creatures compared with mammals such as whales (see page 165), elephants (see page 168) or rats (see page 172) is undergoing significant revision. Bird brains, it is being discovered, may be wired very differently from those of mammals, but some species have at least an equal capacity for intelligence, communication and culture. The crow's brain, relative to its body, is among the largest in the animal kingdom, equivalent in size to that of a chimpanzee. They are even referred to by some experts today as flying monkeys.

Recently observed behaviour certainly supports that view. Crows, like humans and some primates, habitually manufacture tools to make feeding easier. American crows have been seen to craft splints out of wood to skewer insects lodged in crevices. Meanwhile, New Caledonian crows, challenged by problem-solving exercises in a research laboratory, were found to be capable of bending wires into hooks to retrieve a small bucket of food. On another occasion, a raven in Maine, USA, untwisted ten separate wires, one by one, to open a squirrel-proof feeder.

Some ornithologists believe crows exhibit their own distinctive culture by acquiring and passing on knowledge, a skill normally credited only to humans. The recently observed antics of carrion crows living in Sendai, northern Japan, vividly illustrate this point. These crows wait for cars to stop at traffic lights, position nuts in front of the car wheels, and return to eat the kernels once the cars have moved off and crushed the shells. A similar pattern of behaviour has been found to have spread out over a range of several miles during the last few years, demonstrating that these crows can pass on acquired knowledge through cultural exchange.

Are crows that treat cars as nut-crackers examples of creatures developing superior intelligence as a result of living in a modern world dominated by humans? Perhaps natural selection is now acting on crows, foxes and other non-human species to sharpen their intelligence, honing their survival skills in ways that may one day rival the ingenuity of mankind. A speculative zoological study on the possible course of evolution in a world 'after man' lists crows as evidence that: '*... intelligence as high as man's may evolve once more ... After all, mammals were scurrying about the feet of dinosaurs for a good hundred million years before they came to anything ...*'

Birds, like mammals, survived the catastrophe that wiped out the dinosaurs (see page 126). Who knows? Crows, once venerated by humans but now mostly regarded as scavenging menaces, could be man's intellectual successors in the event of another extraterrestrial strike.

Fruit Fly

ORDER: DIPTERA
SPECIES: DROSOPHILA MELANOGASTER
RANK: 15

An ancient way of life that has prospered with the rise of people and farming, morphing human history along the way.

IF A NOBEL PRIZE for medicine could be awarded to a non-human species then a serious contender would have to be the common fruit fly. Thomas Morgan, an American Nobel Prize winner, was able to verify nineteenth-century monk Gregor Mendel's theory about genetics and inheritance only thanks to staring into the whites of a mutant fruit fly's eyes in 1910.

In the year 1900, following the rediscovery of Mendel's pioneering work on pea plants, a race began between European and American scientists to discover the precise mechanism by which physical (and behavioural) features are passed on from one generation of living things to the next. The mechanics of inheritance had fascinated biologists even before Charles Darwin proposed his theory of evolution by natural selection in 1859.

Morgan, professor of experimental zoology at Columbia University in Ohio, had the inspiration to choose the fruit fly (*Drosophila melanogaster*) as a model species on which to base his research. This fly had a host of features that made it ideal for studying genetics. It was fast to breed, quick to incubate, easy to observe, a good size to handle and, best of all, because females are significantly larger than males, it was easy to distinguish one gender from the other – an important factor when embarking on a programme of selective breeding.

Milk bottles were additional aids in Morgan's experiments, carried out inside his top-floor 'Fly Room', beginning in 1908. Working with his team of students, Morgan attempted to identify individual flies with some form of observable mutation. By breeding a mutant fly with an apparently normal partner he hoped to see how the mutation manifested itself in future generations. In 1910 Morgan struck the geneticist's equivalent of gold – a mutant male fly with white eyes ...

Into the bottle went this precious creature, accompanied by a single female with standard-issue red eyes. Once nature had taken its course, Morgan could trace the pattern of the white eyes in the mutant's family tree as the gene cascaded down several generations. Before long, Morgan and his students were tracing all kinds of different mutations – genes that gave flies pale yellow bodies, wrinkled hairs or short wings, each time confirming Mendel's theories concerning the pattern of genetic inheritance. Morgan's book *The Mechanism of Mendelian Heredity*, published in 1915, explained how genes responsible for different physiological traits were placed in fixed locations on sexually transferred chromosomes. At last biologists had a concrete system of inheritance to underpin Darwin's theory. What's more, Morgan's flies earned him the Nobel Prize for Medicine in 1933 for identifying chromosomes as the vector for the inheritance of genes.

Thomas Morgan studying fruit flies under a microscope in his US laboratory.

Morgan's fruit-fly revelations had other consequences. Genetic research like this led to the revival of the concept of selective breeding in humans (practised in ancient Sparta), placing it on a firm scientific footing. An increasing number of American politicians, scientists and commentators began calling for a eugenics programme to purify the human race of its own 'mutant' genes. To what extent the fruit fly can be blamed for the launch of such a programme in Nazi Germany is debatable, but it is a plausible unintended consequence of Morgan's rational research.

Another American scientist, Seymour Benzer, followed Morgan's work, using fruit flies to track the inheritance of behaviours such as sleep patterns, cognitive skills and sexual courtship routines. His 'white-eyed moment' struck in 1971 when he found a fruit fly with an irregular sleeping pattern, which he traced, through selective breeding, to a specific mutation in the fruit-fly gene pool. Similar research in the late 1980s led to the identification of a fruit-fly gene for long-term memory.

So what? What's the big deal about scientists being able to spot genes in fruit flies responsible for white eyes, wrinkly legs, sleeplessness or a photographic memory? The true significance was revealed when it was discovered that humans and flies share many of the same genes. These include the crucial genes responsible for embryonic body design (see Hox genes, page 61) as well as many that cause hereditary disease. Up to 75 per cent of known human disease genes are similar or identical to those found in fruit flies, allowing scientists to explore potential treatments for human afflictions such as Alzheimer's, Huntington's and Parkinson's disease simply by experimenting on flies. Such research has led to yet more scientists winning Nobel Prizes, thanks to their work on easy-to-propagate fruit flies.

If fruit flies sound like an unlikely source of clues leading to cures for disease, then maggots (fly larvae) as a way of combating antibiotic-resistant bacteria such as MRSA will probably sound even more far-fetched. However, before the development of antibiotics in the mid 1940s (see page 314) one of the most successful treatments for infected open wounds was the deliberate placement of maggots on diseased flesh.

Maggot biotherapy was originally deployed by the Aborigines in Australia and the Mayans in Mexico, yet the practice was rediscovered in the West only during Napoleonic times and used later during the American Civil War (1861–5). Confederate medic Dr J. F. Zacharias wrote about the miraculous healing powers of maggots, which *'in a single day would clean a wound much better than the agents we had at our command ...*' During World War I it was found that if the open wounds of soldiers left on the battlefield became infested with maggots they remained infection-free, typically healing within six weeks.

Live maggots kill bacteria by secreting their own antibiotic juice. They also feed solely on dead tissue, leaving healthy flesh intact. So effective are maggots at killing bacteria that larvae from the greenbottle fly species *Phaenicia sericata* are now being used in place of penicillin and other antibiotics to fight modern resistant bacterial strains such as MRSA. Maggot medicine, now in use in as many as twenty countries, received approval from US and UK medical authorities as recently as 2004 as an alternative to antibiotics (see also use of viruses as antibiotic treatments, pages 43–4). Maggot-laced dressings, despite being a little tickly for the patient, have one other major advantage over antibiotic drugs: they do not encourage the selection of mutant antibiotic-resistant bacterial strains.

Positive feelings about the impact of everyday flies, past and present, go against normal human sensibilities. Since Biblical times, various cultures have regarded flies as evasive insects that congregate only around excrement and rotting flesh. Often they have been depicted as incarnations of death and the devil. Beelzebub, the Lord of the Flies, took his place alongside Lucifer and Leviathan next to Satan in a seventeenth-century French Catholic witch-hunter's depiction of Hell.

Mythological omens in the minds of many turned into grim reality shortly before the turn of the twentieth century. During the Spanish–Mexican War (1898) it was confirmed by a US government report that flies were the most likely transmission route for the typhoid bacterium, a major cause of death of American soldiers in the four-month-long conflict (see page 401). Today more than one hundred diseases are known to be transmitted by flies to humans. Their maggots hatch from eggs laid in animal dung, itself infested with pathogenic bacteria, viruses, protozoa and other disease-causing agents. The maggots pupate and metamorphose in adults, which transfer these disease organisms to humans when they walk on their food, or regurgitate digestive saliva on it.

One type of fly (*Muscina stabulans*) uses its antennae to detect buried bodies (human or animal) on top of which it lays its eggs. Once hatched, its larvae instinctively burrow down and invade the corpse, feeding on the rotting flesh. Forensic scientists sometimes use the presence (or absence) of maggots like these to determine the time and location of death.

Human population booms and the conversion of land into pasture for dung-dolloping farm-animals have ensured that everyday flies, of which there are thought to be about 240,000 species, continue to thrive all over the world. Their evolution since the age of dinosaurs – amber fossils containing trapped flies have been found dating back to the mid Triassic Period – continues unabated today. While bloodsuckers such as mosquitoes (see page 134) and the tsetse fly (see page 140) have had their own specific impact, the Nobel-Prize-winning fruit fly *Drosophila*, adopted by researchers in their dedicated fly rooms, earns its rightful place alongside those species that have most changed the world. It qualifies not just by virtue of the genetic insights it has provided, but also for its duplicitous role as both transmitter of and cure for diseases in humans.

Tick

FAMILY: IXODIDAE
SPECIES: IXODES SCAPULARIS
RANK: 45

A miniature disease-spreading arthropod that used to feed on dinosaur blood and is thriving today owing to climate change.

IT IS TEMPTING to to think that the most severe effects of global warming are still some way off – at least on a human timescale. Perhaps it is inevitable that at some point there will be no polar ice caps and sea levels will have risen by between five and twenty-five metres. But most scientists predict that this scenario cannot realistically happen until towards the end of the twenty-first century at the earliest. And extreme events such as the imminent melting of the Wilkins Ice Shelf, an ice sheet the size of Jamaica which is currently breaking off Antarctica, although sounding dramatic, don't really affect sea levels since most of this ice is already floating in water.

Unfortunately there is another, much more urgent reason why global warming matters today. Small bloodsucking arthropods called ticks are highly sensitive to changes in atmospheric temperature. Because of global warming and the fact that winters all over the world are getting milder, tick numbers are currently booming and their range is extending northwards. These insects harbour a number of dangerous diseases that are now sweeping through the human populations of North America, Europe and Asia on an epidemic scale.

Many people are ignorant of the hazards of tick bites. It was only relatively recently that Swiss-born American scientist Willy Burgdorfer first identified the type of bacterium that causes Lyme disease, an infection transmitted by ticks to humans. This pathogen, discovered in 1982 and named *Borrelia burgdorferi* in honour of its discoverer, is a missile-shaped spirochaete that swims around in the bloodstream propelled by a coil of twisting tail-like flagella. These Prokaryotic organisms, related to the very earliest living cells, hide away from the oxygenated atmosphere (see page 46) by seeking refuge in the bodies of multicellular creatures be they reptiles, birds, mammals, humans – or ticks.

Since Burgdorfer's discovery, a host of frequently recurring symptoms such as fever, headache, fatigue, depression and arthritis have been diagnosed as Lyme disease. If left untreated the spirochaetes reproduce in the soft tissues, sometimes causing irreversible damage to eyes, heart and brain. Although few individuals are known to have died from this disease, the quality of life in many sufferers has been severely compromised. Thousands of new cases of Lyme disease are being diagnosed each year in North America and Europe, making this infection one of the fastest emerging illnesses in the world.

Ticks are arachnids (the class that includes spiders), arthropods that evolved more than a hundred million years ago, initially feeding opportunistically on the blood of dinosaurs. They thrived in the increasingly grassy world that emerged in the last fifty million years with its vast populations of grazing mammals. Different tick species have different animal hosts, the most popular today being deer (see page 372) and sheep (see page 218). After several days gorging on the blood of an animal, the tick loosens its grip and falls into the long grass where the female lays up to 10,000 eggs. These hatch in the springtime, waiting for an unsuspecting animal to brush past, allowing the baby tick to latch on and gorge for itself. Disease-causing pathogens pass from the tick's saliva into the bloodstream of its host.

Lyme disease is not the only potential check on human wellbeing transported by hard-bodied arthropods such as *Ixodes scapularis*. Other types of bacteria, viruses and protozoa are similarly well served by these tiny bloodsucking vectors. Rocky mountain spotted fever is an American disease spread by ticks that range from Peru, throughout the United States and into Canada. The disease is notoriously hard to diagnose and if left untreated can be fatal in up to 5 per cent of cases. *Rickettsia rickettsii*, the offending bacterium, lives in the male tick's semen and is passed on to the female during copulation. The bacterium then infects her eggs. Once hatched, whatever creature the baby ticks bite also becomes infected by the disease, making the tick population a persistent bacterial reservoir.

Another disease, tick-born encephalitis (TBE), is a viral pathogen that attacks the central nervous system. Although preventable by vaccination, TBE can cause crippling neurological damage in unprotected people. Up to 20 per cent of those infected are left with lifelong complications, sometimes losing their ability to walk. As many as 12,000 cases are reported in Europe and Russia each year.

The impact of these hard-to-diagnose, tricky-to-treat diseases is only just beginning to be felt. Ticks die off in harsh winter climates, but owing to a succession of mild winters, their range is increasing, especially in highly populated areas of the northern hemisphere. People living or holidaying on the 220,000 islands off the east coast of Sweden have to endure one of the most tick-infested environments in the world. Since records of viral tick-borne infections began in 2004, cases have increased from 224 to 584 in just four years, despite the increasing availability of vaccines. Recent research suggests these infections are now spreading to western parts of the country.

'Boat clinics' visit the islands offering free vaccinations against TBE, and holidaymakers are told to be far more concerned about the hazards of tick bites than of sunburn. Meanwhile, a third of Swedish ticks also carry Lyme disease (which, although treatable, has no vaccine) and cause infections in as many as 10,000 people a year – a big number in a country with a population of just nine million.

My eldest daughter contracted Lyme disease when she was five years old. The tick must have brushed on to her from the long grass at the end of our garden. We knew nothing about the hazards of tick bites at the time, but when the tell-tale bulls-eye rash appeared it was clear that spirochaete infection was a possibility. Several blood tests later and the diagnosis was confirmed. Fortunately catching it early on meant an eight-week course of antibiotics stopped these spiralling bacteria in their tracks.

Local doctors were dumbfounded. Incidents of Lyme disease in south-east England were almost unheard of. Not any more. As the viable habitat for tiny bloodsucking creatures creeps ever northwards, people who live in parts of the world unused to diseases traditionally called 'tropical' will have to reassess the risks of walking through innocent-looking countryside or woodland. The arrival of disease-infected ticks is among the early-warning signs of a very different world, soon to come.

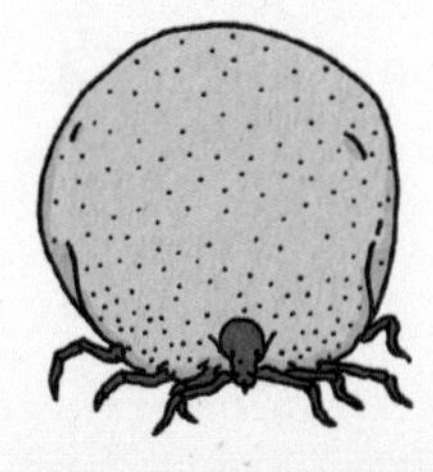

Lactobacillus

CLASS: BACILLI
GENUS: LACTOBACILLUS
RANK: 5

Gutsy bacteria that have, in their own small way, profoundly boosted the fortunes of man.

ASTRONOMERS like to think BIG. While the number of galaxies in the observable universe is thought to be in the region of about a hundred billion, the number of stars (like our Sun) in our galaxy, the Milky Way, is at least twice that number. Actually, those who enjoy *really big* numbers are better served by looking inwards rather than outwards. Take the number of living cells in an average human body – approximately ten trillion. Even more spectacular is the number of bacteria that live inside the human gut – *ten times as many as there are cells in the body* ... In terms of discrete living components, you and I are ten times more bacterium than human. It is only reasonable that the historic contributions of such organisms should be well understood.

Perhaps it is a by-product of Louis Pasteur's revelation of the role of bacteria in disease or Alexander Fleming's discovery of antibiotics (see page 314) that most people today associate bacteria with a negative impact on humanity. That is unfair – witness the contribution of cyanobacteria (see page 34), which gave birth to atmospheric oxygen, or *Rhizobia*, which fix nitrogen in the air to nourish plants (see page 40). Indeed the evolution of all higher forms of life is, essentially, a by-product of bacterial processes. The same as goes on inside our guts night and day.

The huge growth in the human population over the last 2,000 years – along with its communities of billions of sheep (see page 218), cattle (see page 222), pigs (see page 215) and other farm animals – provided a huge boost for bacteria that survive only in oxygen-free environments. People (and animals) are excellent hosts for such bacteria. Not only do our intestines provide a safe, oxygen-free habitat for them but they also get very well fed. It should come as little surprise then that many bacteria that have made people and animals their home also seek to improve their adopted host's health.

Such were the thoughts of Ilya Ilyich Mechnikov (1845–1916), a Russian microbiologist, best remembered for discovering the process of phagocytosis in the human immune system (see pages 49–50), for which he won a Nobel Prize in 1908. During his research he began to suspect that certain bacteria that live inside the intestines are responsible for human fitness and longevity. He was especially intrigued by a genus of bacterium called *Lactobacillus*, which converts lactose (a sugar found in dairy products) into lactic acid. Until the end of his life, Mechnikov was convinced these micro-organisms held the key to healthy living – so he wrote a book about it called *The Prolongation of Life: Optimistic Studies*.

Minoru Shirota (1899–1982), a Japanese scientist, picked up on Mechnikov's ideas a few years after the Russian's death. During the 1930s he experimented by mixing a new strain of *Lactobacillus* with skimmed milk to make a probiotic drink that he called Yakult. His aim was to stimulate the cultivation of healthy bacteria inside the gut to prevent infection and boost the immune system – vital if Japanese children of the time were to avoid malnutrition and disease. When Yakult came on to the market in 1935, the drink was an instant hit, especially in Japan and Taiwan. It is still sold in more than thirty countries today.

Rod-like *Lactobacillus bulgaricus*, the strain found in probiotic bacteria, which secretes lactic acid, inhibiting the growth of pathogens.

Strains of *Lactobacillus* that live inside the colon perform at least two functions that have been of great assistance to humans throughout history. First they help compensate for lactose intolerance, a genetic deficiency in many people. *Lactobacillus* is responsible for the fermentation process that turns milk into cheese, breaking down the sugars in milk (lactase) into digestible lactic acid. This process continues inside the human gut, helping humans tolerate a milk-based diet. Over time, such symbiotic assistance has become incorporated into the genomes of many people in milk-drinking cultures (see page 195).

A second benefit of *Lactobacillus* is that the process of creating lactic acid protects against less beneficial bacteria and other microbial pathogens (viruses, protozoa or fungi) that cannot thrive in acidic environments. The presence of lactic-acid-producing *Lactobacillus* is therefore one of the body's defences against infections spreading via the gut. It is also what stops fermented cheese from going off quickly. Indeed, were it not for the natural preservative effects of a strain of bacterium called *Lactobacillus plantarum*, which occurs naturally in plants and fruits, merchants in ancient Greece would never have been able to trade their precious crops of olives around the Mediterranean without them turning sour in the sun (see page 210).

The impact of bacteria on human history is woefully underwritten – healthy bacteria particularly. Pathogenic diseases caused by other types of bacterium (such as typhus) have left far more tangible evidence in their wake than the therapeutic benefits of probiotic good health. Yet boosts to immunity, higher acidity levels, lactose tolerance and food preservation make an impressive list of benefits. It is ironic to think that these bacteria may have been just as instrumental in supporting the massive rise in the human population as Alexander Fleming's bacteria-snuffing penicillin.

Herpes

FAMILY: HERPESVIRIDAE
SPECIES: HERPES SIMPLEXVIRUS
RANK: 47

A highly infectious genetic agent that has colonized most of the human population and remains an elusive, persistent threat.

VIRUSES HAVE had a fundamental yet little-understood impact on the history of life on Earth and on humanity itself, as the decoding of the human genome has shown (see pages 13–14). Despite the eradication of smallpox (see page 27) or the present containment of avian influenza (see page 17), viral diseases are continually reshaping the lives of countless individuals, nations and cultures.

It is therefore appropriate to end our survey of one hundred species that have changed the world with a virus family that is ubiquitous in all humanity – past and present. The name 'Herpes' comes from the Greek *erpis*, meaning 'to crawl', so-called by the father of European medicine Hippocrates (460–370 BC). It was a good name for a disease in which labial blisters and cold sores creep across the surface of infected human flesh.

Eight types of herpes virus infect humans, ranging from chickenpox and shingles to glandular fever (Epstein-Barr virus). The two most significant are the closely related *Herpes simplexvirus* (HSV) 1 and 2. The first lives inside the mouth, causing cold sores around the lips, while the second causes similar blistering below the waistline, to the genitalia. These two viruses are not limited to any particular region, race or gender. HSV 1 – the oral form – is estimated to be carried by a staggering 70 per cent of the human population – that's more than *four billion* individuals.

Although not all infected people develop symptoms, each one is a carrier. As with HIV/AIDS (see page 18), humans infected with herpes carry the virus for life. Normally, symptoms recede after several weeks, and the virus retreats into the sensory ganglia in tissues around the ear, where it lies in a latent state, unable to be detected by the human immune system. Occasionally (for reasons not yet fully understood) it rears back into life, hijacking cells and reproducing its own viral matter to be spread through saliva, mucus or pus.

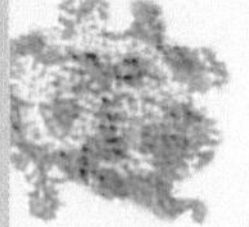

While most humans harbour herpes in a latent form, the infection does not generally appear to be fatal (except occasionally in some individuals where it can attack the central nervous system causing encephalitis). However, if the human herpes virus happens to skip across the species barrier to infect, say, a gibbon, it is almost always lethal. Similarly, when the Simian Herpes B virus, common in macaque monkeys, infects a human there is a 70 per cent chance of death.

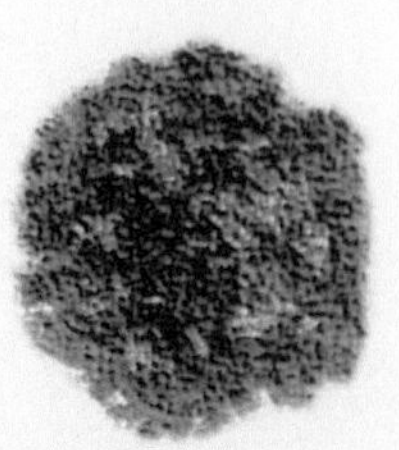

Herpes is a typical example of how viruses can establish a kind of evolutionary equilibrium with their host species. But if the host changes, the consequences for the evolution of other species can be catastrophic. It is even thought by some experts that a cause of the demise of the Neanderthal species of humans, the last to survive apart from *Homo sapiens*, was the leap of the herpes virus across the species barrier from *Homo sapiens* to *Homo neanderthalensis*. Such an interpretation would explain the selective demise of that single species while leaving *H. sapiens* unharmed. This theory is supported by the fact that there is so little evidence of interbreeding between the *H. sapiens* and *H. neanderthalensis* genomes.

While the spread of oral or genital herpes throughout the human population is not presently regarded as a major health threat, there are ominous signs that this irremovable, incurable virus may carry a nasty sting in its tail. Recent research suggests that the exponential rise in Alzheimer's disease (or senile dementia) may be due to HSV 1 and 2 attacking people who have reduced immunity in old age. When the virus wakes up, as it often does, it is able to penetrate the central nervous system and enter the brain.

Here it wreaks havoc, damaging brain cells, which degenerate, leading to mood swings, memory loss, a breakdown of communication skills, disorientation and eventually death. Alzheimer's disease (first identified in 1906 by German psychiatrist Alois Alzheimer) is currently the most expensive disease to treat in the developed world. The annual costs of care are estimated to run to $100 billion in the United States alone. Approximately twenty-six million people suffer dementia worldwide. Owing to the world's ageing population the number is expected to rise dramatically by 2050.

Herpes simplex about to latch on to the surface of a soon-to-be-infected cell.

There are no known strategies for vaccinating against persistent viruses such as herpes and HIV. Such viruses represent the genuine front-line in humanity's struggle to wrest control of the evolutionary process for its own ends – whether for food production (see page 191), material wealth (see page 237), longevity and recreation (see page 268), companionship (see page 318) or in the pursuit of beauty (see page 336).

Another way of picturing this contest is to travel full circle, back to the beginning of life itself. All life, ever since its original incarnation roughly four billion years ago, can be viewed as an ongoing struggle for evolutionary supremacy between loose strands of genetic code (viruses) and cellular forms of life (bacteria). Both have continued their quest for immortality through higher forms of life from the leafy tissues of plants to the bloody bodies of creatures. As part of this conflict, elaborate survival strategies have evolved, such as the biodiversity of sexual reproduction (see page 47) or the gene-splicing systems mastered by genomic retroviruses (see page 14).

Humanity's attempt over the last 12,000 years to craft nature to suit its purposes is superimposed uneasily on top of this age-old struggle. Exposure to viruses may therefore present the greatest threat to mankind and its precariously balanced artificial construct. As humans demonstrated so effectively with the release of myxomatosis in 1950, a virus can systematically all but wipe out a single species, while leaving an almost identical genetic line unscathed (see page 332).

What if such an event were to strike humanity? Would a world without people return to the state of nature as it was four million years ago before the first *Australopithecus* pivoted up on to two feet? Or would some other form of intelligent life eventually evolve and similarly try to superimpose its own rational way of thinking on to the self-organizing systems of nature? Who knows what on Earth would evolve next?

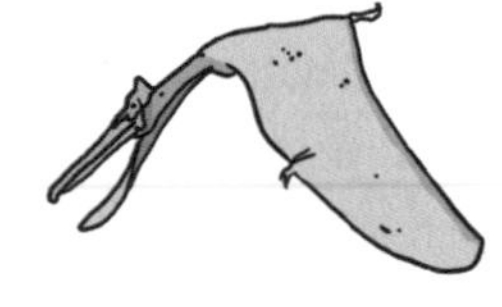

The Ladder of Life

The top 100 species ranked in order of overall impact

WHICH LIVING THINGS within our survey of '100 species that changed the world' have had most overall impact? The following table presents one interpretation. Five simple (but entirely arbitrary) criteria, worth 100 points each, have been devised to assess the impact of each selected species.

Evolutionary impact

An example of a species that has done well according to this criteria is cyanobacteria, which ejected oxygen into the Earth's atmosphere, poisoning many other bacteria but provoking the rise of higher forms of life. Modern humans have also had a big evolutionary impact owing to their deployment of artificial selection. Corals in the seas and oaks on land provide habitats that have promoted biodiversity. Meanwhile *Archaeopteryx*, thought to have been the pioneer of feathered flight, is an example of significant evolutionary innovation.

Impact on human history

Winners according to this criteria include *Lactobacillus*, bacteria that thrive in the anaerobic environments of mammal stomachs. Domesticated animal species (horses, dogs, sheep) and crops (cotton, maize, rice and potatoes) also fare well. Poor performers include those that became extinct before the arrival of humanity (such as *Tiktaalik*) or have never been pliable to domestication (such as slime mould and dragonflies).

Environmental impact

This assesses the impact of living things on the land, sea and sky. Species that have done well according to this measure include *Rhizobia* (nitrogen-fixing bacteria that fertilize the soil), algae (absorbs CO_2 and emits O_2), cows (which emit methane) and cotton (responsible for drying out the Aral Sea). Species with little discernible contribution include bats, olives, Rhyniophytes, HIV/AIDS and potyvirus.

Global reach

This score is determined from the extent of a species' range when at its height. Those that score well have dominated the land (such as *Lystrosaurus*, smallpox, earthworm), the air (*Quetzalcoatlus*) or the seas (sperm whale). Those that fare less well are local niche growers such as the durian and vanilla.

Longevity

Even extinct species such as *Lepidodendron* can score highly (150 million years compares very well against the 160,000 years of *Homo sapiens*), whereas more recent-to-evolve species – especially those domesticated by and dependent on humans for their survival – fare worse.

The top spot ...

The humble earthworm takes pride of place in the top 100 with 431/500 points, even though it doesn't come top in any single category. It is a suitable winner, being described by Charles Darwin in memorably glowing terms: '*It may be doubted whether there are many other animals which have played so important a part in the history of the world, as have these lowly organized creatures.*' (See page 97.) Agriculturally based human civilizations could never have emerged without the constant recycling of these subterranean wrigglers, which also rank high up on the list of nature's most robust, enduring and environmentally beneficial creatures.

What about humans?

Our species, *Homo sapiens*, comes sixth, and is the only mammal featured in the top 10. We lose points chiefly as a result of our recent evolutionary emergence. Human ancestors also fare well because of crucial evolutionary innovations such as increasing brain size (*Homo erectus*) and learning how to stand up and walk on two feet (*Australopithecus*). Without these the platform for artificial selection, which has propelled our species *Homo sapiens* so high up the table, might never have become established.

The 'top 10'

Others in the top 10 divide into two groups. The first are those that emerged early on in the evolution of life on Earth and have contributed dramatically to the Earth's environment, such as cyanobacteria, algae, *Rhizobia* and stony corals. The second are those that have made a profound difference to human history, such as *Penicillium*, yeast, *Lactobacillus* and influenza.

Other high achievers

Thanks to the variety of the five criteria, extinction is no barrier to success; witness the *Lepidodendron*, that pioneering tree of the Carboniferous Period that left so much fine quality coal for human posterity. Meanwhile, the survival skills of eucalyptus, with its recent global spread, utility to man and devastation of other species, are reflected in its elevated position, as are the environmental changes wrought by cows, predominantly in terms of bolstering human populations levels and as a premier source of greenhouse gas emissions.

Group selection

Each species in the table has been categorized into one of twelve groups: *virus, bacterium, fungus, protoctist, invertebrate, fish, non-flowering plant, flowering plant, amphibian, reptile, bird and mammal.* By far the most dominant and influential evolutionary force on Earth, according to this analysis, are the flowering plants (with 32 entries) followed by mammals (17), invertebrates (16), fish (6), viruses (5), non-flowering plants (5), bacteria (5), fungi (4), reptiles (4), birds (3), protoctists (2) and amphibians (1). Despite their predominance in the top 100, there are no flowering plants at all in the top 10 species, suggesting that flowering plants, although the most dominant life-force, are more the foot-soldiers of evolutionary change than its pioneers.

The top 100 species numbers 1–25

Overall rank	Species	Evolutionary impact: Score	Evolutionary impact: Rank	Impact on human history: Score	Impact on human history: Rank	Environmental impact: Score	Environmental impact: Rank	Global reach: Score	Global reach: Rank	Longevity: Score	Longevity: Rank	Total score	Group	Page number
1	**EARTHWORM**	87	17	93	8	95	4	82	11	74	16	431	*Invertebrate*	*97*
2	**ALGAE**	95	3	41	68	89	7	100	2	95	2	420	*Non-flowering plant*	*54*
3	**CYANOBACTERIA**	100	1	13	83	100	1	90	6	100	1	403	*Bacterium*	*34*
4	**RHIZOBIA**	65	28	85	25	87	8	74	36	85	5	396	*Bacterium*	*40*
5	**LACTOBACILLUS**	74	24	89	17	56	22	100	1	76	13	395	*Bacterium*	*386*
6	**HOMO SAPIENS**	98	2	100	1	98	2	83	10	14	78	393	*Mammal*	*182*
7	**STONY CORALS**	91	13	74	40	85	10	63	56	78	11	391	*Fish*	*66*
8	**YEAST**	69	27	92	11	32	42	100	3	82	7	375	*Fungus*	*293*
9	**INFLUENZA**	42	55	90	16	34	35	100	4	85	6	351	*Virus*	*16*
10	**PENICILLIUM**	93	8	95	3	13	65	60	60	75	15	336	*Fungus*	*314*
11	**LEPIDODENDRON**	65	29	74	41	68	14	80	17	45	32	332	*Non-flowering plant*	*90*
12	**SEA SQUIRT**	91	14	11	100	76	12	72	37	80	8	330	*Fish*	*76*
13	**HOMO ERECTUS**	84	19	92	12	67	16	66	47	20	74	329	*Mammal*	*179*
14	**MOSQUITO**	85	18	82	29	23	54	72	38	66	22	328	*Invertebrate*	*134*
15	**FRUIT FLY**	62	34	89	18	23	55	82	12	68	20	324	*Invertebrate*	*381*
16	**EUCALYPTUS**	45	50	73	44	87	9	76	28	42	36	323	*Flowering plant*	*264*
17	**COW**	40	56	89	19	98	3	78	24	7	86	312	*Mammal*	*222*
18	**FLEA**	78	22	82	30	13	66	76	29	62	26	311	*Invertebrate*	*137*
19	**WHEAT**	60	36	92	13	59	21	70	41	24	67	305	*Flowering plant*	*199*
20	**ROUNDWORM**	80	21	13	84	35	34	95	5	80	9	303	*Invertebrate*	*68*
21	**OAK**	72	25	84	26	30	47	66	48	45	33	297	*Flowering plant*	*143*
22	**NORWAY SPRUCE**	28	70	69	49	64	20	68	43	68	21	297	*Non-flowering plant*	*95*
23	**AUSTRALOPITHECUS**	93	9	94	4	54	25	32	79	23	71	296	*Mammal*	*176*
24	**GRASS**	35	63	67	52	67	17	85	9	37	41	291	*Flowering plant*	*358*
25	**ANT**	55	40	43	65	45	28	82	13	66	23	291	*Invertebrate*	*154*

Domestic success

The chances of making it into the top 50 are substantially greater for species whose genes have responded vigorously to the demands of artificial selection. Sheep, rice, sugarcane, maize, cotton, dogs, potatoes, cats and pigs are, as a result, global players in the few super-species that have come to dominate life on Earth.

Extinction still no barrier

Despite their demise *c.* 250 million years ago trilobites (ranked 31, see also page 70) were once so dominant that they are amongst the most common fossils found in the world today. Their pioneering system for predation (underwater vision) was what probably triggered such extraordinary success. The genes of *Tyrannosaurus*, just three places behind at number 34, created the ultimate theropod, a lineage that lasted from the first appearance of dinosaurs 230 million years ago through to their demise *c.* 170 million years later. Birds, such as *Archaeopteryx* (ranked 41) are theropod offshoots, so the *Tyrannosaurus*'s design lives on in the world's favourite food, chicken (see page 397), albeit in a much reduced and modified form. *Quetzalcoatlus* just makes it into the top 50, representing the first flying vertebrates, the pterodactyls – creatures able to circumnavigate the world.

Top trumps

How come elephants are so high up the table? After all, these are creatures limited in range (Africa and Asia) and with markedly less impact on human history compared to, say, other transporters such as the horse (ranked 58) or camel (ranked 87). On reflection, however, it is their ecological contribution that propels these majestic creatures up the ladder. Before the arrival of Neolithic farming *c.* 12,000 years ago, these were nature's premier lumberjacks, which, along with fire, were responsible for clearing forests in ways that allowed vital new growth. At the other end of the scale their dung provided essential habitats for seeding the nests of termites, creatures essential for the process of decomposing wood. Horses and camels, both beneficiaries of human domestication, have far more limited environmental credentials.

The biggest loser?

It must be the dragonfly, in at number 50. If this table reflected life on Earth 300 million years ago, there is little doubt that these aggressive carnivorous creatures would have made it into the top 10. Their declining relevance is reflected in their smaller size and the ecological niches in which they live. However, they are still successful, unlike many creatures of that era, such as the pioneering amphibian *Eryops* (see page 108), which became extinct aeons ago and therefore doesn't make it into today's top 100.

The top 100 species **numbers 26–50**

OVERALL RANK	SPECIES	EVOLUTIONARY IMPACT		IMPACT ON HUMAN HISTORY		ENVIRONMENTAL IMPACT		GLOBAL REACH		LONGEVITY		TOTAL SCORE	GROUP	PAGE NUMBER
		SCORE	RANK	SCORE	RANK	SCORE	RANK	SCORE	RANK	SCORE	RANK			
26	**SHEEP**	40	57	94	5	68	15	78	25	8	84	288	*Mammal*	*218*
27	**RICE**	50	46	93	9	65	19	64	55	7	87	279	*Flowering plant*	*230*
28	**RABBIT**	35	64	64	54	77	11	68	44	33	53	277	*Mammal*	*330*
29	**SUGARCANE**	55	41	93	10	55	24	62	58	4	95	269	*Flowering plant*	*202*
30	**WATER MOULD**	25	71	87	23	44	31	71	40	40	38	267	*Protoctist*	*51*
31	**TRILOBITE**	94	5	13	85	13	67	78	26	69	19	267	*Invertebrate*	*70*
32	**RAT**	46	49	82	31	24	50	80	18	33	54	265	*Mammal*	*172*
33	**HONEY BEE**	60	37	91	14	25	49	42	71	44	34	262	*Invertebrate*	*152*
34	**TYRANNOSAURUS**	70	26	13	86	33	40	82	14	61	29	259	*Reptile*	*125*
35	**MAIZE**	50	47	88	21	48	26	66	49	6	91	258	*Flowering plant*	*234*
36	**ACACIA**	65	30	62	56	32	43	57	63	42	37	258	*Flowering plant*	*146*
37	**TSETSE FLY**	65	31	74	42	19	63	34	76	65	24	257	*Invertebrate*	*140*
38	**SHARK**	75	23	27	77	13	68	65	51	76	14	256	*Fish*	*78*
39	**COTTON**	30	69	86	24	92	5	42	72	5	93	255	*Flowering plant*	*252*
40	**BAMBOO**	89	15	68	50	21	61	34	77	39	40	251	*Flowering plant*	*150*
41	**ARCHAEOPTERYX**	92	11	13	87	45	29	80	19	21	73	251	*Bird*	*122*
42	**DOG**	40	58	96	2	24	51	80	20	9	82	249	*Mammal*	*323*
43	**POTATO**	50	48	89	20	21	62	62	59	24	68	246	*Flowering plant*	*206*
44	**ELEPHANT**	20	77	76	34	76	13	42	73	32	57	246	*Mammal*	*168*
45	**TICK**	60	38	65	53	13	69	45	67	62	27	245	*Invertebrate*	*384*
46	**CAT**	45	51	88	22	24	52	79	23	8	85	244	*Mammal*	*327*
47	**HERPES**	56	39	84	27	13	70	78	27	12	81	243	*Virus*	*388*
48	**PIG**	35	65	77	33	56	23	68	45	7	88	243	*Mammal*	*215*
49	**QUETZALCOATLUS**	45	52	13	88	30	48	90	7	63	25	241	*Reptile*	*120*
50	**DRAGONFLY**	65	32	13	89	13	71	76	30	72	17	239	*Invertebrate*	*103*

Just slipping out of the top 50

The lobe-finned fish would have featured more highly in the table had it been drawn up in the pre-human era, but today only a few lobe-fins survive (for example, the beautiful blue coelacanth, which was 'rediscovered' in a South African fishing catch in 1933). Lobe-fins cling on at the margins of the oceanic ecosystem, hardly reflective of their momentous evolutionary importance as the lineage that developed much of the apparatus required for fish to learn to live on land (see page 112).

Why aren't horses higher?

Surely horses could do better than a mere 58? After all, they have dominated the history of humanity for the last 4,000 years from being mankind's pre-eminent weapons system to its most ubiquitous form of transportation. Yet, despite their impact on human history (where they rank sixth) domestic horses have relatively little, if any, environmental impact. Their numbers are nothing like on the scale of cows or sheep and being a domesticated species (wild horses became extinct several hundred years ago) their longevity score inevitably suffers. Horses are not generally farmed for consumption, putting the populations of cows (which provide meat and transport) and sheep (a source of wool and meat) on a far more numerous trajectory in the human-centric world. Finally, the age of the horse as premier battle system came to an end with the rise of the petrol engine, putting these creatures into a group of those falling into steady decline.

A tale of two opposites

You couldn't get two species much more contrasting that *Lystrosaurus* (an extinct early mammal) and tobacco (a flowering plant), yet they cosy up alongside each other at numbers 66 and 67 in the table. The former was a lifeline for the lineage of vertebrate land-life that evolved to become mammals by surviving the perilous Permian Extinction 252 million years ago. The latter is a plant that has been responsible for the death of hundreds of millions, if not billions, of humans since it was globalized by European explorers who 'discovered' the Americas in the late fifteenth century. Both have had dramatic impacts on life as it is today, but for completely different reasons.

Sounds like sense

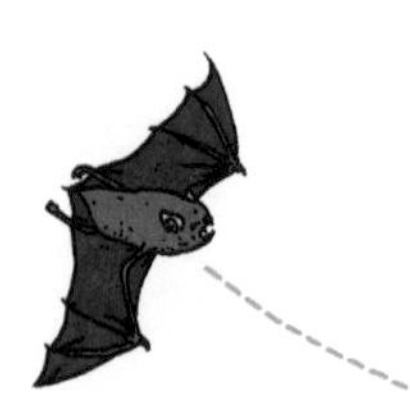

Bats, the world's only flying mammals, squeaked into the top 75 thanks only to their extraordinary capability to see in the dark using sound waves. This survival edge is matched by their unrivalled diversity (there are more than 1,000 species, representing 20 per cent of all mammal species), meaning they are well equipped for future turmoil if some virus or environmental catastrophe unfolds. My bet is that in some future table, bats might rise even higher.

The top 100 species numbers 51–75

OVERALL RANK	SPECIES	EVOLUTIONARY IMPACT		IMPACT ON HUMAN HISTORY		ENVIRONMENTAL IMPACT		GLOBAL REACH		LONGEVITY		TOTAL SCORE	GROUP	PAGE NUMBER
		SCORE	RANK	SCORE	RANK	SCORE	RANK	SCORE	RANK	SCORE	RANK			
51	LOBE-FINNED FISH	94	6	13	90	34	36	23	91	72	18	236	*Fish*	*112*
52	PROTOTAXITES	94	7	13	91	34	37	70	42	24	69	235	*Fungus*	*86*
53	APPLE	22	76	76	35	33	41	72	39	31	58	234	*Flowering plant*	*344*
54	CANNABIS	45	53	64	55	13	72	75	35	37	42	234	*Flowering plant*	*286*
55	DEER	52	44	45	63	34	38	76	31	25	65	232	*Mammal*	*372*
56	VELVET WORM	95	4	13	92	13	73	31	82	80	10	232	*Invertebrate*	*72*
57	COD	35	66	68	51	42	33	65	52	20	75	230	*Fish*	*212*
58	HORSE	38	61	94	6	13	74	76	32	7	89	228	*Mammal*	*242*
59	SPERM WHALE	20	78	57	60	23	56	87	8	40	39	227	*Fish*	*165*
60	AZOLLA	13	88	18	80	90	6	44	68	61	30	226	*Non-flowering plant*	*93*
61	CHICKEN	35	67	75	38	32	44	76	33	6	92	224	*Bird*	*226*
62	ROSE	20	79	58	58	45	30	63	57	37	43	223	*Flowering plant*	*341*
63	SMALLPOX	19	87	94	7	18	64	80	21	9	83	220	*Virus*	*24*
64	CROW	55	42	27	78	23	57	80	22	31	59	216	*Bird*	*378*
65	RED FOX	45	54	42	67	23	58	81	16	25	66	216	*Mammal*	*375*
66	LYSTROSAURUS	84	20	13	93	24	53	82	15	13	79	216	*Reptile*	*118*
67	TOBACCO	20	80	91	15	32	45	32	80	31	60	206	*Flowering plant*	*289*
68	DUNG BEETLE	13	89	43	66	32	46	54	65	62	28	204	*Invertebrate*	*100*
69	ERGOT	34	68	54	61	13	75	65	53	37	44	203	*Fungus*	*304*
70	ANTHRAX	13	90	18	81	13	76	66	50	90	3	200	*Bacterium*	*36*
71	TEA	13	91	76	36	44	32	29	84	37	45	199	*Flowering plant*	*282*
72	SPONGE	20	81	15	82	13	77	60	61	90	4	198	*Invertebrate*	*58*
73	GRAPE	13	92	84	28	13	78	58	62	28	61	196	*Flowering plant*	*296*
74	SLIME MOULD	65	33	13	94	13	79	27	86	78	12	196	*Protoctist*	*49*
75	BAT	55	43	26	79	13	80	65	54	35	51	194	*Mammal*	*163*

Flower power

Dominating the lower ranks of the top 100 species are myriad flowering plants that have transformed human history and, with it, the planet and other life. Material wealth comes from rubber trees; health from cinchona; mind numbing from poppies and coca; gastronomic pleasure from olives, cacao, oranges and bananas; while the lotus and lavender represent symmetrical beauty and perfumed fragrance. Evolutionary ingenuity and genetic pliability gather at their absolute best, it seems, in species rooted to the ground – perhaps because they rely most on other species for services such as pollination and spreading seeds while having evolved an awesome array of chemical defences against being eaten.

Fastest riser

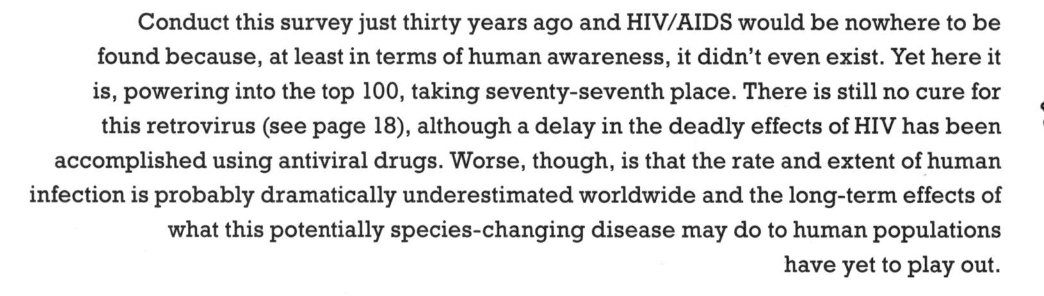

Conduct this survey just thirty years ago and HIV/AIDS would be nowhere to be found because, at least in terms of human awareness, it didn't even exist. Yet here it is, powering into the top 100, taking seventy-seventh place. There is still no cure for this retrovirus (see page 18), although a delay in the deadly effects of HIV has been accomplished using antiviral drugs. Worse, though, is that the rate and extent of human infection is probably dramatically underestimated worldwide and the long-term effects of what this potentially species-changing disease may do to human populations have yet to play out.

Future prospects

Of all the species in the top 100 there is just one that earns its place for its *potential* impact. Pseudomonas (see page 38) makes it only because of its ability to create biodegradable plastic out of artificial pollution. How quickly such alchemy could contribute towards reducing the enormous problem of non-biodegradable plastic waste that pollutes the oceans is as much dependent on humans as on bacteria. But the potential for organic solutions to the material needs of humans is so enormous that this species scores highly in terms of environmental impact just by being possible.

The bottom spot

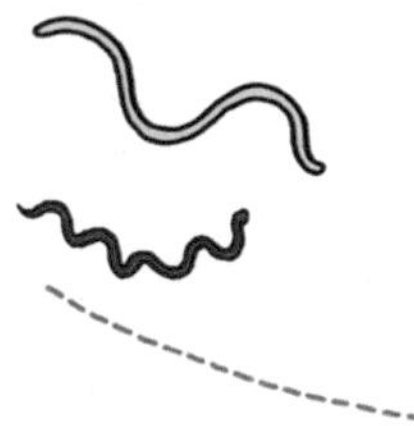

Being amongst the top 100 is what really counts, but as in any league table something has to take bottom place. The potyvirus squeezes in after provoking the first financial crash in modern history and, in the process, propelling its host species, the tulip, to pre-eminence in the human-dominated world. Without notable other accomplishments, this virus languishes down at number 100 and I would not be surprised if, in the event of future (or rival) surveys, it fell out altogether.

The top 100 species numbers 76–100

OVERALL RANK	SPECIES	EVOLUTIONARY IMPACT SCORE	EVOLUTIONARY IMPACT RANK	IMPACT ON HUMAN HISTORY SCORE	IMPACT ON HUMAN HISTORY RANK	ENVIRONMENTAL IMPACT SCORE	ENVIRONMENTAL IMPACT RANK	GLOBAL REACH SCORE	GLOBAL REACH RANK	LONGEVITY SCORE	LONGEVITY RANK	TOTAL SCORE	GROUP	PAGE NUMBER
76	RUBBER TREE	25	72	72	45	34	39	27	87	34	52	192	*Flowering plant*	*256*
77	HIV/AIDS	20	82	76	37	13	81	67	46	13	80	189	*Virus*	*18*
78	SEA SCORPION	52	45	13	95	23	59	55	64	44	35	187	*Invertebrate*	*74*
79	POPPY	20	83	72	46	13	82	46	66	33	55	184	*Flowering plant*	*308*
80	OLIVE	40	59	72	47	13	83	27	88	28	62	180	*Flowering plant*	*209*
81	CACAO	25	73	71	48	23	60	34	78	27	64	180	*Flowering plant*	*274*
82	DIMETRODON	89	16	13	96	13	84	35	74	23	72	173	*Reptile*	*116*
83	RHYNIOPHYTE	92	12	13	97	13	85	35	75	20	76	173	*Non-flowering plant*	*88*
84	CINCHONA	24	75	78	32	13	86	26	90	28	63	169	*Flowering plant*	*316*
85	COFFEE	13	93	54	62	48	27	28	85	24	70	167	*Flowering plant*	*278*
86	SILKWORM	20	84	75	39	13	87	43	70	7	90	158	*Invertebrate*	*260*
87	CAMEL	36	62	74	43	13	88	27	89	5	94	155	*Mammal*	*248*
88	TIKTAALIK	93	10	13	98	13	89	13	95	18	77	150	*Amphibian*	*114*
89	DURIAN	62	35	13	99	13	90	13	96	48	31	149	*Flowering plant*	*148*
90	HAMSTER	20	85	34	71	13	91	76	34	1	100	144	*Mammal*	*334*
91	BLACK PEPPER	13	94	62	57	13	92	19	92	37	46	144	*Flowering plant*	*351*
92	CHILLI PEPPER	39	60	29	76	13	93	13	97	37	47	131	*Flowering plant*	*356*
93	LAVENDER	13	95	32	72	13	94	32	81	37	48	127	*Flowering plant*	*348*
94	PSEUDOMONAS	13	96	31	75	66	18	13	98	3	97	126	*Bacterium*	*38*
95	ORANGE	25	74	39	69	13	95	44	69	4	96	125	*Flowering plant*	*361*
96	COCA	13	97	32	73	13	96	18	94	37	49	113	*Flowering plant*	*300*
97	VANILLA	13	98	36	70	13	97	13	99	37	50	112	*Flowering plant*	*346*
98	BANANA	20	86	44	64	13	98	31	83	3	98	111	*Flowering plant*	*364*
99	LOTUS	13	99	32	74	13	99	19	93	33	56	110	*Flowering plant*	*354*
100	POTYVIRUS	13	100	58	59	13	100	13	100	2	99	99	*Virus*	*22*

Postscript

Thirty species that nearly made it!

PAINFUL DECISIONS about what to include or leave out in the top 100 species have inevitably had to be made. Here I console myself with a list of those I would like to have featured, but could not because of overlaps, repetition, or for fear of creating a list that was insufficiently representative of the diversity of species on Earth. Please feel free to comment, debate, agree or disagree with any of my choices on the forum at the website: www.whatonearthevolved.com. All contributions are welcomed, responded to and gratefully received.

	COMMON NAME	SPECIES (UNLESS STATED)	SUMMARY
1	**COCCOLITHOPHORE**	*Emiliania huxleyi*	One of the most widely distributed single-celled micro-organisms in the seas, it is thought to have had a significant impact on the Earth's biosphere (see pages 55–6).
2	**CYCAD**	Cycadales (Order)	The first plants to develop seeds as a way of protecting their embryos (see page 84).
3	**SPIDER**	*Parasteatoda tepidariorum*	Evolving approximately 380 million years ago from a crab/scorpion-like ancestor (see page 74), there are about 40,000 spider species alive today. *Parasteatoda tepidariorum* is especially common in America.
4	**EDAPHOSAURUS**	*Edaphosaurus boanerges*	A sail-back pelycosaur from the same family as *Dimetrodon* (see page 116) but with a significant claim to fame as being one of the first (if not the first) herbivorous land animal.
5	**ICHTHYOSAURUS**	*Ichthyosaurus breviceps*	An extinct marine reptile and one of the first creatures able to give birth to live young (viviparous).
6	**CROCODILE**	*Crocodylus acutus*	Large semi-aquatic reptile that somehow survived the extinction event that led to the death of the non-avian dinosaurs, 65.5 million years ago.
7	**COMMON DAISY**	*Bellis perennis*	One of the few non-cultivated flowers (or weeds) to be admired for their bright colours and perfect symmetry (see also page 133).
8	**TERMITE**	*Mastotermes darwiniensis*	Descended from cockroaches, these small, highly social, wood-eating insects build enormous nests, sometimes taller than a house, usually containing several million individuals.
9	**NEANDERTHAL**	*Homo neanderthalensis*	An extinct member of the human genus that evolved *c.* 500,000 years ago, with the last individuals dying out in Europe about 25,000 years ago.
10	**GOLDEN EAGLE**	*Aquila chrysaetos*	Well-known birds of prey that have been instrumental in controlling populations of insects, mammals and reptiles, including mice, rats and snakes.
11	**SNAKE**	*Titanoboa cerrejonensis*	Carnivorous reptiles descended from lizards that 'lost' their legs, adapting to life on land in a manner that lets them coil up in trees and under long grasses. Titanoboa can grow up to thirteen metres long!
12	**GREY WOLF**	*Canis lupus*	Wolves can survive in almost any habitat, including forests, deserts, mountains and urban areas.
13	**GORILLA**	*Gorilla gorilla*	Vegetarian, knuckle-walking primates that survive in only two African species but have an extremely close genetic relationship with humans.

14	**SUGAR BEET**	*Beta vulgaris*	Sugar beet now accounts for 30 per cent of the world's sugar production, so although the small brother of sugarcane (see page 202), it plays a significant part in modern agriculture.
15	**GOAT**	*Capra aegagrus*	One of the first animal species to be domesticated in south-west Asia and eastern Europe. Being such a close cousin of the sheep precluded it from the top 100 species.
16	**MOUSE**	*Mus musculus domesticus*	As the most common laboratory animals, mice are key to the search for cures for human diseases.
17	**SOYA BEAN**	*Glycine max*	The major impact of this crop on modern human history brought it to within a whisker of inclusion in the top 100 – its exclusion was mainly due to its relatively recent rise to global status.
18	**FLAX**	*Linum usitatissimum*	An extremely versatile crop grown since ancient times around the Mediterranean and the Middle East for its seeds, fibres and oil. Flax is not grown in such large quantities today, which is why it only nearly made it.
19	**FROG**	Anura (Order)	Frogs are one of the most ancient and diverse orders of amphibians (about 5,000 species are known, and fossils date back 250 million years) and have adapted to live in regions ranging from the tropics to the Sub-Arctic.
20	**KOLA NUTS**	*Cola* (Genus)	The other principal constituent in the world's 'favourite drink' (see page 300).
21	**PLASMODIUM**	*Plasmodium falciparum*	These microscopic parasitic protozoa that cause malaria have already featured in the story of the mosquito (see page 134) and cinchona (see page 316).
22	**TYPHOID**	*Salmonella typhi*	A disease-causing bacterium that spreads by contaminated food and water, or is carried on the legs of flying insects (see fruit fly, page 381), and survives phagocytosis by the human immune system (see slime mould, pages 49–50).
23	**CHOLERA**	*Vibrio cholerae*	Historically this bacterium has been a significant human killer, causing death by dehydration from acute diarrhoea. If left untreated mortality rates can reach as high as 60 per cent.
24	**TUBERCULOSIS**	*Mycobacterium tuberculosis*	A bacterial disease that attacks the lungs and other parts of the body. It is established in up to a third of the world's population – mostly in Asia and Africa – although only a minority develop the full-blown disease.
25	**POLIO**	*Poliovirus*	The virus that causes the disease poliomyelitis, resulting in wasted limbs, paralysis and sometimes death. A vaccine was pioneered by American virologist Jonas Salk in 1952 and mass vaccination programmes began five years later.
26	**SUNFLOWER**	*Helianthus annuus*	Another asterid (see also common daisy, opposite) but with a much larger flowering head and seeds that can be turned into a valuable source of nutrition, such as cooking oil and margarine, as well as bio-diesel.
27	**RUST**	*Puccinia triticina*	A fungal parasite that causes widespread economic damage to crops such as wheat and rye (see picture on page 83).
28	**LEMON**	*Citrus limon*	Genetically so similar to oranges (see page 361) that to include it in the top 100 would have been over-generous to the genus *Citrus*.
29	**TOMATO**	*Solanum lycopersicum*	Native to South America, yellow tomato varieties were the first to be domesticated (by *c.* 500 BC). They began to be grown in Spain, Italy and England from the 1600s. More than one hundred million tonnes of tomatoes are harvested annually today.
30	**PINWORM**	*Enterobius vermicularis*	A type of roundworm (see page 68) that has thrived with the growth of human populations and infects an estimated 200 million people worldwide.

Further Reading

If you'd like to read the endnotes for this book, they can be found at www.whatonearthevolved.com.

General

- *On the Origin of Species* by Charles Darwin (John Murray, 1859)
- *The Descent of Man* by Charles Darwin (John Murray, 1871)
- *Principles of Geology* by Charles Lyell (John Murray, 1872)
- *Genome* by Matt Ridley (Fourth Estate, 1999)
- *Diversity of Life* by Lynn Margulis, Karlene Schwartz and Michael Dolan (Jones and Bartlett, 1999)
- *The Selfish Gene* by Richard Dawkins (Oxford University Press, 1976)
- *The Crucible of Creation* by Simon Conway Morris (Oxford University Press, 1998)
- *The Variety of Life* by Colin Tudge (Oxford University Press, 2002)
- *The Rise of Life* by John Reader (Collins, 1986)
- *The Complete Guide to Prehistoric Life* by Tim Haines and Paul Chambers (BBC Books, 2003)
- *The Ancestor's Tale* by Richard Dawkins (Houghton Mifflin, 2004)
- *A Natural History of Domesticated Animals* by Juliet Clutton-Brock (Cambridge University Press, 1999)

1. On Viruses

- *Viruses and the Evolution of Life* by Luis Villarreal (ASM, 2005)
- *Princes & Peasants* by Donald Hopkins (University of Chicago Press, 1983)
- *The Life & Death of Smallpox* by Ian and Jennifer Glynn (Profile, 2004)
- *The Tulip* by Anna Pavord (Bloomsbury, 1999)

2. On Simple Cells

- *The Origins of Life* by John Maynard Smith and Eörs Szathmáry (Oxford University Press, 2000)
- *Enriching the Earth* by Vaclac Smil (MIT Press, 2004)
- *Oxygen: The Molecule that Made the World* by Nick Lane (Oxford University Press, 2002)
- *The Ages of Gaia* by James Lovelock (Oxford University Press, 1988)
- *Acquiring Genomes: A Theory of the Origins of Species* by Lynn Margulis and Dorion Sagan (Basic Books, 2002)
- *The World Without Us* by Alan Wiseman (Viking, 2007)

3. On Symbiosis

- *Emergence, Connected Lives of Ants, Brains, Cities & Software* by Steve Johnson (Allen Lane, 2001)
- *The Red Queen* by Matt Ridley (Viking, 1993)
- *Propitious Esculent* by John Reader (Heinemann, 2008)

4. On Sea Life

- *Trilobite!* by Richard Fortey (Flamingo, 2001)
- *The Theory of Evolution* by J. M. Smith (Cassell, 1962)
- *Acquiring Genomes* by Lynn Margulis and Dorion Sagan (Basic Books, 2002)
- *In the Blink of an Eye: How Vision Kick-started the Big Bang of Evolution* by Andrew Parker (Free Press, 2004)
- *Old Fourlegs* by J. L. B. Smith (Longman, 1938)
- *Endless Forms Most Beautiful* by Sean Carroll (Weidenfeld & Nicolson, 2006)

5. On Pioneers of the Land

- *Why Size Matters* by John Bonner (Princeton University Press, 2006)
- *The Secret Lives of Plants* by David Attenborough (BBC Books, 1995)
- *The Earth Moved* by Amy Stewart (Frances Lincoln, 2004)
- *Fungi* by Roy Watling (Natural History Museum, 2003)
- *Amber* by Andrew Ross (Natural History Museum, 1998)
- *Evolution of Insects* by David Grimaldi and Michael Engels (Cambridge University Press, 2005)
- *Fossil Plants* by Paul Kenrick and Paul Davis (Natural History Museum, 2004)

6. On Fish that Came Ashore

- *Your Inner Fish* by Neil Shubin (Allen Lane, 2008)
- *The Emerald Planet* by David Beerling (Oxford University Press, 2007)
- *Extinctions in the History of Life* edited by Paul Taylor (Cambridge University Press, 2004)

7. On Biodiversity

- *The Insect Societies* by E. O. Wilson (Harvard University Press, 1974)
- *Fleas, Flukes & Cuckoos* by Miriam Rothschild (Collins, 1952)
- *The Diversity of Life* by E. O. Wilson (Allen Lane, 1993)

8. On the Rise of Reason

- *The Origins of Virtue* by Matt Ridley (Viking, 1996)
- *Bats in Question* by D. E. Wilson (Smithsonian Institution Press, 1997)
- *The Mating Mind* by Geoffrey Miller (Heinemann, 2000)
- *Leviathan: The History of Whaling in America* by Eric Dolin (Norton, 2007)
- *Coming of Age with Elephants* by Joyce Poole (Hodder & Stoughton, 1996)
- *The Elephant's Secret Sense* by Caitlin O'Connell (Free Press, 2007)
- *The Story of Rats* by S. Anthony Barnett (Allen & Unwin, 2001)
- *After Man: A Zoology of the Future* by Dougal Dixon (Granada, 1981)

9. On Agriculture

- *Food, A History* by Felipe Fernández-Armesto (Macmillan, 2001)
- *Plants, Man & Life* by Edgar Anderson (Andrew Melrose, 1954)
- *Sugar: The Grass that Changed the World* by Sanjida O'Connell (Virgin, 2004)
- *The Potato* by Larry Zuckerman (Macmillan, 1999)
- *A History of the British Pig* by Julian Wiseman (Duckworth, 1986)
- *Sheep* by Alan Butler (O Books, 2006)
- *Cattle* by Valeria Porter (The Crowood Press, 2007)
- *China: Land of Discovery and Invention* by Robert Temple (Patrick Stephens, 1986)
- *People, Plants & Genes* by Denis Murphy (Oxford University Press, 2007)
- *Planet Chicken* by Hattie Ellis (Sceptre, 2007)
- *Cod: A Biography of the Fish that Changed the World* by Mark Kurlansky (Jonathan Cape, 1998)

10. On Material Wealth

- *War Horse* by Louis Dimarco (Westholme Yardley, 2008)
- *Inside Your Horse's Mind* by Lesley Skipper (J. A. Allen, 1999)
- *The Behaviour of Horses* by Marthe Kiley-Worthington (J. A. Allen, 1987)
- *The Camel and the Wheel* by Richard Bulliet (Harvard University Press, 1975)
- *The Camel in Australia* by Tom McKnight (Melbourne University Press, 1969)
- *The Fibre that Changed the World* edited by Douglas Farnie (Oxford University Press, 2004)
- *Tears of the Tree* by John Loadman (Oxford University Press, 2005)
- *Jacquard's Web* by James Essinger (Oxford University Press, 2004)

11. On Drugs

- *On Drugs* by David Lenson (University of Minnesota Press, 1995)
- *Tastes of Paradise* by Wolfgang Schivelbusch (Vintage, 1992)
- *Matters of Substance* by Griffin Edwards (Allen Lane, 2004)
- *Consuming Habits* edited by Jordan Goodman, Paul E. Lovejoy and Andrew Sherratt (Routledge, 1995)
- *The Pursuit of Pleasure* by Rudi Matthee (Princeton University Press, 2005)
- *Forces of Habit* by David Courtwright (Harvard University Press, 2001)
- *Fungi* by Brian Spooner and Peter Roberts (Collins, 2005)
- *A Short History of Wine* by Rod Phillips (Allen Lane, 2000)
- *Phylloxera: How Wine was Saved for the World* by Christy Campbell (HarperPerennial, 2004)
- *Narcotic Culture: A History of Drugs in China* by Frank Dikötter et al (Hurst & Company, 2004)
- *The Alchemy of Culture* by Richard Rudgeley (British Museum Press, 1998)

12. On Companionship

- *Man and the Natural World* by Keith Thomas (Allen Lane, 1983)
- *Dog* by Susan McHugh (Reaktion Books, 2004)
- *Cat* by Katherine Rogers (Reaktion Books, 2006)
- *Rabbits and their History* by John Sheail (David & Charles, 1971)
- *Hamsters & Guinea Pigs* by David le Roi (Nicholas Vane, 1955)

13. On Beauty

- *A History of the Fragrant Rose* by Allen Patterson (Little Books, 2007)
- *The Botany of Desire* by Michael Pollan (Bloomsbury, 2002)
- *Vanilla* by Tim Ecott (Penguin, 2004)
- *Lavender (The Genus Lavandula)* by Maria Lis-Balchin (CRC Press, 2002)
- *Out of the East* by Paul Freedman (Yale University Press, 2008)
- *The Encyclopaedia of Grasses* by Rick Drake (Timber Press, 2007)
- *The Grass is Greener* by Tom Fort (HarperCollins, 2000)
- *The Citrus Industry* by Herbert John Webber et al (University of California, 1967)
- *Curry* by Lizzie Collingham (Chatto & Windus, 2005)
- *Bananas: An American History* by Virginia Scott Jenkins (Smithsonian Institution Press, 2000)

14. On Rivalry

- *Something New Under the Sun* by J. R. McNeill (Allen Lane, 2000)
- *Canids* by Claudio Sillero-Zubiri, Michael Hoffmann and David W. Macdonald (Union Internationale pour la Conservation de la Nature et de ses Ressources, 2004)
- *In the Company of Crows & Ravens* by John Marzluff and Tony Angell (Yale University, 2005)
- *After Man: A Zoology of the Future* by Dougal Dixon (Granada, 1981)
- *Time, Love, Memory* by Jonathan Weiner (Faber & Faber, 1999)
- *Fly* by Steve Connor (Reaktion Books, 2006)
- *Maggots, Murder and Men* by Zakaria Erzinclioglu (Colchester Books, 2000)
- *Deer of the World* by G. Kenneth Whitehard (Constable, 1972)

Index

Page numbers in *italics* refer to illustrations.

A

B

C

D

G

H

M

N

O

P

Q

R

S

Acknowledgements

EVERY LIVING THING that I have encountered whilst writing this book has been a unique privilege and joy to research, interpret and describe. What extraordinary stories they all have to tell! Ultimately, therefore, it is only thanks to nature's unfathomable genetic diversity that such a survey has been possible.

However, discovering and weaving such tales into a connected narrative would have been completely impossible were it not for the scholarship of literally thousands of people – too numerous to mention by name. Their many books, articles and conversations have been integral to my attempt to turn a patch-work quilt into what is hopefully a meaningful, informative and interconnected illumination of the story of life on Earth and the place of humanity within it.

When considering the many people who have worked tirelessly to turn this text into such a beautifully illustrated book, things get personal. **Richard Atkinson**, my superlative commissioning editor, has been a tower of support and inspiration from start to finish. **Natalie Hunt**, editor-in-chief of the *What on Earth ... ?* series, has worked with unbounded professionalism, often seven days a week, to ensure that despite our ambitious deadlines this project has been safely delivered without missing a beat. Meanwhile, illustrative genius **Andy Forshaw** has yet again sprinkled his magic pictures on just about every page while **Will Webb**, designer, has laid out each spread with total skill, love and care. Thanks also to **Richard Emerson** who scrutinized the text, making many useful comments, suggestions and corrections throughout and to **Andrew Lownie**, my literary agent, for his ongoing advice, encouragement and support.

My deep gratitude also goes to blue-badge tour-guide **Anne Marie Ehrlich** (on pictures), Bond-girl **Penelope Beech** (on marketing), cockroach-hating **Anya Rosenberg** (on publicity) and Turkish delights **Ruth Logan** and **Katie Mitchell** (on rights – responsible for masterminding the translation of *What on Earth Happened?* into fourteen languages so far ...). Thanks also to **Polly Napper** (on production), **Sarah Barlow** (on final polish), expert mathematician **Erica Jarnes** (on fact-checking), sales supremo **David Ward** (on sales) and the entire Bloomsbury sales force (including the very tall and super-diligent **Andrew Sauerwine**). Finally, my sincere thanks go to **Richard Charkin** (on management) for his continued faith in and support for the *What on Earth ... ?* concept.

I dedicate this book to my wonderful wife **Gins**, our two lovely daughters **Matilda** and **Verity** and also to little **Flossie**, who has rightly come out top dog (see page 322). It is entirely thanks to your constant inspiration, love, patience, company and support that my relentless wrestling with all this stuff about the planet, life and people has been in the least bit possible.

Comments and suggestions about this book can be posted on the discussion forum at **www.whatonearthevolved.com**. All contributions and feedback are most welcome.

Picture Credits

The Advertising Archives: pages 190, 214, 229, 258, 276, 295, 301, 309, 350, 363. **Alamy Images:** pages 40, 94, 140, 160, 231, 361, 362, 372. **Ann & Bury Peerless Picture Library:** page 175. **The Art Archive:** pages 102, 207, 209, 232, 235, 243, 246, 249, 260, 261, 272, 275, 292, 296, 297, 313, 316, 319, 328, 346, 353, 354, 359, 364, 365. **Associated Press:** page 37. **The Bridgeman Art Library:** pages 129, 132, 145, 195, 206, 210, 219, 239, 259, 269, 288, 321, 373. **The British Library:** page 284. **Corbis Images:** pages 23, 105, 117, 123, 126–7, 153, 171, 176, 185, 199, 213, 217, 247, 255, 257, 281, 283, 298, 302, 305, 335, 341, 342, 355, 356, 366, 375, 379, 382. **Getty Images:** pages 137, 345. **Glasgow City Council:** page 90. **Mary Evans Picture Library:** pages 215, 227, 234, 323, 325, 326, 338. **NASA:** pages 6–7. **Natural History Museum:** pages 61, 66–7, 70, 111, 121, 160, 166, 181. **Photolibrary:** pages 79, 348–9. **The Royal Society:** page 29 (right). **Science Photo Library:** pages 10, 15, 17, 35, 53, 55, 58, 63, 68, 71, 73, 81, 83, 88, 97, 130, 131, 140, 143, 183, 203, 253, 265, 306, 315, 337, 387, 388–9. **The Wellcome Institute:** pages 25, 27, 29 (left), 48, 100, 108, 134, 135, 138, 162, 168, 173, 174, 223, 250, 251, 317, 331.

First published in Great Britain in 2009

Bloomsbury Publishing Plc, 36 Soho Square, London W1D 3QY
Bloomsbury Publishing, London, New York and Berlin

A CIP catalogue record for this book is available from the British Library

Design: willwebb.co.uk
Illustrations: andyforshaw.com
Picture Research: Anne-Marie Ehrlich
Index: Helen Smith

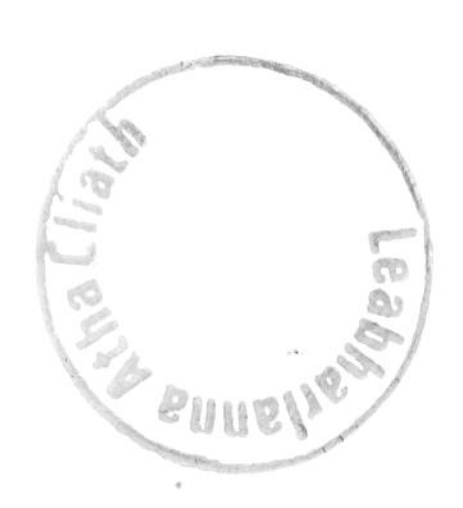

ISBN 978 0 7475 9962 3

10 9 8 7 6 5 4 3 2 1

Printed and bound in Spain by Cayfosa S. A.

www.bloomsbury.com/christopherlloyd
www.whatonearthevolved.com

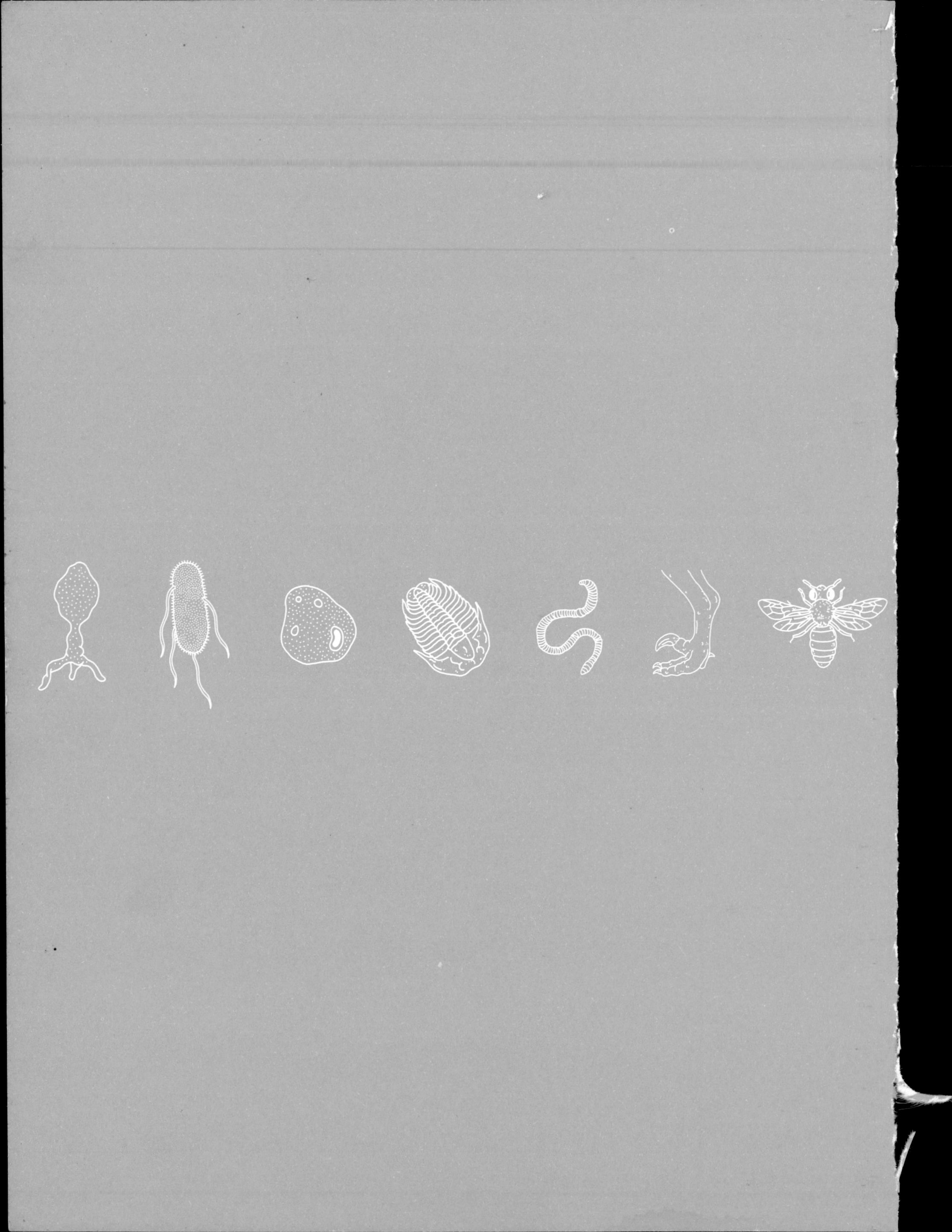